AF700319

An Introduction to Quantum Computing for Computer Engineers

Marcus Stephen Edwards

An Introduction to Quantum Computing for Computer Engineers

Springer

Marcus Stephen Edwards
Vancouver, BC, Canada

ISBN 978-3-032-03649-0 ISBN 978-3-032-03650-6 (eBook)
https://doi.org/10.1007/978-3-032-03650-6

© The Editor(s) (if applicable) and The Author(s), under exclusive license to Springer Nature Switzerland AG 2026

This work is subject to copyright. All rights are solely and exclusively licensed by the Publisher, whether the whole or part of the material is concerned, specifically the rights of translation, reprinting, reuse of illustrations, recitation, broadcasting, reproduction on microfilms or in any other physical way, and transmission or information storage and retrieval, electronic adaptation, computer software, or by similar or dissimilar methodology now known or hereafter developed.
The use of general descriptive names, registered names, trademarks, service marks, etc. in this publication does not imply, even in the absence of a specific statement, that such names are exempt from the relevant protective laws and regulations and therefore free for general use.
The publisher, the authors and the editors are safe to assume that the advice and information in this book are believed to be true and accurate at the date of publication. Neither the publisher nor the authors or the editors give a warranty, expressed or implied, with respect to the material contained herein or for any errors or omissions that may have been made. The publisher remains neutral with regard to jurisdictional claims in published maps and institutional affiliations.

This Springer imprint is published by the registered company Springer Nature Switzerland AG
The registered company address is: Gewerbestrasse 11, 6330 Cham, Switzerland

If disposing of this product, please recycle the paper.

This work is dedicated to the people who made themselves central to my journey from an inexperienced undergrad interested in research to the Ph.D. researcher who has written this book. In particular, without the inspiration and mentorship of Dr. Shohini Ghose whose courses sparked my interest in quantum computing, the support of Dr. Robert Mann who co-supervised me in my Masters with Dr. Ghose, and Dr. Joseph Salfi whose leadership and guidance helped me to mature into a more capable researcher, I would not have found my way in a career in quantum computing research.

Thank you.

Preface

As I myself have studied and read a number of textbooks and technical books on the topic of quantum computing over the last 7 years I have not found one that presents theoretical understanding to the reader in terms that are most familiar to a computer engineer. Instead, I always find myself reading books that take a computer-science-based approach or a physics-based approach. In part this is because the field is only now beginning to employ engineers. For a long time, quantum computers have belonged only in the experimental research laboratory. Only now do we begin to see companies like IBM working on quantum computers that are being built and re-built, improved iteratively in a way that mimics Moore's law. Having studied quantum computing at six universities and worked for three quantum computing companies in industry, I wish to present an integrated picture of the problem of quantum computer engineering that will give the reader a foot in the door.

The book will get the reader's feet wet in virtually every aspect of building a quantum computer (with a bias toward the types of quantum computers I have worked on, i.e. spin qubit-based computers). The reader will not be pulled into exhaustive academic treatments of abstract topics, but they will instead learn how relevant theory relates immediately to practice with examples, diagrams and tutorials. I will draw from 7 years of relevant experience and knowledge from over 10 related courses, so that the reader can benefit without the same amount of time commitment.

The hope is to provide something I have not seen elsewhere in one place: a book that gives the reader a solid idea of what goes into building quantum computers from the basic physics of the qubit to the engineering of the control systems and distributed cloud systems that orchestrate quantum program execution, and everything in between. The book aims to be broad and practical rather than focused and full of detailed minutiae, though it is certainly at times thorough and mathematically somewhat rigorous (mathematicians might disagree). By the end, I hope you feel comfortable talking about and working with quantum computers, if not building one for yourself in a lab or industry setting.

Vancouver, BC, Canada Marcus Stephen Edwards
December 2024

Acknowledgements Many thanks to Joan Weng for her thorough reading as well as thoughtful suggestions and feedback on early drafts of this work. Thanks also to Debopriyo Biswas for a very informative interview on ion trap quantum computers, which is his area of research, and ion trap quantum computer architectures.

Competing Interests The author declares that he is a member of the Quantum Science and Technology Lab at the University of British Columbia, a Summer Software Resident at Xanadu Quantum Technologies and a former employee of Photonic Inc.

Contents

Physical Constants

h	Planck's constant
$\hbar$	Reduced Planck's constant
L	de Broglie wavelength
k	Boltzmann's constant
c	Speed of Light
μ_B	Bohr magneton
Φ_0	Flux quantum
ε_0	Vacuum permittivity
μ_0	Vacuum permeability

Acronyms

ADC	Analog to Digital Conversion
ALU	Arithmetic Logic Unit
APS	American Physical Society
BFS	Breadth-First Search
CAD	Computer-Aided Design
CNOT	Controlled NOT
CPB	Cooper Pair Box
CS	Computer Science
CVQC	Continuous Variable Quantum Computing
DAC	Digital to Analog Converter
DDS	Direct Digital Synthesis
DFT	Density Functional Theory
DFT	Discrete Fourier Transform
DQD	Double Quantum Dot
DUT	Device Under Test
EBL	Electron Beam Lithography
ECC	Error Correction Code
EDSR	Electric Dipole Spin Resonance
ERC	Electrical Rules Checker
FOSS	Free Open Source Software
GCP	Google Cloud Platform
GHZ	Greenberger–Horne–Zeilinger
GKE	Google Kubernetes Engine
GPU	Graphics Processing Unit
HEMT	High Electron Mobility Transistor
IQC	Institute for Quantum Computing
JSON	JavaScript Object Notation
KE	Kinetic Energy
LO	Local Oscillator
MPS	Matrix Product State
MVC	Model-View-Controller

NMR	Nuclear Magnetic Resonance
OOP	Object Oriented Programming
PE	Potential Energy
POVM	Positive Operator Valued Measure
PPU	Pulse Processing Unit
PSB	Pauli Spin Blockade
QASM	Quantum Assembly
QCS	Quantum Control System
QFP	Quantum Flux Parametron
QFT	Quantum Fourier Transform
QMASM	Quantum Macro Assembler
QND	Quantum Non-Demolition
QPC	Quantum Point Contact
QPU	Quantum Processing Unit
QST	Quantum State Tomography
QSTL	Quantum Science and Technology Laboratory
QUBO	Quadratic Unconstrained Boolean Optimization
RBM	Restricted Boltzmann Machine
SEM	Scanning Electron Microscope
SET	Single Electron Transistor
SNR	Signal to Noise Ratio
SOI	Spin-Orbit Interaction
SPS	Single Photon Source
SRB	Streamlined Randomized Benchmarking
TQD	Triple Quantum Dot
VNA	Vector Network Analyzer
VQE	Variational Quantum Eigensolver

Chapter 1
From Computer Engineering to Quantum Computing

Abstract Quantum computer engineering is a bit like classical computer engineering. There are some wacky concepts to wrap your head around, however, including superposition, entanglement and spooky action at a distance. In this chapter we explain these concepts in terms that should be familiar to every computer engineer, with examples using familiar computer electronics.

1.1 Learning Objectives

You might be familiar with the appearance of a typical quantum computer: a cylindrical "drum" suspended by a frame, connected to a control computer and pump system. Maybe you have seen the drum opened to reveal the chandelier-like temperature stages inside that allow a QPU at the bottom to reach tens of milliKelvin. Does it make you wonder what all goes into making a quantum computer? Well, we are going to explore exactly what goes into the quantum computing stack in these chapters!

Depicted in Fig. 1.1 are the pieces that go into building a quantum computer. By the end of this book we will have addressed all of the items, but for now we will focus on the first: quantum basics. Our learning objectives for this chapter will include the following:

- Understand the difference between a bit and a qubit.
- Understand superposition and related experiments.
- Understand entanglement and what it means for theories of the foundations of quantum mechanics.

We will revisit this diagram with each chapter. Now, let's dive into it!

© The Author(s), under exclusive license to Springer Nature Switzerland AG 2026

M. S. Edwards, *An Introduction to Quantum Computing for Computer Engineers*,
https://doi.org/10.1007/978-3-032-03650-6_1

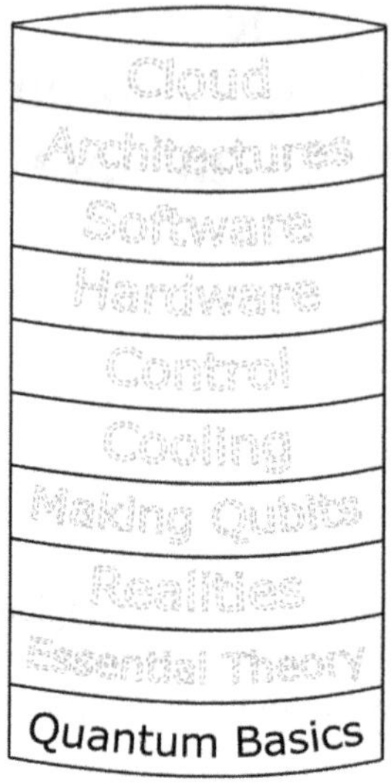

Fig. 1.1 Learning objective for this chapter: quantum basics

1.2 Motivation

> If you think you understand quantum mechanics, you don't understand quantum mechanics.—Dr. Feynman [1]

Quantum computing is a computing paradigm that has been dreamed of by physicists since the early 1900s [2]. It is only now that these dreams are becoming a practical reality.

On-boarding a quantum computer programmer is not as simple as on-boarding a full-stack developer or software engineer. A quantum computer programmer will benefit from a PhD or Masters degree in a related field. Even then, quantum computer programming is not a topic covered in Quantum Information Science (QIS) programs.

IBM is doing a notably good job with their graphical interactive programming interfaces, visual execution explanations and friendly-looking web tools. Still, these tools are targeted at graduate students, post-docs and PhDs.

Many tools more useful to an engineer are emerging as well. In 2024 these include unique quantum computer assembly languages, circuit-level programming languages somewhat analogous to hardware description languages, algorithm level languages, full-stack libraries and a handful of interoperable "universal languages". These are being developed and offered by a wide range of companies, including IBM, Rigetti, DWave, Xanadu, Google, Microsoft and more. A versatile quantum computer programmer in the coming noisy intermediate-scale quantum (NISQ) era will need to know them all.

While there have been a number of good books on this subject, we think that there is a need for an accessible technical introduction that is written specifically for students with a computer engineering (as opposed to physics) background. There is no reason that an enthusiastic engineer or developer can't prepare him or herself for the NISQ era. Sure, there are some tricky concepts in quantum physics, but we know from experience that engineers deal with a lot more complexity on a daily basis than is recognized in many academic circles, and we believe the engineers can handle it!

1.3 Conventions

For this first chapter, we will make use of conventions that will be familiar to engineers in order to make a transition to thinking like "a quantum mechanic" easier.

We will sometimes use the engineering notation in mathematical equations. That means vectors will be represented by bolded symbols, derivatives by dots, etc. This might make physicists squeamish but should be more familiar to those with an engineering background.

$$\dot{\mathbf{A}} = \frac{\partial \mathbf{A}(r,t)}{\partial t} = e_i A_0 j\omega \ \exp(j\omega t)/\exp(j\mathbf{k}\cdot\mathbf{r})$$

We will also use Dirac "Bra-Ket" notation in the description of quantum systems after introducing the notation shortly.

$$|+\rangle = \frac{1}{\sqrt{2}}(|0\rangle + |1\rangle)$$

1.4 Why Quantum Computing?

> ...scientists need to establish a quantum-programming community to nurture an ecosystem of software. This community must be interdisciplinary, inclusive and focused on applications.—Zeng et al. First Quantum Computers Need Smart Software [3]

The interested reader likely has some concept of quantum computing, whether it is vague or specific. We guarantee you that whatever depth of understanding you may have, there is always more to it than you think! That is the wonderful thing about quantum computing. It is a species of physical computing which is very versatile. Being versatile means being widely applicable. So, the language of quantum computing is not just useful for solving systems of quantum particles. The paradigm can be aptly applied to describe and solve systems of many kinds.

Consequently, learning the language of quantum computing equips you with a powerful tool. It is a tool that, when appropriately applied, enhances understanding and utility. This is why companies are applying quantum computing in the areas of chemistry, machine learning, optimization, finance, communications, aerospace and more.

A single, isolated quantum particle is a much more valuable resource than a pile of electrons that flows through modern digital electronic circuitry. Without going deeply into the resource theory that (might) justify the use of quantum computers, we will explain in a somewhat unique way how quantum computers work, starting from computer engineering concepts that should be familiar to our particular audience.

1.5 Classical and Quantum Registers

At the low-level, quantum computer programmers deal with the same primitive constructs as classical programmers. Assembly languages like Intel's X86 are designed based on families of microprocessors with unique register architectures. In the case of X86, the instruction set is designed for the 64-bit Intel 8088 and its variants. A low-level programmer should be familiar with the various registers of at least one register architecture: the data registers, address registers, program counters, stack pointers, frame pointers, status registers, etc. Quantum computers don't usually make use of such a variety of purpose-built registers (yet), though there are proposed schemes (Quantum Abstract Machines (QAMs) [4]) in which they do so [5–8]. So, a quantum computer programmer only (usually) deals with data registers.

However, some of the data registers a quantum computer programmer uses are arrays of qubits (quantum bits), and some of them are arrays of regular bits.

One promising technique for creating arrays of qubits is ion trapping. Since an ion is a particle with a charge, it can be manipulated by charged electrodes. For an ion to be trapped, it must first float unknowingly into a point between a trap's electrodes. With the ion in place, the electrodes on all sides of the ion begin oscillating their charges in sync, so that the ion is kept in a saddle point at the center.

In one type of ion trap called a Paul trap, four electrodes are used. Two are configured to attract the ion at a time, while the other two will repel the ion. This is illustrated by the electromagnetic potential contour graphed below (Fig. 1.2).

Instantaneously, the ion will be confined to the center by the electrode charges along one dimension and begin to be expelled from the center by the charges in the other dimension. The electrodes rapidly oscillate their charges in sync so that the

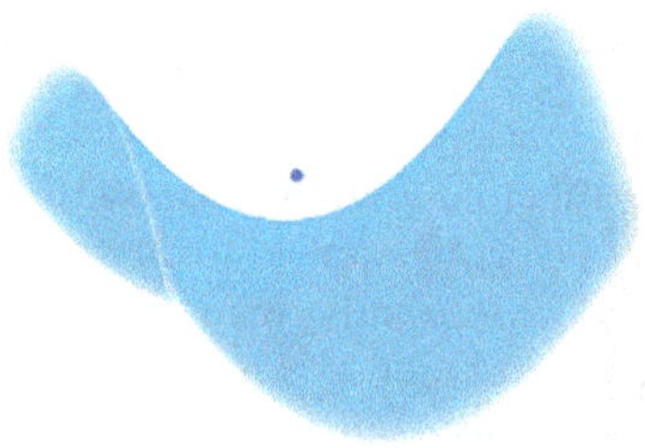

Fig. 1.2 Ion trap instantaneous potential approximation

confinement of the ion is equally effective in each dimension overall, and the ion experiences an average of zero electromagnetic potential energy at the center.

Multiple ions can be trapped in this manner in a way that they are individually addressable, controllable and can also be made to interact. A quantum computer needs at least one low-level mechanism for controlling individual ions, and one for facilitating interaction between the ions. Controlled laser pulses help us in both cases.

A digital microprocessor must be able to apply a minimal set of primitive operations to its registers' bits. This set could be set, clear, measure; set, flip, measure; or clear, flip, measure. Like a bit, a qubit has two computational basis states: 0 and 1. Also, like a digital microprocessor, a Quantum Processing Unit (QPU) needs to be able to perform at minimum one of set, clear, measure; set, flip, measure; or clear, flip, measure to its qubits.

1.6 Bits Versus Qubits by Analogy

The appeal of quantum computers is really based entirely in the difference between qubits and bits. So, let's look closely at what a qubit is that a digital bit is not.

First of all, what is a digital bit? The theorists like to simply say that it is a Boolean value, 0 or 1, and leave it at that. However, the engineers that work with hardware or embedded systems know this is only an idealization of the real world. Yes, just like there are now many types of qubits being proposed every day, there have been many exotic schemes for making digital bits. One of the author's undergraduate computer science professors remembered working with rod cell memory, which used the writable and readable polarities of magnetically charged metal rods to store bits of information [9] (Fig. 1.3).

By electronically inducing currents in the wires coiled around the rods, one could clear or set the polarities of the magnetic fields. By electronically measuring the

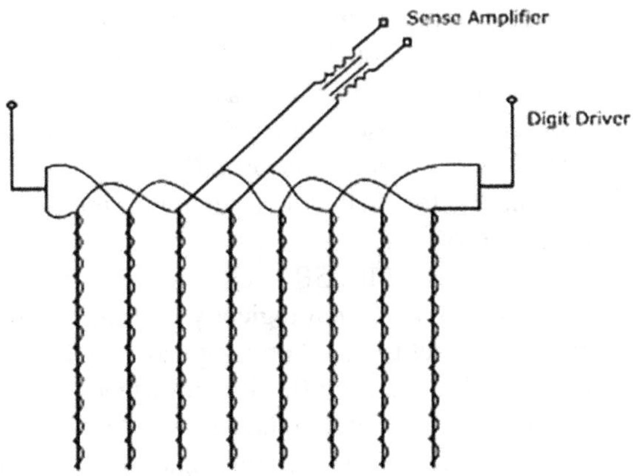

Fig. 1.3 1966 thin film rod memory

currents induced by the polarized rods, one could measure the bit value, retrieving it for use by the computer.

This works because, as we all learned in high school science, the electric field corresponding with a magnetically charged rod is oriented in a circular direction about the rod's length. The direction of this electric field corresponds with the direction of the current induced in a nearby conductor in this circular dimension (the coiled wire).

The length of the rods plays two roles. First of all, the strength of the electric field at any point outside a rod is determined by the sum of all of the rod's constituent point charges' contributions. So, the electric field strength depends on the number of these contributing point charges and, equivalently, on the length of the rod.

Second, the rods have length in only one linear dimension. This makes it easy to magnetize them in that dimension. It also makes any magnetization in other dimensions negligible.

Imagine we shrunk the rods' lengths down to almost nothing, and then made them symmetric in every dimension. We would lose both the ability to measure the electric fields and to charge them reliably in only the desired dimension. This would seem to render them rather useless to us!

However, extremely tiny magnets are loosely analogous to (charged) ions. Just like rod memory, the polarity of the ion's spin is used in a QPU to represent its value (there's a bit more to this—we'll cover it in Chap. 4). A linear axis is chosen so that a spin polarization pointing in one direction along this axis is taken to mean a value of 0, and the other direction is taken as 1. A significant difference between ions and magnetized rods is that the ions can be polarized in any direction.

Since our measurement apparatus is still based on a single axis, called an orthonormal basis, the measurement of any polarization will yield either a 0 or 1 result, which the system will assume as its new state. In quantum jargon, this is called the "collapse of the state" to an "observable value". This is generally presented as one of the wackiest properties of quantum systems. However, something similar should not really seem all that unfamiliar to engineers with some knowledge of classical memories.

We have moved on from polarized rods and now use solid-state memories in our computers. Dynamic random-access memory (DRAM) type memories use arrays of small capacitors to store bits of information. Static random-access memory (SRAM) type memories use digital latches to achieve the same. Non-volatile memory (NVM) memories typically use floating-gate metal-oxide-semiconductor field-effect transistors (MOSFETs). Let's consider a (loose... bear with me) analogue of the quantum measurement effect that can be seen with the most simple of all digital latches: the SR latch [10] (Fig. 1.4).

The truth table of the SR latch is in Fig. 1.5.

The idea is that when a logically high digital pulse is applied to the set input S, the output Q is set to 1. When a logically high pulse is applied to the reset input R, the output Q is cleared to 0. The cross-feedback loop between the two NOR gates allows the latch to maintain the value of Q after the pulse responsible for setting or clearing its value has finished propagating—the value is "latched".

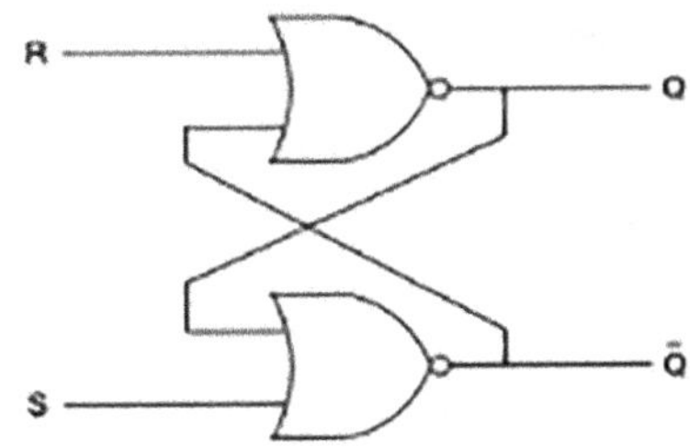

Fig. 1.4 NOR gate SR latch

Fig. 1.5 SR latch truth table

S	R	Q	$\bar{Q}$
0	0	Latched	
0	1	0	1
1	0	1	0
1	1	Metastable	

If a high pulse is applied to both inputs at the same time, a race condition is produced. If both pulses could really be exactly simultaneous, and both gates identically manufactured, the latch would be in an undefined or metastable state for an indefinite period of time. In reality, whichever of the gates' outputs change first will cause feedback to the other gate and assert its value as the winner of the process that is the race condition. The other pulse will not affect Q.

Before the winning gate does assert itself, we cannot be certain which will win. We can only model the circuit's state with a model that has our uncertainty built-in. When we use a probability to describe the (un)certainty of our belief that something will occur, this is called a Bayesian probability. We can describe the system using a pbit, which is essentially a record of how we bet on the individual gates' relative manufacturing qualities to prevail (electrons will choose the path of least resistance as per the principle of least action). A pbit is a weighted sum of bits. That is, it is a state defined by association with the basis states 0 and 1. Our bets on either gate become the weights in the sum. So, if we knew the clearing gate was new and the setting gate had been found in a proverbial dumpster, we might bet that the output would most likely be cleared to 0. The resistance to motion that an electron will experience in the clearing gate is likely less. If we also knew that the input signal on the reset input was 5 V, while the setting signal on the reset input was only 2.5 V, we could factor this into our bet as well. With both of these pieces of information, we might bet that Q is cleared with 80% probability. The electrical pressure that the electron experiences to take the clearing path is higher than the pressure to take the

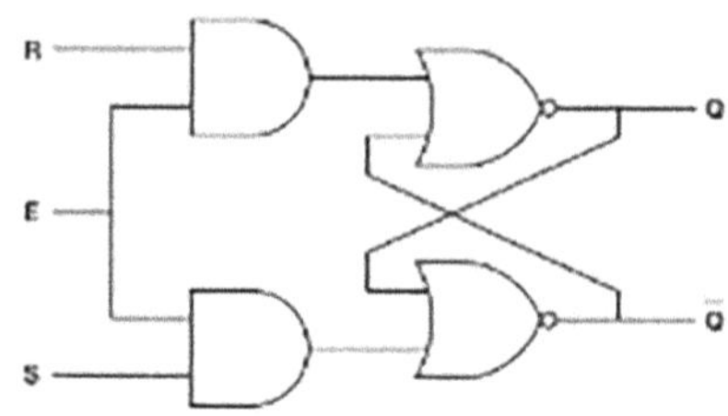

Fig. 1.6 Gated SR latch

setting path. In this sense, our weights are like estimates of the ratio of the voltages that we would find across each path.

$$Q = 0.8 \cdot \mathbf{0} + 0.2 \cdot \mathbf{1}$$

Once one value asserts itself, Q "collapses" (probably to 0) and our prediction seems to have been sufficiently accurate. Since the value is now fully latched, there is no longer any uncertainty about the state of the stored bit.

Another simple latch is the gated SR latch. It adds an enable control signal E to the SR latch, so that the S and R signals only propagate when this control signal E is high (Fig. 1.6).

With S and R both high, what the enable signal allows us to do is to control exactly when a latching race-condition competition may occur. This bounds the evaluation and resolution of any uncertainty in the value Q in time. After we set E high, we can be certain that within the time of two gate propagation delays, Q will have assumed an entirely certain Boolean value. This would be referred to as the device's "settling time".

When E is low, the SR latch is decoupled from its input lines. When E is set high, the circuit couples the SR latch to its input lines, and the circuit is closed. This transition initiates the competition to determine which of the opposed gates "wins", via the principle of least action. The competition lasts at most as long as the fixed period of the "settling time", after which the system assumes the value of the Boolean result.

If you, as a computer engineer, can understand the metastable state of the SR latch and how to describe it with Bayesian probability, then you can also understand quantum computing.

Qubits are isolated two-level systems, like the decoupled SR latch. When a qubit is measured, the qubit is coupled to an external control system. This initiates a competition to determine the "winner" of the opposing collapsing transitions. Such a competition lasts at most as long as the fixed period of the "decoherence time", after which the system assumes the value of the Boolean result.

In the case of the SR latch, the tug-of-war occurs between two opposing voltages: logical high and logical low. It is a truly one-dimensional competition. Neither gate can influence the outcome to become an imaginary value, for example.

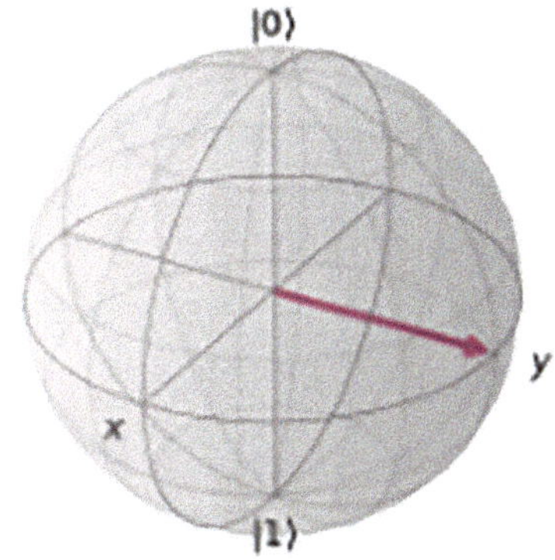

Fig. 1.7 Bloch sphere

In the case of a qubit, the measurement axis is imposed entirely by the measurement apparatus and is in some sense arbitrary. The qubit is not limited to tug-of-war competitions between two specific opponents. Rather, a qubit's value can be pulled in a theoretically infinite number of directions. The qubit's value will collapse to 0 or 1 after the decoherence time, just like the SR latch. However, our uncertainty about the state of a qubit before decoherence is not captured by a pbit. Rather, the state of a qubit is better represented by a vector in 3 dimensions. This vector can be pulled in any direction and is often depicted with spherical coordinates θ and ϕ on a sphere called the "Bloch Sphere"[1] (Fig. 1.7).

Describing the uncertain state of the SR latch was equivalent to making an educated guess about what the state Q would be after it was measured. The way people usually describe a quantum state is equivalent to making an educated guess about where its state vector is before it is measured. You could also express a quantum state using a pbit. This would be the same as guessing what the state would be after measurement. But this pbit would be a projection of the state vector, and state information would be lost. The loss of information follows simply and directly from the fact that we are using a lower-dimensional measurement apparatus to probe a higher-dimensional system.

In the case of the SR latch, a large number of electrons each navigate the pair of NOR gates. What each electron is doing is choosing its path based on the ratio of the path conductances. The better, more conductive gate will admit more electrons per unit time. However, as long as there is some conductivity in each path, there will always be some number of electrons in each. This is Kirchhoff's Current Law of Conservation of Charge in action and is a lot like the case of a basic current divider (Fig. 1.8).

$$\sum i_{\text{in}} = \sum i_{\text{out}}$$

[1] Originally Swiss, Felix Bloch worked on atomic energy at Los Alamos National Lab during World War II. He eventually became the first head of CERN [11].

Fig. 1.8 Current divider

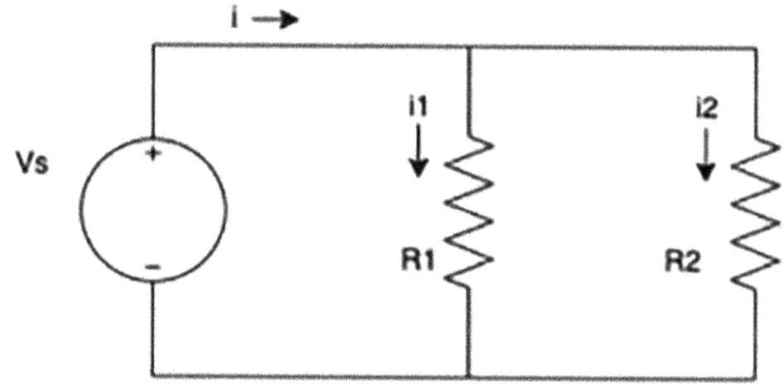

$$\sum i_s = \sum i_1 + i_2$$

$$i_1 = \frac{R_2}{R_1 + R_2} \cdot i_s$$

$$i_2 = \frac{R_1}{R_1 + R_2} \cdot i_s$$

What it means when we measure any electrical current in any wire with an ammeter is that there are enough electrons that have chosen that wire as their path to register.

When a lot of electrons want to take the same path, this creates electrical pressure or voltage. So, like an ammeter tells us about the logical state of a circuit after enough electrons have made their path decisions to create a detectable signal, a voltmeter makes an educated guess about where the electrons will go before they make the journey.

Describing a quantum state before measurement is like using a voltmeter. Describing it after measurement is like using an ammeter. There is a misconception that quantum computers simply deal with individual particles where classical computer circuits deal with the flow of many. This is true, but there is more to say. The comparison can be made, but only very carefully.

Quantum computers give their computation results as histograms. A histogram is a frequency plot. A histogram of electron frequency in a current divider would be a tally of the number of electrons that are seen traveling each path, over a certain period of time. This concept is actually already built into the current, since current is the amount of charge seen over time, and charge comes in the quantized unit of the charge of an electron q_e.

$$i = \frac{n \cdot q_e}{t}; \; n \in \mathbb{Z}$$

Incident over time is exactly the definition of frequency, so the current can be rewritten trivially in terms of the electron frequency.

$$i = f_e \cdot q_e$$

You might have heard people try to introduce quantum computing by asking you what would happen if you put just one electron through an electrical circuit. They might have tried to blow your mind by explaining that the electron's state is still given in terms of both possible paths! How could one electron go both ways?!

At this point, it is important to be careful not to forgot that there is a difference between the Bayesian description of our uncertainty about what an electron will do, and the frequentist measurement of what an electron has done. There are also more practical considerations to be had as well.

If someone insists that the electron would somehow travel both paths at once, and that your intuitions are clearly insufficient to deal with quantum mechanics, you can gently correct them. We suspect your intuition would tell you that a single electron would have insufficient charge to create any measurable voltage or current, having done nothing interesting at all. You would be right. Most of the time.

Real quantum effects in an electron's path choice would only occur for a free electron under the right conditions, and electrons in wires are not free electrons. Also, these quantum effects in particle's path choices only come into play when considering a length scale of the thermal de Broglie wavelength L. Here h is Planck's constant, m is the electron's mass, k is Boltzmann's constant and T is the ambient temperature.

$$L = \frac{h}{\sqrt{2\pi mkT}}$$

At room temperature, the relevant length scale would be around 10 picometers.

But that's not the end of the story. We can still describe our uncertainty about the single electron's path, despite never being able to experimentally verify our guess with just one electron. We can go further by trying to verify our guess about the single electron by using a lot of electrons.

Say each electron traveling through our original SR latch is identical, and we bct that they are each likely to travel through the clearing gate with 80% probability, as we bet before. Then if we put 100 of these identical electrons through the circuit, we would expect the circuit to latch low 80 times, and this behavior would suggest that we were right about the individual electrons.

Unfortunately, we can't do this since one electron isn't enough to operate the circuit. So, we have to just put all 100 electrons through at once, and we'll likely never see the setting gate win.

This is where quantum computing really comes into the picture. In a quantum computer, the circuitry is miniaturized, cooled and isolated to such a degree that one particle can operate the circuit on its own, in a measurable way. The so-called quantum state is still just our guess about what path it will take. We still have to put all 100 particles through our circuit in order to verify our guess. Each particle is still only seen to take one path. The (huge) difference is simply that we don't have to put them all through at once. This means we can tally exactly how many times each path is taken.

The sheer coldness of a quantum circuit means that we can actually determine the “voltage” and “current” due to one particle.[2] If we have just one particle, we can use and re-use it. It still needs to run the circuit 100 times, but now we only need one physical particle to furnish our experiments. Note that while this seems at first like an incredible resource-saving effect, it isn't. The cost of keeping a quantum particle isolated and cooled is very high.

In quantum physics, the pressure that a particle experiences to take each path is given as a weight in the sum that defines its “quantum state”. Let's find the quantum states that would describe our Bayesian pbit from before.

$$Q = o.2 \cdot \mathbf{0} + 0.8 \cdot \mathbf{1}$$

Since this is a pbit, it doesn't fully describe a quantum state. It could however accurately describe what our expectation would be for the value of a quantum state after it is measured.

We can generally describe a quantum state on the Bloch sphere in spherical coordinates [12]. This format is called a polar qubit. Here $0 \leq \theta \leq \pi$ and $0 \leq \phi \leq 2\pi$.

$$Q = \cos\left(\frac{\theta}{2}\right) \cdot \mathbf{0} + \sin\left(\frac{\theta}{2}\right) \exp(i\phi) \cdot \mathbf{1}$$

Note how with a pbit, our weights are normalized so that they add to 1. Here, they are clearly not. An obvious issue is that one of these weights is imaginary! We can recover a type of normalization, however, if we treat both numbers as imaginary and consider the squares of their absolute values.

$$\alpha = \cos\left(\frac{\theta}{2}\right) \exp(i0)$$

$$\beta = \sin\left(\frac{\theta}{2}\right) \exp(i\phi)$$

$$\alpha^2 = \cos^2\left(\frac{\theta}{2}\right)$$

$$\beta^2 = \sin^2\left(\frac{\theta}{2}\right) \exp(2i\phi)$$

[2] It is strange that you get statistics out of one particle in quantum mechanics. But even in the quantum realm, the great Canadian hockey player Wayne Gretzky's words hold true: “You miss 100% of the shots you don't take”. We still need at least one “shot”!.

$$|\alpha|^2 + |\beta|^2 = 1$$

In our case, subbing in θ leaves us with some freedom in ϕ. This designates a set of valid points on the Bloch sphere which would correspond with our 80/20 pbit.

$$\cos\left(\left(\frac{\pi}{2} - \sin^{-1}(0.6)\right)/2\right) = \sqrt{0.8}$$

$$\sin\left(\left(\frac{\pi}{2} - \sin^{-1}(0.6)\right)/2\right) = \sqrt{0.2}$$

$$Q = \sqrt{0.8} \cdot \mathbf{0} + \sqrt{0.2} \cdot \exp(i\phi) \cdot \mathbf{1}$$

You can see immediately in this equation that our probability of measuring **0** is $0.8 = |\alpha|^2$. By rearranging for ϕ in $0.8 + (\sqrt{0.2}\exp(i\phi))^2 = 1$ we can solve for $\phi = \pi n, n \in \mathbb{Z}$ as well.

This describes a set of quantum states that would correspond with our pbit.

> **Difference between a bit and a qubit?**
> A bit is a single Boolean unit of information, physically represented by controlling the measurable flow of many electrons through an electronic device in one path or another. A qubit is a unit of information that can describe what "path"[3] a *single* particle might take to produce a measurable outcome.

1.7 But What Is a Qubit Really?

If the way that probabilities were encoded into the squares of the absolute values of the imaginary weights in our polar qubit equation didn't seem intuitive, that is because it isn't intuitive. The counterintuitive nature of the accepted qubit representation is almost unanimously agreed upon.

Ions used for quantum computing are used for their spin-1/2 particles, which are essentially 2-level systems. Spin-1/2 particles include the proton, neutron, electron, neutrino, quark and photon. These particles can be measured using a measurement axis with two outcomes: 0 and 1—these are the two "levels". There are particles with many more than two levels, which cannot be as easily used to create qubits, but can be more easily used to create higher-dimensional units of information that fall into the more general bucket of qudits.

[3] This "path" terminology is often used in Feynman's path integral formulation of quantum mechanics and is also used in the formulation of some interesting quantum compilation strategies [13]

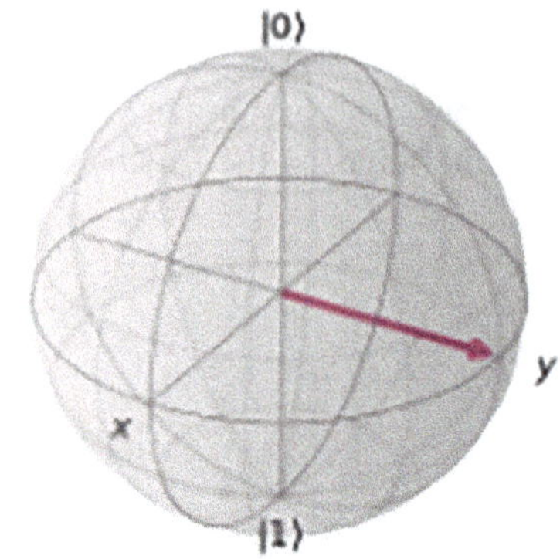

Fig. 1.9 Bloch sphere graphed using IBM's QisKit library [17]

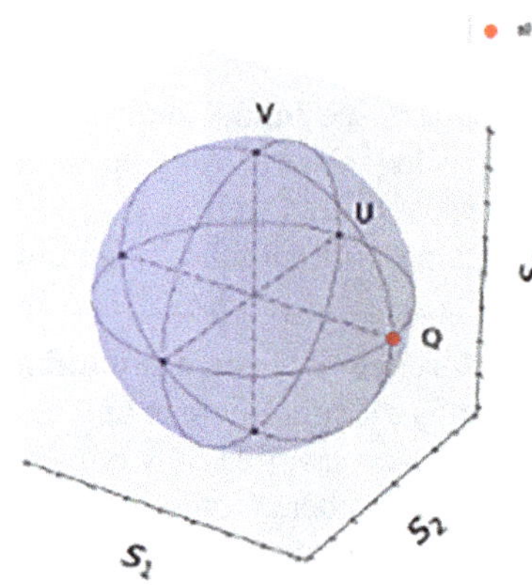

Fig. 1.10 Poincaré sphere graphed using python polarization library [18]

> **What is a qubit?**
> Qubits are 2-level systems often encoded into the spins of spin-1/2 particles. These can be measured to yield a Boolean outcome.

What this means is that if one understands one type of spin-1/2 particle, he/she can understand them all. This allows us to draw accurate analogies between many fields, including the classical optics that most undergraduate students are taught. It is also why people are making quantum computers using everything from photonic optical qubits to chemicals like trichloroethylene [14, 15].

Someone who has taken a university optics course may have thought the Bloch sphere presented in the last section vaguely familiar. The Bloch sphere of quantum mechanics is the Poincaré[4] sphere of classical optics in disguise (Figs. 1.9 and 1.10).

There is no complicated mapping between the Bloch and Poincaré spheres; they are simply identical. The Poincaré sphere visualizes the Jones calculus of fully polar-

---

[4] Poincaré is also credited with discovering chaoticity in the famous three-body problem of orbital mechanics [16].

ized light and Mueller calculus of partially polarized or incoherent light. Fully polarized light is represented by vectors that point onto the surface of the sphere. Partially polarized/incoherent light is represented by vectors that point somewhere inside the sphere.

The Bloch sphere visualizes the mechanics of pure quantum states (which turns out the same as Jones calculus) and the density matrices of mixed quantum states (the same as Mueller calculus). Pure quantum states are exactly equivalent to fully polarized states of light and are represented by vectors that point onto the surface of the Bloch sphere. Mixed states are represented by vectors that point somewhere inside the sphere.

Students may remember that light is an electromagnetic field that can be polarized linearly, circularly or elliptically. In every case, the magnetic field B and the electric field E are orthogonal to one another and to the direction of the wave's motion k (Fig. 1.11).

The points on the surface of the Poincaré sphere which intersect with each of the x-, y- and z-axes are labeled based on their horizontal, vertical, left, right, diagonal and anti-diagonal orientations as $|H\rangle$, $|V\rangle$, $|L\rangle$, $|R\rangle$, $|D\rangle$, $|A\rangle$, respectively. Any Jones vector which points to a place on the surface not equal to $|R\rangle$, $|L\rangle$ and not on the circle that intersects $|H\rangle$, $|V\rangle$, $|D\rangle$ and $|A\rangle$ is representative of an elliptical polarization.

Perfectly circular polarizations are represented by vectors that point at $|R\rangle$ and $|L\rangle$. Vectors pointing onto the circle that intersects $|H\rangle$, $|V\rangle$, $|D\rangle$ and $|A\rangle$ represent linearly polarized EM fields.

In vector form, these EM field polarizations are represented according to the following identities:

$$|V\rangle = \begin{bmatrix}1\\0\end{bmatrix},\ |H\rangle = \begin{bmatrix}0\\1\end{bmatrix},\ |D\rangle = \frac{1}{\sqrt{2}}(|H\rangle + |V\rangle),\ |A\rangle = \frac{1}{\sqrt{2}}(|H\rangle - |V\rangle)$$

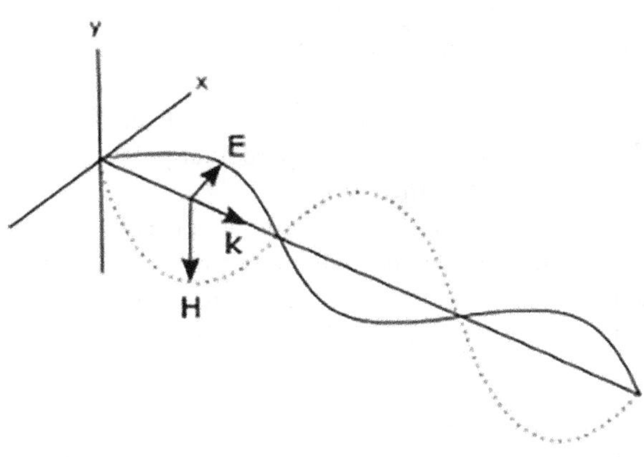

Fig. 1.11 Electromagnetic plane wave field distribution

Fig. 1.12 Adding two polarized electromagnetic field distributions

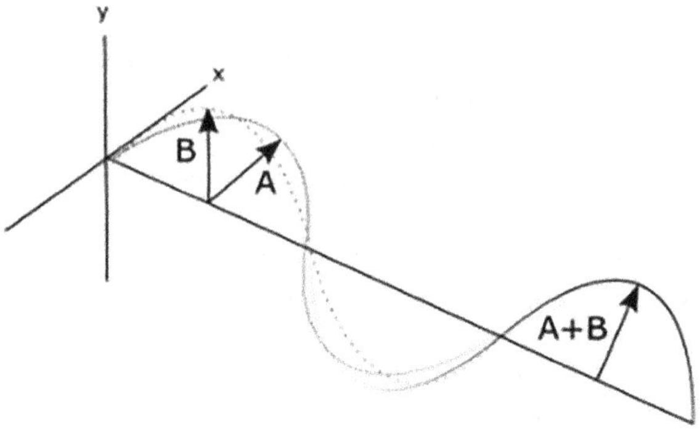

$$|R\rangle = \frac{1}{\sqrt{2}}(|H\rangle - i\,|V\rangle),\ |L\rangle = \frac{1}{\sqrt{2}}(|H\rangle + i\,|V\rangle).$$

The convention is to define $|H\rangle$ and $|V\rangle$ explicitly as linearly independent unit vectors, and the rest of these vectors as combinations of these. However, the assignment of the vectors to specific dimensions is arbitrary and any can be described additively in terms of others (Fig. 1.12).

When we talk about the Bloch sphere, we are simply generalizing these ideas beyond the domain of photonics. The Bloch sphere (and equivalently the Poincaré sphere) are not applicable only to the description of light waves. They are applicable to describe the physics of many systems, including that of tiny magnets like ions and many other phenomena.

1.7.1 Dirac's Three-Filter Experiment

So, what makes a qubit quantum? Influential quantum physicist Paul Dirac[5] described an intuition-building experiment that you can do at home in the 1930s [20]. The experiment is very simple and can be done with three polarizing filters and natural light.

Recall that a polarizing filter doesn't actually polarize light. Rather, it filters out the photons that don't have a particular polarity. This is a polarity determined by the orientation of the filter.

There are three filters and three corresponding phases to the experiment. After natural light passes through a first filter, the remaining photons can be described as entirely vertically polarized. This means they are in the state represented as $|V\rangle$, or equivalently $|0\rangle$ (Fig. 1.13).

[5] Dirac was described as reticent to the point that scientific experiment and theory were delayed by his keeping things to himself [19].

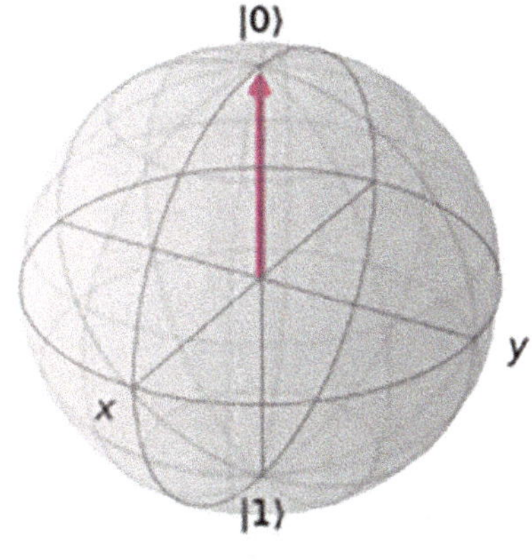

Fig. 1.13 Vertically polarized light on the Bloch sphere

Then a second filter, oriented at a $\pi/2$ angle from the first, is placed on top of the first. This would only let horizontally polarized light ($|H\rangle = |1\rangle$) through. Since all horizontally polarized light has already been filtered out by the first filter, no light should get through both filters.

Finally, a third filter is inserted between the first two, at a $\pi/4$ angle. The result is that a fraction of the original light passes through the three filters. What is happening here helps us draw the distinction between what are considered quantum and classical physics.

The application of each filter is not unlike the closing of a circuit. Earlier we looked at a digital circuit that resolved an uncertain Bayesian probability to a Boolean bit value when the circuit became closed. Here we are looking at a quantum circuit that resolves an uncertain Bayesian quantum state to a pure quantum state when a filter is applied.

Consider this. In the SR latch example, why did the circuit resolve itself to a Boolean value, specifically? Why did it not resolve to a decimal value, or something else? To answer this, we have to go deeper than the hand-wavy logic of digital primitives like the NOR gates that made up the circuit. Recall the SR latch from earlier (Fig. 1.14).

The depiction below is designed to give us an intuitive picture of the circuit's intended behavior. We know, however, that it does not give us an intuitive understanding of some behaviors, like the race conditions that occur when the circuit is

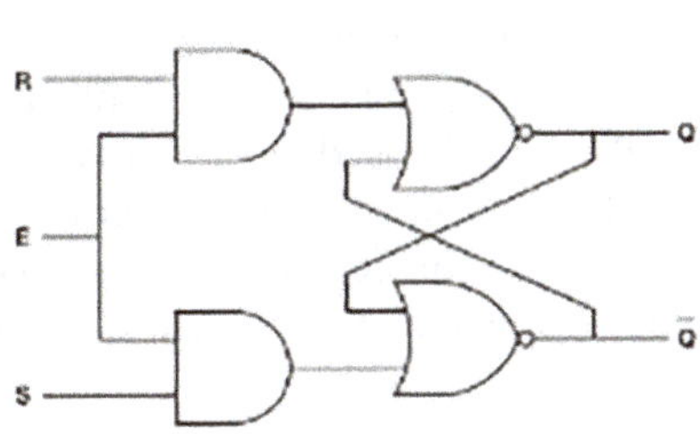

Fig. 1.14 Gated SR latch

Fig. 1.15 TTL NOR

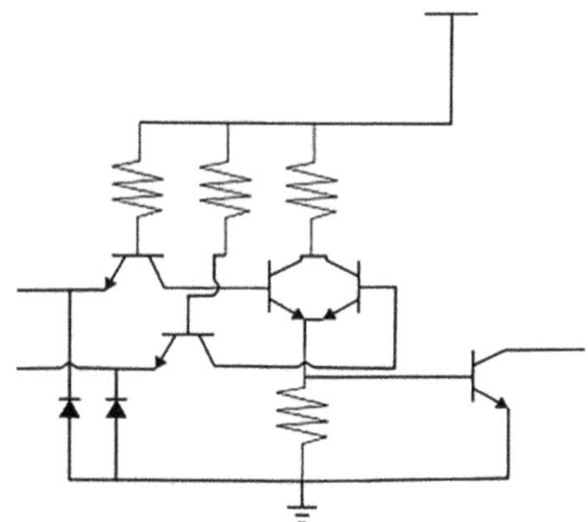

in a metastable state. In order to understand the circuit from the perspective of a physicist or computer engineer, we must realize that the digital circuit is a convenient abstraction for intuitive design that summarizes the behavior of a more complex network of discrete electrical components. When we do this, the circuit moves from the domain of math and CS (Computer Science) to the domain of physics and ECE (Electrical/Computer Engineering). Consider a TTL (Transistor-Transistor-Logic) implementation of a NOR gate (Fig. 1.15).

TTL denotes a family of digital logic circuits that are built using BJTs (Bipolar Junction Transistors), including the 7400 series ICs (Integrated Circuits) that are used in many undergraduate ECE (Electrical and Computer Engineering) labs.

The BJT is the active switching element used in all TTL circuits [10]. A BJT is a two-junction semiconductor with three biased terminals. The three terminals' materials are made either positive- or negative-biased via doping, which involves inducing a permanent extra positive or negative charge on the materials, respectively. The middle terminal is always made to have the opposite bias polarity of the other two. The two BJT types are named "npn" and "pnp" transistors, labeling the collector, base and emitter terminals by their doping polarities, respectively.

The behavior of the well-known BJT is simple. Consider the npn transistor. When a voltage > 0.7 V is found between the base and emitter, the base is considered saturated and current is allowed between the collector and the emitter, treating the transistor like a closed switch between the collector and emitter. A voltage of $< 0.7V \doteq \frac{1}{\sqrt{2}}V$, for example, means current will not be found between the collector and emitter, and the transistor acts like an open switch between the collector and emitter.

We can also take a closer look at the discrete electronics that make up a quantum emulator [21] based on similarly well-known electrical components that the reader should be familiar with in order to gain an appreciation of how quantum mechanics works from an ECE perspective. Indeed, relatively simple circuits can be used to emulate quantum mechanical systems such as Bose-Einstein condensates, imaginary

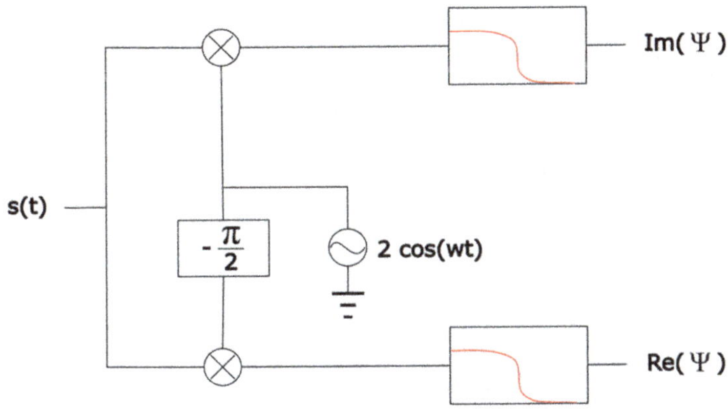

Fig. 1.16 Demodulation state extraction

Schrödinger potentials, or Gross–Pitaevskii ground states, and this is an active area of research [22].

The emulation approach we will briefly consider (called quadrature modulated tonals) uses a sequence of modulating signals to represent a single n-qubit state in terms of all 2^n complex amplitudes. The quantum state is represented by a complex baseband signal. This baseband signal modulates a carrier wave with frequency ω_c by a tonal of frequency ω. The real and imaginary parts of the qubit state are represented in the quadrature and in-phase components of the representative signal $s(t)$ (Fig. 1.16).

$$s(t) = \psi_R(t)\cos(\omega_c t) - \psi_I(t)\sin(\omega_c t)$$

The state can be extracted from $s(t)$ by signal demodulation. Note that the circuit makes use of four-quadrant multipliers, a phase shifter and two low-pass filters.

An inner product between two states is defined as the time average over a time of $T \in \frac{2\pi}{\omega} N$ of a complex product. The complex product is obtained by another circuit (Fig. 1.17).

$$\langle \phi | \psi \rangle = \frac{1}{T} \int_0^T \phi(t)^* \psi(t) \mathrm{d}t$$

This gives us what we need to construct a Hilbert space for our qubit (something you will learn about in the next chapter!) Defining a Hilbert space is essentially what lets us do quantum computing. This is because the probabilities we have been talking about which are the squares of our complex coefficients look like this according to quantum mechanics.

$$|\psi(x,t)|^2 = \psi^*(x,t)\psi(x,t) = P(x,t)$$

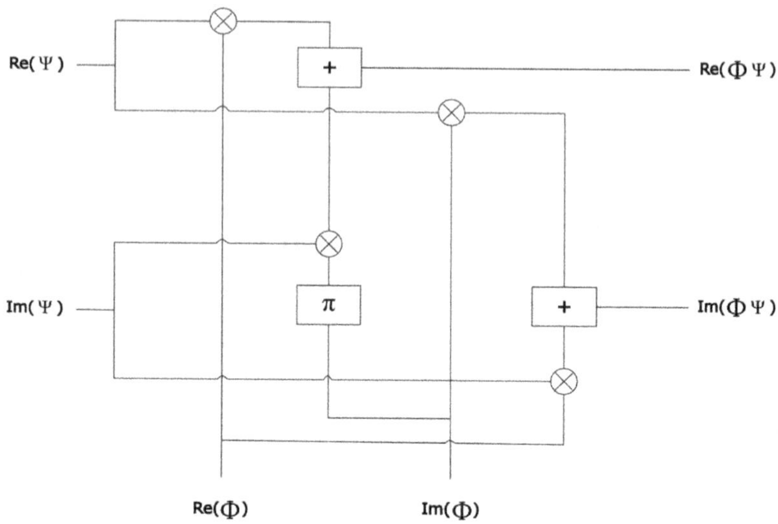

Fig. 1.17 Complex product circuit

Therefore, we can understand the determination of the probability of finding a quantum state at a particular place and time as the outcome of a calculation involving this kind of product. Using the time-average approach, we can extract the components of the state. From there we can calculate the probabilities. Unfortunately, this doesn't scale very well. Since we have a frequency-based encoding, more qubits are accommodated by increasing the frequency resolution. The frequency resolution can be increased by increasing the time T. Our emulator doesn't look very good in this department. A T of 1 year can only help us to resolve 60 qubits, and it would take an amount of time on the order of the age of the universe to represent 95 qubits.

See the difference between this and the NOR gate SR latch? The outcome of the metastable race condition of the SR latch was given qualitatively because we did not have enough information to assign specific probabilities to each outcome. Whereas we cited finding one gate in a proverbial dumpster as an imprecise contributing factor to our probability estimates in the case of the SR latch, here in the case of the emulator circuit we give our probabilities in terms of the specific, calculable electrical signals in our circuit.

Since NOR gates are made of specific discrete components, we could also apply a more rigorous quantitative approach to the SR latch, if we first made a domain transition from math/CS to physics/ECE. This would mean changing the level of description we use to describe our circuit from that of the convenient abstraction of digital logic, to that of discrete electronics. If we made this transition, we could get to work characterizing each of the five BJT transistors that make up the TTL NOR gate and then find that the race condition outcomes which could seem random are actually explainable.

We need to follow similar steps to make Dirac's three-filter experiment explainable. It's just like popping the hood on your car and getting your hands dirty. That's what physics is all about!

So, at the classical level of description, the mystery seems to be that we have vertically polarized light somehow passing through a horizontal filter if it first encounters a $\pi/4$ filter. At this level of description, we have difficulty explaining the system. So, we need to transition to the quantum level of description. Only, we do not have a nice quantum state fully accessible to us in the form of amplitude-modulated tonals which we can understand. We don't have access to the underlying physical system. We only know that it works *like* our emulation. That, and it doesn't take us 13.77 billion years to accommodate 95 qubits, somehow nature is more efficient (something for us to exploit!)

The first fundamental phenomenon characteristic of quantum physics is superposition. A "superposition" is just a fancy term for a linear combination, a sum of vectors. We have already discussed it, actually. We stated earlier in this section that any vector on the Bloch sphere can be described additively in terms of others. The change to this statement that occurs when we switch to the quantum domain is that we realize any vector on the Bloch sphere is a linear combination of others. We have switched our perspective from that of one describing to explaining. Rather than saying how a vector can be described, we are saying what it is. We have switched from math to physics, from CS to ECE.

When we say that something *is*, it is called assigning ontology to that thing. Quantum physicists debate back and forth on what is ontological about, or what makes up, a quantum state. Quantum mechanics is a statistical theory, as we've already seen and will discuss in greater depth throughout the book. A statistical theory typically involves probabilities which are defined with respect to large numbers of objects. We retain some of the standard statistical terminology when referring to "expectations", for example, in quantum mechanics. However, it is the exception rather than the rule in quantum physics that we deal with more than one quantum object at a time. More often than not, we will find ourselves working with a single qubit defined in the spin state of a single isolated particle somewhere in a trap. The only theory that predicts this object's behavior however still looks statistical. So, what are the ontological objects and what is extrapolative mathematics? This has been a debate for some time. Quantum systems have been described in terms of ensembles as far back as the 1930s [23], and ensemble interpretations have existed since Max Born's statistical interpretation won him the Nobel Prize in 1954. Ensemble interpretations have been argued by Einstein[6] [25] and continue to be argued for in new papers as recently as 2022 [26]. It seems that our statistical equations should describe an ensemble of objects that differ by some hidden variable. Indeed, hidden variable theories have been proposed. However, the 2022 Nobel Prize in physics was awarded to Zelinger, Aspect, and Clauser for experimentally carrying out experiments which ruled these out conclusively called "loophole-free" Bell tests. In a nutshell, there is no hidden

[6] It is well known that Einstein was critical of the prevailing interpretation of quantum mechanics, the Copenhagen interpretation [24]. It is still one of the most frequently taught interpretations today.

variable or microstate to make our description of reality more fundamental than the quantum state, even though our state remains statistical in nature. If this problem interests you, there is still work going on in the realm of interpretations of quantum mechanics!

These interpretations of quantum mechanics define quantum physicists' perspectives, and really all of the conclusions that follow from their domain of science. It is very important to recognize that their results can be applied in other domains (via in-betweens like the effective theories of quantum chemistry [27] and quantum-classical correspondence [28, 29]), but only after a proper domain transition has occurred. As we have already seen, a domain transition changes the proper wording of scientific statements in meaningful ways. It has been conclusively proven that quantum physics is not a fundamental theory of the universe on its own [30], and virtually all modern Theories of Everything (ToEs) incorporate domain transitions between locally consistent theories like quantum physics, classical electrodynamics and general relativity [31–36]. An example of why we need different theories is because scales decouple so that the large-scale physics is not dependent on all the details of our small-scale models [37]. The quantum domain is the smallest of them all.

In summary, when we pop the hood and switch to the quantum domain, we find the following mathematical identity is basically physically true (under normalization):

$$|V\rangle = |D\rangle - |A\rangle = \frac{1}{\sqrt{2}}(|H\rangle + |V\rangle) - \frac{1}{\sqrt{2}}(|H\rangle - |V\rangle).$$

So, our vertical light seems to "split" into its diagonal and anti-diagonal components. This is why it is frequently said that quantum particles can be "in many places at once". Half of these diagonal photons can pass through the final filter.

This is supposed to blow your mind! However, it's similar to saying the decoupled SR latch is in many states at once when it is described probabilistically. Just like the SR latch resolved to a single Boolean value as soon as it was coupled to external signals, the qubit will resolve to a single vector as soon as it is coupled to an external system.

Here's what is happening here. Datasheets for SR latches say the metastable state of the SR latch is an invalid state and the combination of two high inputs is an invalid input because in the domain of Boolean algebra (math) and digital logic (CS), there is no such thing as uncertainty or competitive outcomes like in a race condition. This is why the D latch was invented, to make sure users don't accidentally find themselves in the spooky domain of electrical engineering by accident.

In just the same way, classical physics says that a photon should have a trivial polarization distribution. It should not be hard to guess how a photon's polarization is oriented, in other words. But unfortunately, as we know, there are metastable states for photons too!

When we insert the $\pi/4$ filter between the other two, we put each photon into a metastable state. Classical physics tells us this is an illegal state and is of no use to us, just like SR latch datasheets. Quantum physics tells us that when a photon gets into a metastable state, we can only describe it using Bayesian statistics [38, 39], until it becomes coupled to a measurement device at which point it is resolved to a single value. This is just like what we did when we formulated our first pbit.

What many quantum physicists will talk about is an "instantaneous collapse" of the state. However, there is growing evidence that the collapse is not instantaneous. Quantum Darwinism is a theory that quantum information is actually fundamentally a description of competitive processes [40] and has recently passed experimental validation [41]. People are also beginning to realize that quantum states do not collapse instantaneously [42, 43].

In the case of electronics, measurement devices we like to use include voltmeters, ammeters and LEDs. In quantum optics, we use filters, photo-diodes and the like. Really, anything that constrains a state to a value that we can use to record an observation counts.

What people mean when they make a distinction between a classical photonic apparatus and a quantum photonic apparatus is the same as the distinction between a room-temperature electronic circuit and a superconducting one cooled to 1.5 K. The results of particles' path choices that can only be seen on aggregate in the first regime can be seen on an individual basis in the second.

In this sense, the three-filter experiment is not a quantum photonic system. The experiment is enough to prove that the results on aggregate are due to an uncertain competitive process, just like the SR latch was enough to demonstrate that the final result was due to an uncertain competitive process. It is however not sufficient to demonstrate that the same competitive process applies to each individual particle, and not just on aggregate.

> Some time before the discovery of quantum mechanics people realized that the connexion between light waves and photons must be of a statistical character. What they did not clearly realize, however, was that the wave function gives information about the probability of one photon being in a particular place and not the probable number of photons in that place.—Paul Dirac, The Principles of Quantum Mechanics [20]

In general, a diagonal $\pi/4$ filter forces each incident photon state to resolve to a diagonally polarized state, if it can. If an incident photon is in the diagonal state, it passes with 100% likelihood. If an incident photon is in an orientation that is not equivalent to any linear combination involving the $|D\rangle$ state, it cannot pass through the diagonal filter. If it is in a state like $|V\rangle$ that is equivalent to a sum of states involving $|D\rangle$, then the weights of the terms in this sum determine the probability of that photon's passing.

In the case of Dirac's experiment, only vertically polarized light hits the diagonal filter. Vertically polarized light is equal to a uniform sum of $|D\rangle$ and $|A\rangle$. So, the probabilistically normalized sum is the following:

$$|V\rangle = \frac{1}{\sqrt{2}}|D\rangle - \frac{1}{\sqrt{2}}|A\rangle$$

$$|\alpha| = |\beta| = \frac{1}{\sqrt{2}}$$

$$|\alpha|^2 = |\beta|^2 = \frac{1}{2}.$$

This means the probability that each photon will resolve to the $|D\rangle$ state is exactly 1/2 . This is why exactly 1/2 the photons that hit the diagonal filter pass through.

The same phenomenon occurs when the $|D\rangle$ states hit the final horizontal filter, since $|D\rangle$ is a uniform linear superposition of $|V\rangle$ and $|H\rangle$. So, the final filter cuts the total number of passing photons in half again. That concludes Dirac's three-filter experiment.

Remove the diagonal filter in the middle, and you have vertical light hitting the horizontal filter. Vertical light is not equal to any superposition involving horizontal light, so there can be no compatible resolution.

To experimentally prove that these competitive processes apply to each individual particle, and not just on aggregate, we can refer to the famous double-slit experiment that most student's learn in high school physics. The setup is that one photon at a time is shot toward a barrier with two slits: the two paths that the photon may follow to get to the other side. They can be labeled 0 and 1. The place that each photon hits the wall on the far side of the barrier is observed. The experiment shows concretely that while each individual photon is only ever observed to hit the wall in one place, the distribution of locations that the photons hit shows evidence of an interference effect. Since the individual photons are sent through the gauntlet one at a time, they can't be interfering with one another. So, the conclusion is that each photon must create interference between the two paths on its own. This means that the photon resolves a single spatial result directly from the set of its possible paths. No single path is chosen until the photon is measured by incidence with the wall. Sometimes one path will ultimately win, sometimes the other will, but the distribution shows that the mechanism of resolution depends on the set of possible paths right until the end, when the photon is coupled with the measurement device that is the wall.

What is superposition?

Until the last moment, the best description of the particle's path is a *superposition*, a sum of complex weighted basis states and an expression of uncertainty. That final position of the particle depends on the superposition as a whole and not on a single trajectory.

This can be slightly confusing. You might think "wait, why is the wall a measurement device, while the slits are not? Don't they both constrain the photon's state to a value we could record?" The answer is yes! They are all arguably measurement devices! However, the slits are not as precise or constraining as the wall. The two slits constrain the overall photon state Q to a linear combination of two equally likely states.

$$Q = \frac{1}{\sqrt{2}} \cdot \mathbf{0} + \frac{1}{\sqrt{2}} \cdot \mathbf{1}$$

This state is not fully resolved! There is still a probability distribution in the state. However, it is partially resolved. The slits did constrain the state somewhat. This is because it is possible to couple a quantum particle to another system that is less constraining than a linear filter or a wall. It is also possible to couple a quantum system like a qubit to another qubit. There is an entire range in between the extremes.

The most powerful quantum systems use networks of many qubits that are coupled to one another, before finally coupling the entire system to a digital measurement system when it is time for recording outputs, or observations. The most "quantum" thing that can happen when you interact multiple qubits together is called "quantum entanglement". It can bend one's perception of reality, but since you are comfortable with accepting the nuanced ontology of superpositions, maybe you will also be prepared to accept that a multi-qubit state can be in a configuration such that it cannot be expressed as a product of two separable single-qubit states anymore. This phenomenon gives rise to strange physics such as "spooky action at a distance" wherein non-local correlations can be observed between entangled particles—no matter the distance that separates them.

What is entanglement?
An entangled state is a state of two or more qubits that cannot be expressed as a product of individual separable single-qubit states anymore.

1.7.2 Bell States

Some of the most talked-about quantum entangled states are the Bell states. See how in each one, it is impossible to multiplicatively factor out a $|0\rangle$ or a $|1\rangle$. These special superpositions can be created using a $\pi/2$ rotation of the state and an operation that interacts two qubits called a controlled-NOT (CNOT) gate. The CNOT does exactly what it says, it flips the state of the second qubit if the first qubit's state is a $|1\rangle$, without collapsing either state. We will look closer at the Bell states and how to create them in the next chapter!

$$|\Phi^+\rangle = \frac{|0\rangle\,|0\rangle + |1\rangle\,|1\rangle}{\sqrt{2}}$$
$$|\Phi^-\rangle = \frac{|0\rangle\,|0\rangle - |1\rangle\,|1\rangle}{\sqrt{2}}$$
$$|\Psi^+\rangle = \frac{|0\rangle\,|1\rangle + |1\rangle\,|0\rangle}{\sqrt{2}}$$
$$|\Psi^+\rangle = \frac{|0\rangle\,|1\rangle - |1\rangle\,|0\rangle}{\sqrt{2}}$$

> The best possible knowledge of a whole does *not* include the best possible knowledge of its parts – and this is what keeps coming back to haunt us.—Erwin Schrödinger, 1935 [44]

Some of the oldest and most important measures of the "quantumness" of a system are Bell's inequalities. A violation of these inequalities indicates that a system violates the principle of locality, meaning that there is room for non-local quantum effects or "spooky action at a distance", as Einstein called it. Bell originally proved that no hidden-variable theory is able to restore the principle of locality for quantum mechanical systems by introducing these inequalities in 1964 [45].

Some of the first equations you will come across in an introductory quantum computing course are Bell's Inequalities. These inequalities are usually introduced early on, in order to help students establish that quantum entanglement is really unintuitive.

At the time Bell published the first Bell Inequality [45], there was a debate in the physics community around whether the violations of *causality* and *locality* that had been observed in some quantum systems could be resolved by introducing "hidden variables".

The idea was that since quantum particles were more complex than we could capture with any single use of a measurement device, perhaps the results we observed could be explained by some more complex "hidden" structure that we couldn't measure directly.

A lot of philosophical implications hung on this debate. If there were hidden variables, it seemed that quantum mechanics could be considered deterministic. If not, it seemed that measurement outcomes were truly random. Physicists didn't know what the hidden structure in a quantum particle could be, so they left it as generic as possible.

Implications for foundations of physics.
What Bell achieved in his 1964 publication was proving that no hidden variable theory could possibly exist that would restore the principle of locality for entangled spin-$\frac{1}{2}$ particles.

The setup for hidden variables was the following. Here σ_1, σ_2 give the orthonormal components of two qubits in the singlet spin state ψ, and a, b are unit vectors.

$$|\psi\rangle = \frac{1}{\sqrt{2}}\left(\left|\frac{1}{2}, -\frac{1}{2}\right\rangle - \left|-\frac{1}{2}, \frac{1}{2}\right\rangle\right) \tag{1.1}$$

In the absence of a hidden variable theory, we would expect the measurement of $\sigma \cdot a$ to be dependent on the initial quantum state a.

$$A(a) = \sigma \cdot a \tag{1.2}$$

Introducing hidden variables is very straightforward. Bell represented all the necessary hidden variables as the set of variables λ. It was not necessary to define $A(a, \lambda)$ directly, only to analyze λ's effect on the expectation value for the product of A and B.

It was known from empirical observations that $< A\ B > = -a \cdot b$, since whenever $\sigma_1 \cdot a$ was measured to be 1, $\sigma_2 \cdot a$ would be -1.

$$< A\ B >= \sum_{A,B} p(A, B)\ A(a)\ B(b) = -a \cdot b \tag{1.3}$$

However, with the introduction of variables λ that *determine* both A and B, the expectation value becomes:

$$P(a, b) = \int \mathrm{d}\lambda\ p(\lambda)\ A(a, \lambda)\ B(b, \lambda). \tag{1.4}$$

Bell proved that $P(a, b)$ cannot be equal to $-a \cdot b$ by deriving the first Bell inequality.

$< A\ B >$ are normalized probability distributions, meaning that all the outcome probabilities add up to 1. Since outcomes A, B are determined by λ, we can say:

$$\int \mathrm{d}\lambda\ p(\lambda) = 1. \tag{1.5}$$

It should be possible for $< A\ B > = -a \cdot b$ to equal -1, since we know that A must equal $-B$ when one of the two is observed to be equal to 1. This means we can re-state $< A\ B >$.

$$\begin{aligned} < A\ B > &= \int \mathrm{d}\lambda\ p(\lambda)\ A(a, \lambda)\ B(b, \lambda) \\ &= -\int \mathrm{d}\lambda\ p(\lambda)\ A(a, \lambda)\ A(b, \lambda) \end{aligned} \tag{1.6}$$

Bell introduced a third unit vector c in order to obtain the difference $P(a,b) - P(a,c)$:

$$\begin{aligned} P(a,b) - P(a,c) &= -\int d\lambda\ p(\lambda)\ [A(a,\lambda)A(b,\lambda) - A(a,\lambda)A(c,\lambda)] \\ &= \int d\lambda\ p(\lambda)\ A(a,\lambda)A(b,\lambda)[A(b,\lambda)A(c,\lambda) - 1]. \end{aligned} \tag{1.7}$$

Since $1> = \int d\lambda\ p(\lambda)\ A(a,\lambda)A(b,\lambda)> = -1$,

$$\begin{aligned} |P(a,b) - P(a,c)| &\le \int d\lambda\ p(\lambda)\ [1 - A(b,\lambda)A(c,\lambda)] \\ &= 1 - \int d\lambda\ p(\lambda)\ A(b,\lambda)A(c,\lambda) \\ &= 1 + P(b,c). \end{aligned} \tag{1.8}$$

This gives the first Bell Inequality:

$$1 + P(b,c) \ge |P(a,b) - P(a,c)| \tag{1.9}$$

Bell then introduced ϵ to represent the difference between the two $P(a,b)$ and $-a \cdot b$.

$$|P(a,b) + a \cdot b| \le \epsilon \tag{1.10}$$

He then showed that this ϵ could not be made arbitrarily small. Hopefully this gives you an idea of how these inequalities work.

References

1. Feynman, R.P.: "Surely You're Joking, Mr. Feynman!": Adventures of a Curious Character. W. W. Norton & Company (2010). Google-Books-ID: 7papZR4oVssC
2. Hawking, S.: The Dreams that Stuff is Made of: The Most Astounding Papers on Quantum Physics and How They Shook the Scientific World. Running Press (2011)
3. Zeng, W., Johnson, B., Smith, R., Rubin, N., Reagor, M., Ryan, C., Rigetti, C.: First quantum computers need smart software. Nature **549**(7671), 149–151 (2017). https://doi.org/10.1038/549149a. https://www.nature.com/articles/549149a
4. Nunez-Corrales, S., Matteo, O.D., Dumbell, J., Edwards, M., Giusto, E., Pakin, S., Stirbu, V.: Productive Quantum Programming Needs Better Abstract Machines (2025). https://arxiv.org/abs/2505.00718
5. Wang, Q., Ying, M.: Quantum random access stored-program machines. J. Comput. Syst. Sci. **131**, 13–63 (2023)
6. Zhang, Z., Ying, M.: Quantum Register Machine: Efficient Implementation of Quantum Recursive Programs (2024). https://arxiv.org/abs/2408.10054
7. Yuan, C., Villanyi, A., Carbin, M.: Quantum control machine: The limits of control flow in quantum programming. Proc. ACM Program. Lang. **8**(OOPSLA1), 1–28 (2024). https://doi.org/10.1145/3649811

8. Valiron, B.: On Quantum Programming Languages. arXiv preprint arXiv:2410.13337 (2024)
9. Higashi, P.: A thin-film rod memory for the NCR 315 RMC computer. IEEE Trans. Electron. Comput. **EC-15**(4), 459–467 (1966). https://doi.org/10.1109/PGEC.1966.264353. https://ieeexplore.ieee.org/document/4038819
10. Floyd, T.L.: Digital Fundamentals, 11th edn. Pearson Education India (2014)
11. Felix Bloch: Facts (2025). https://www.nobelprize.org/prizes/physics/1952/bloch/facts/
12. Mermin, N.D.: Quantum Computer Science: An Introduction. Cambridge University Press (2007). Google-Books-ID: q2S9APxFdUQC
13. Amy, M.: Towards Large-scale Functional Verification of Universal Quantum Circuits. Electron. Proc. Theor. Comput. Sci. **287**, 1–21 (2019). https://doi.org/10.4204/EPTCS.287.1
14. Nielsen, M.A., Knill, E., Laflamme, R.: Complete quantum teleportation using nuclear magnetic resonance. Nature **396**(6706), 52–55 (1998). https://doi.org/10.1038/23891. https://www.nature.com/articles/23891
15. Laflamme, R., Wilson, C.: Quantum Information Processing Devices At University of Waterloo (2019). Graduate course
16. Gray, J.J.: Henri Poincaré | French Mathematician & Physicist | Britannica. Encyclopædia Britannica (2025). https://www.britannica.com/biography/Henri-Poincare
17. plot_bloch_vector. https://docs.quantum.ibm.com/api/qiskit/qiskit.visualization.plot_bloch_vector
18. py_pol package — Python polarization 1.1.3 documentation. https://py-pol.readthedocs.io/en/master/py_pol.html#module-py_pol.drawings
19. Close, F.: Paul Dirac: a physicist of few words. Nature **459**(7245), 326–327 (2009). https://doi.org/10.1038/459326a. https://www.nature.com/articles/459326a
20. Dirac, P.: Principles of Quantum Mechanics. Clarendon Press (1930)
21. Cour, B.R.L., Ott, G.E.: Signal-based classical emulation of a universal quantum computer. New J. Phys. **17**(5), 53017 (2015)
22. Pomorski, K., Pluszyński, Hałubek, E.: Towards the Construction of an Analog Solver for the Schrödinger and Ginzburg–Landau Equations Based on a Transmission Line. Condens. Matter **9**, 35 (2024). https://doi.org/10.3390/condmat9040035
23. Einstein, A.: Physics and reality. J. Franklin Inst. **221**(3), 349–382 (1936). https://doi.org/10.1016/S0016-0032(36)91047-5. https://www.sciencedirect.com/science/article/pii/S0016003236910475
24. Ballentine, L.E.: Einstein's Interpretation of Quantum Mechanics. American Journal of Physics **40**(12), 1763–1771 (1972). https://doi.org/10.1119/1.1987060
25. Schilpp, P.A. (ed.): Albert Einstein: philosopher-scientist, 3rd edn. 6. pr edn. No. 7 in The library of living philosophers. Open Court, La Salle, Ill (1995)
26. Hance, J.R., Hossenfelder, S.: The wave function as a true ensemble. Proc. R. Soc. A: Math. Phys. Eng. Sci. **478**(2262) (2022). https://doi.org/10.1098/rspa.2021.0705
27. Szabo, A., Ostlund, N.S.: Modern Quantum Chemistry: Introduction to Advanced Electronic Structure Theory. Macmillan, New York (1982)
28. Bolivar, A.O.: Quantum-Classical Correspondence: Dynamical Quantization and the Classical Limit. Springer Science & Business Media (2004). Google-Books-ID: xTafy9uEZvgC
29. Olavo, L.S.F.: Foundations of quantum mechanics (II): equilibrium, Bohr–Sommerfeld rules and duality. Physica A: Statist. Mechan. Appl. **271**(3), 260–302 (1999). https://doi.org/10.1016/S0378-4371(99)00216-2. https://www.sciencedirect.com/science/article/pii/S0378437199002162
30. Einstein, A., Podolsky, B., Rosen, N.: Can quantum-mechanical description of physical reality be considered complete? Phys. Rev. **47**(10), 777–780 (1935). https://doi.org/10.1103/PhysRev.47.777. https://link.aps.org/doi/10.1103/PhysRev.47.777
31. Hugh Everett: The Theory of the Universal Wave Function. Princeton University (1956). http://archive.org/details/TheTheoryOfTheUniversalWaveFunction
32. Wilczek, F.: A Beautiful Question: Finding Nature's Deep Design. Penguin Books Limited (2015). Google-Books-ID: Oh3ICAAAQBAJ

33. Fuchs, C.A.: QBism, the Perimeter of Quantum Bayesianism (2010). https://doi.org/10.48550/arXiv.1003.5209. http://arxiv.org/abs/1003.5209. ArXiv:1003.5209 [quant-ph]
34. The New Yorker: A Different Kind of Theory of Everything. https://www.newyorker.com/science/elements/a-different-kind-of-theory-of-everything
35. Wolfram, S.: A Class of Models with the Potential to Represent Fundamental Physics. Complex Syst. **29**(2), 107–536 (2020). https://doi.org/10.25088/ComplexSystems.29.2.107. http://arxiv.org/abs/2004.08210. ArXiv:2004.08210 [gr-qc, physics:hep-th, physics:math-ph]
36. Hossenfelder, S.: Lost in Math: How Beauty Leads Physics Astray, First, edition Basic Books, New York (2018)
37. Hossenfelder, S.: Existential Physics: A Scientist's Guide to Life's Biggest Questions. Viking, New York (2022)
38. Fuchs, C.A.: Qbism, the perimeter of quantum bayesianism (2010). https://arxiv.org/abs/1003.5209
39. von Baeyer, H.C.: QBism: The Future of Quantum Physics. Harvard University Press (2016). http://www.jstor.org/stable/j.ctv24w633w
40. Zurek, W.H.: Quantum darwinism. Nat. Phys. **5**(3), 181–188 (2009). https://doi.org/10.1038/nphys1202. https://www.nature.com/articles/nphys1202
41. Ball, P.: Quantum Darwinism, An Idea to Explain Objective Reality, Passes First Tests (2019). https://www.quantamagazine.org/quantum-darwinism-an-idea-to-explain-objective-reality-passes-first-tests-20190722/
42. Ball, P.: Quantum Leaps, Long Assumed to Be Instantaneous, Take Time (2019). https://www.quantamagazine.org/quantum-leaps-long-assumed-to-be-instantaneous-take-time-20190605/
43. Minev, Z., Mundhada, S., Shankar, S., Reinhold, P., Gutiérrez-Jáuregui, R., Schoelkopf, R., Mirrahimi, M., Carmichael, H., Devoret, M.: To catch and reverse a quantum jump mid-flight. Nature **570**(7760), 200–204 (2019). https://doi.org/10.1038/s41586-019-1287-z
44. Trimmer, J.: A Translation of Schroedinger's "Cat Paradox" Paper. https://unicamp.br/~chibeni/textosdidaticos/schrodinger-1935-cat.pdf
45. Bell, J.S.: On the einstein podolsky rosen paradox. Physics Physique Fizika **1**, 195–200 (1964). https://doi.org/10.1103/PhysicsPhysiqueFizika.1.195

Chapter 2
Essential Quantum Mechanics and How to Code It

Abstract In this chapter, we cover the bases when it comes to quantum mechanics and maintain the perspective of an engineer by including code snippets that should allow one to build their own Python scripts. We build up to and describe the basics of quantum computing in a way that shows how it follows from the quantum mechanics of particles. To be more specific, we supplement selected topics covered in upper-level quantum mechanics and quantum computing courses (such as [1–4] at the author's alma-maters) with more detailed background information as well as symbolic code meant to make this highly theoretical topic more practical for the reader who is more familiar with engineering systems that modeling the physics of them. The hope is that you follow along with the physics and the code in order to grasp the concepts in the language most accessible to you at an introductory level. Then, we refer the reader to [5] for the full quantum mechanical derivations which we sample here, and to texts such as [6, 7] for a deeper dive into the software side of things.

2.1 Learning Objectives

In this chapter we will aim to understand the essential theory behind quantum computing, which will lay a groundwork for future chapters. These essentials include (Fig. 2.1):

- Understand wavefunctions and how to work with them.
- Understand conjugate variables beyond position and momentum.
- Understand how to apply commutators.
- Understand the measurement postulate of quantum mechanics.
- See how uncertainty relations arise and know how to use them.
- Understand how to discretize quantum information with linear algebra.
- See how quantum information arises from quantum physics.
- Know a handful of quantum algorithms and how they work.

We will make this chapter more practical with an applied example of how these concepts in action yield something useful in quantum key distribution protocols.

© The Author(s), under exclusive license to Springer Nature Switzerland AG 2026

M. S. Edwards, *An Introduction to Quantum Computing for Computer Engineers*,
https://doi.org/10.1007/978-3-032-03650-6_2

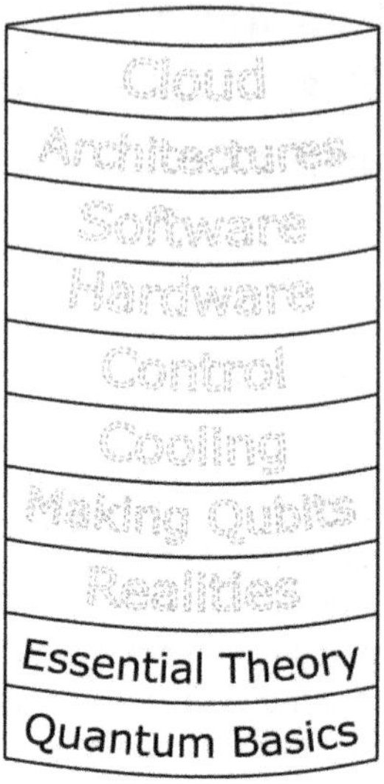

Fig. 2.1 Learning objective for chapter: essential theory

2.2 Background

Modern physics is founded on the understanding of particles like photons, electrons, neutrons, protons, etc., and their properties. One of the most important properties of matter in our case is the particle-wave duality. From classical physics we know the basic relationships.

$$\text{Waves} : c = f\lambda$$
$$\text{Particles} : p = mv$$
$$\text{Photons} : E = hf$$
$$\text{Waves} : p = \frac{h}{\lambda}$$

Here we have that the frequency f and wavelength λ of light multiplied must equal the (constant) speed of light 3×10^8 m/s. We also see that the energy of a photon is Planck's constant h times its frequency f. For particles, momentum p is given by the particle's mass m times its velocity v. For waves, momentum p is given by Planck's constant divided by the wavelength.

De Broglie,[1] who was a member of the French military through World War 1 and for a large part of his career as a physicist before being inducted into the Académie des Sciences in 1933, used Einstein's relativity equations to unify matter and energy [9].

[1] De Broglie once worked during his military service in a radio station in the Eiffel Tower [8].

$$E^2 = p^2c^2 + (mc^2)^2$$
$$m = 0 \quad \text{for a photon}$$
$$E = pc \rightarrow p = \frac{E}{c}$$
$$p = \frac{h}{\lambda}$$

Bohr then explained that since electrons' orbits don't decay, they don't follow the laws of classical physics. Instead, their energies are quantized, i.e. for quantum number $n \in \mathbb{Z}$ (the energy level), $E_n = \frac{-13.6\ \text{eV}}{n^2}$. Sommerfeld added to this by accounting for non-spherical orbiting bodies and adding two more variables. The first is the angular momentum quantum number $l < n, \in \mathbb{Z}$ which gives the orbital angular momentum L of the particle by $L^2 = \hbar^2 l(l+1)$. The second is the magnetic quantum number $m < l, \in [-l \ldots l]$ which gives the orbital angular momentum along a particular axis by $L_z = m_l \hbar$. The allowed energy states became combinations of these variables: l, m and n. Pauli took this further to say that for every value of l, m and n there is a fourth quantum number s, the spin. The spin is the particle's *intrinsic* angular momentum and works similarly to angular momentum in ways that we will see shortly. These quantum numbers identify the particles that orbit the nuclei in atoms of Hydrogen, Helium, Lithium and so on adding one electron at a time [10] (Table 2.1).

Before we dive into the purely quantum mechanical description of matter via the wavefunction let's stop to think about how what we've described thus far behaves. We can use a specific type of python package, called symbolic computing libraries, to explore these relationships. Sympy in particular comes equipped with a quantum mechanics package. We use Sympy to define symbols such as p for momentum and m for mass. These symbols allow us to create equations without immediately evaluating them. This will allow us to manipulate equations algebraically in the future.

Let's calculate the momentum of a particle and use it to calculate the wavelength in the wave-picture of the same matter. We'll make use of the *lambdify* function to turn our symbolic expressions into callable functions. The first parameter to *lambdify* specifies which symbols we want to substitute values for. The second is the symbolic expression in question and for the third we specify to use the Numpy backend because it's fast.

Table 2.1 Quantum numbers of electrons in atoms

Element	Electrons	n	l	m	s
H	1	1	0	0	$\frac{1}{2}$
He	2	1	0	0	$\pm\frac{1}{2}$
Li	3	1	0	0	$\pm\frac{1}{2}$
		2	0	0	$\frac{1}{2}$
Be	4	1	0	0	$\pm\frac{1}{2}$
		2	0	0	$\pm\frac{1}{2}$

Example: Momentum and Wavelength with Sympy

```
from sympy import lambdify, symbols, Integer, Float
from sympy.physics.quantum.constants import hbar
from math import pi

m, v, p = symbols('m v p')

momentum = lambdify([m, v], m * v, 'numpy')
calculated_p = momentum(10, 25)
print(calculated_p)

wavelength = lambdify(p, hbar * Integer(2) * Float(pi) / p, 'numpy')
calculated_w = wavelength(calculated_p)
print(calculated_w)
```

We get the following output.

```
250
2.6504275632610314e-36
```

2.3 Position and Momentum: Conjugate Variables

In the last section we very quickly discussed the properties of particles and waves. We wrote a few lines of python that allow us to calculate things like the momentum, velocity and mass of a particle or the wavelength of a wave. Now we will discuss a more broad framework for describing the properties of objects such as particles and waves.

In the wavefunction framework, we are concerned with how to describe objects' "observable" properties. This turns out to be a question of how we measure those properties. We also consider properties that evolve in time.

Schrödinger came up with the wavefunction representation when he was on a skiing trip, as the story goes.[2] It is heavily inspired by classical mechanics, but different enough that it produced quantum mechanics. His intuition gave us the following idea. Every object can be represented by a wavefunction ψ that contains all the information about the object's observable properties. For example for a particle moving in time and space we would have $\psi(x, t)$. We can also have a wavefunction of multiple objects $\psi(x_1, x_2, \ldots, t)$. The probability of observing an object at a particular position x at time t is given by [11, p. 15]:

$$|\psi(x, t)|^2 = \psi^*(x, t)\psi(x, t) = P(x, t)$$

To capture this kind of equation we are best served by Sympy's quantum package which has a strong concept of wavefunctions, differential operators and virtually everything else we will need in this section. When working with code we sometimes

[2] Supposedly he was in Arosa, Switzerland, with a mistress, over the Christmas of 1925. The author learned this story by word of mouth.

need to be more concrete. Let's suppose we have a specific wavefunction $\exp(i * (k * x - \omega * t))$ with $k = 1$ and $\omega = 3$. Then we can get our expression of the probability the following way.

Example: Wavefunctions and Probability with Sympy

```
from sympy import I, exp, symbols, conjugate
from sympy.physics.quantum import Wavefunction

k = Integer(1)
o = Integer(3)
x, t = symbols('x t', commutative=False)

psi = Wavefunction(exp(I * (k * x − o * t)), x, t)
prob = conjugate(psi) * psi
```

This all seems arbitrary but as it turns out, using the square root of the probability to describe the wavefunction instead of simply using the probability as in classical mechanics is a key differentiator of quantum mechanics and corresponds with empirical observation. This model predicts what we see in nature. Since $P(x, t)$ is a probability, the sum of all our probabilities must equal to 1 [11, p. 12].

$$\int_{-\infty}^{\infty} \psi^*(x,t)\psi(x,t)dx = 1$$

Example: Assert Normalization with Sympy

```
from sympy import Integral, Eq

Eq(1, Integral(conjugate(psi) * psi, (x, −oo, oo)))
```

> **What is a wavefunction?**
> This yields the following assessment: that only square integrable functions are valid wavefunctions. This space of functions is called a Hilbert space, and it is also an inner product space.

We should have been able to notice that the inner product must be defined for our functions since $\psi^*(x,t)\psi(x,t)$ is an inner product of $\psi(x,t)$ with itself [11, p. 95].

If $\int_{-\infty}^{\infty} \psi^*(x,t)\psi(x,t)dx = \infty$ then our ψ is not a valid wavefunction. If it is simply equal to $N \neq 1$ then we can normalize it [11, p. 13].

$$\psi(x,t) = \frac{\psi(x,t)}{\sqrt{N}}$$

Example: Normalization with Sympy

```
from sympy import sqrt

psi / sqrt(Integral(conjugate(psi) * psi, (x, −oo, oo)))
```

These probabilities are descriptions of our best knowledge about an object's properties and are not the actual properties themselves. What else can we say about ψ? We can draw upon more concepts from classical physics to draw more conclusions. For example, starting with the principle of the conservation of energy: kinetic energy (KE) + potential energy (PE) = total energy (E) we can reach Schrödinger's time-dependent equation [11, p. 1].

$$KE\psi + PE\psi = E\psi$$
$$-\frac{\hbar^2}{2m}\frac{\partial^2\psi(x,t)}{\partial x^2} + U(x,t)\psi(x,t) = \frac{i\hbar\partial\psi(x,t)}{\partial t}$$

Example: Schrödinger's Time-Dependent Equation with Sympy

```
from sympy import Function, Symbol, Derivative

U = Function('U')
m = Symbol('m')

Eq(− (hbar ** 2) / (2 * m) * Derivative(psi, x, x) + U(x, t) * psi, I * hbar * Derivative(psi, t))
```

An expression for the energy of a system that is given in terms of *Operators* is called a Hamiltonian. But where did these terms come from? Let's take a look at kinetic energy first. Recall from classical mechanics that $KE = \frac{1}{2}mv^2$. Here *k* is the wave number from classical mechanics.

$$\begin{aligned} KE &= \frac{1}{2}mv^2 \\ &= \frac{p^2}{2m} \\ &= \frac{h^2}{2\lambda^2 m} \\ &= \frac{h^2k^2}{4\pi^2 2m} \\ &= \frac{\hbar k^2}{2m} \end{aligned}$$

For physical intuition regarding these $\frac{\hbar^2\mathbf{k}^2}{2m}$ terms, we return to the origins of quantum mechanics [12].

Boltzmann[3] suggested in 1877 [14] that the energy levels of a physical system like a molecule could be discrete. Boltzmann's theory followed from the statistical mechanics of the day and led to Planck's theory of quantum physics in 1900.

Planck started [15] by abducing from the observed frequency dependence of energy emitted by black bodies that energy must be quantized. This yielded Planck's law:

$$I(c, T) = \frac{2hv^3}{c^2} \frac{1}{\exp(\frac{hv}{kT} - 1)}$$

Here $I(v, T)$ is the energy per unit time radiated per unit area of emitting surface in the normal direction per unit solid angle per unit frequency by a black body at temperature T; h is the Planck constant; c is the speed of light in a vacuum; k is the Boltzmann constant; ν is the frequency of the electromagnetic radiation; and T is the temperature of the body in kelvins.

In the classical theory, blackbody radiation is modeled as the radiation emitted from oscillating charged particles at the object's surface. Classically, one would expect that the energy emitted would only be determined by the amplitude of the oscillator, not the frequency.

$$I(v) = \frac{8\pi k_B T v^2}{c^2}$$

However, it turns out empirically that this is not the case. The source of the radiation is a set of underlying oscillations that are produced by the thermal motions of charged particles. So, if we treat each particle as a simple harmonic oscillator, we can understand how Planck created his explanation of blackbody radiation.

A simple harmonic oscillator is described by Hooke's law.

$$F = -k(x - x_0)$$

The energy of a simple harmonic oscillator is given by the following.

$$E = \frac{p^2}{2m} + \frac{1}{2}k(x - x_0)^2$$

Planck[4] changed this equation by discretizing E such that $E \propto nhv$, where h is a constant known as "Planck's constant" which allows the model to correspond with the empirical observations.

Planck knew that if a collection of identical systems were at the same temperature T and started from different classical initial conditions, the probability that they would have a particular energy E was proportional to the Boltzmann factor.

---

[3] Boltzmann is also attributed with showing that the second law of thermodynamics is statistical [13].

[4] After making his critical contribution in what would become quantum theory, Planck struggled and failed to show that classical theory was instead sufficient [16].

$$\exp\left(\frac{-E}{k_B T}\right)$$

This led him to the following probability that a system at temperature T would yield a particular energy $E = nhv$.

$$\frac{\exp(\frac{-nhv}{k_B T})}{\sum_{n=0}^{\infty} \exp(\frac{-nhv}{k_B T})}.$$

This is a probability distribution; a geometric series. So, it must be $normalizable$, meaning that when we sum over n, the sum of all the probabilities must equal 1. By normalizing this formula one can derive an average or *expected* energy of $\bar{E}$:

$$\bar{E} = \frac{\sum_{n=0}^{\infty} nhv \exp(\frac{-nhv}{k_B T})}{\sum_{n=0}^{\infty} \exp(\frac{-nhv}{k_B T})}$$

$$\bar{E} = \frac{hv}{\exp(\frac{-nhv}{k_B T}) - 1}.$$

By discretizing the simple harmonic oscillator, Planck invented the "quantum harmonic oscillator" which is the basis of much of quantum mechanics.

$$E_n = \hbar\omega(n + \frac{1}{2}) = (2n + 1)\frac{\hbar}{2}\omega$$

The complexions become known as the "quantum state". The nature of a complexion, or quantum state, is described in terms of properties we consider physical, like the momentum and position of a particle. Einstein interpreted Planck's oscillating particles to be photons, and proposed that the energy of a photon is proportional to its frequency. De Broglie had shown [9] that in Einstein's theory of Special Relativity [17], energy and momentum are related in the same way and frequency and wave number. From this, it followed that the momentum of a photon p would be inversely proportional to its wavelength λ or wave number k.

$$E = hf \; for \; photons$$

$$c = f\lambda \; for \; waves$$

$$E^2 = p^2c^2 + (mc^2)^2$$

$$m \to 0 \; for \; photons$$

$$\therefore \; E = pc \to p = \frac{E}{c}$$

$$\therefore \; p = \frac{h}{\lambda} = \hbar k$$

From here we can describe the kinetic energy of a particle.

$$KE = \frac{1}{2}mv^2$$

$$p = mv \rightarrow KE = \frac{p^2}{2m}$$

$$= \frac{h^2}{2\lambda^2 m}$$

$$= \frac{h^2 k^2}{4\pi^2 2m}$$

$$\therefore \; KE = \frac{\hbar k^2}{2m}$$

$$k = \frac{2\pi}{\lambda}, \hbar = \frac{h}{2\pi}$$

Returning from this aside on intuition from the history of physics, see that when we have $\psi(x, t) = \exp(i(kx - \omega t))$ as per the classical wave equation, we get:

$$\frac{\partial^2 \psi(x, t)}{\partial x^2} = -k^2 \psi(x, t)$$

k has become a differential operator. That is, k is no longer just a constant. Rather, it is a non-commutative operator that takes the partial second derivative of its operand with respect to x. We can represent differential operators using the *DifferentialOperator* class provided in Sympy's quantum package. In the following *Derivative's* first parameter we specify a function f of a symbolic variable x that will be replaced by a *Wavefunction* when the partial derivative is evaluated with an operand. The remaining parameters specify which variable to take the derivative with respect to and how many times. The second parameter to the *DifferentialOperator* constructor specifies which function in the *Derivative* parameters to replace with the expected *Wavefunction*.

Example: Differential Operators with Sympy

```
from sympy.physics.quantum import DifferentialOperator

f = Function('f')
k_squared = - DifferentialOperator(Derivative(f(x), x, 2), f(x))
```

Also, with regard to the total energy term,

$$\begin{aligned}\frac{\partial}{\partial t}\psi(x,t) &= -i\omega\psi(x,t)\\ i\hbar\frac{\partial}{\partial t}\psi(x,t) &= \hbar\omega\psi(x,t).\\ &= hf\psi(x,t)\end{aligned}$$

We can use Sympy to help us with simplifying expressions during such a calculation. See how expressions like $i\hbar(-i\omega\psi(x,t))$ are simplified automatically for us. We can also check the equality of the left-hand side and right-hand side of an equation at any step.

Example: Simplifying Expressions with Sympy

```
h = hbar * 2 * pi
Eq(I * hbar * Derivative(psi, t), I * hbar * −I * o * psi)
print(I * hbar * −I * o * psi)
print(Eq(hbar * o * psi, h * o / (2 * pi) * psi))
```

```
3*hbar*Wavefunction(exp(I*(−3*t + x)), x, t)
True
```

From these a momentum operator can also be derived [11, p. 17].

$$\begin{aligned}KE = \frac{1}{2}mv^2 = \frac{p^2}{2m} \rightarrow p^2 &= -\frac{\hbar^2\partial^2}{\partial x^2}\\ p &= -\frac{i\hbar\partial}{\partial x}\end{aligned}$$

Combining all this, we can calculate the expectation of the momentum [11, p. 16].

$$\begin{aligned}\langle p\rangle &= \int_{-\infty}^{\infty} p(x)P(x)dx\\ &= \int_{-\infty}^{\infty} \psi^*(x,t)\frac{-i\hbar\partial\psi(x,t)}{\partial x}dx\end{aligned}$$

The Fourier[5] transform is very useful in quantum mechanics and quantum computing. Similarly to Planck, Fourier invented the Fourier transform when he was studying how heat propagates. Fourier was interested in how heat diffuses through a metal rod. The classic Fourier transform $\int f(t)\exp(-2\pi i\xi t)dt$ is an expansion of a function

[5] Fourier was a part of Napoleon's expedition to Egypt and was made a baron by Napoleon in 1809 [18].

over a basis of complex exponential eigenfunctions. Fourier considered the Laplacian $\Delta f = \nabla^2 f = \frac{\partial^2}{\partial t^2} f$, which will also come up for us as we consider Schrödinger's equation. The Fourier transform solves the differential equation $\Delta f = 0$.

$$\Delta \exp(-2\pi i \xi t) = \frac{\partial^2}{\partial t^2} \exp(-2\pi i \xi t) = -(2\pi \xi)^2 \exp(-2\pi i \xi t)$$

Taking the Fourier transform is the same as decomposing a function into a weighted linear combination of frequencies, where the weight of one eigenfunction is like the intensity of that frequency. The Fourier transform is therefore said to represent the original function in the frequency domain.

Momentum is encoded into the positional wave equation this way, so the wave equation can be expressed in terms of momentum using a Fourier transform [11, p. 108]. Like the classic Fourier transforms between the time and frequency domains, here we transform from position space to momentum space.

$$\xi(p, t) = \frac{1}{\sqrt{2\pi\hbar}} \int_{-\infty}^{\infty} \exp(\frac{i}{\hbar}(xp))\psi(x, t)dx$$

Example: Wave Equation in Terms of Momentum

```
p = Symbol('p') xi = (1 / sqrt(2 * pi * hbar)) * Integral(exp(I * x
* p / hbar) * psi, (x, -oo, oo))
```

$$P(p_a \leq p \leq p_b) = \int_{p_a}^{p_b} \xi^*(p, t)\xi(p, t)dp$$

Example: Probabilities of Momenta with Sympy

```
# change these parameters to calculate different probabilities
lower = -oo
upper = oo

Integral(conjugate(xi) * xi, (p, lower, upper))
```

We can express the expectation of position in terms of momentum in the same way that we expressed the expectation of momentum in terms of position [11, p. 108]. We transform from momentum space to position space.

$$\langle x \rangle = \int_{-\infty}^{\infty} \xi^*(p, t) \frac{i\hbar \partial \xi(p, t)}{\partial p} dp$$

We can also do a Fourier transform to go from momentum representation to position [11, p. 108].

$$\psi(x,t) = \frac{1}{\sqrt{2\pi\hbar}} \int_{-\infty}^{\infty} \exp(\frac{i}{\hbar}(xp))\xi(p,t)dp$$

Not only can we go back and forth between momentum and position using Fourier transforms, but we can also represent any wavefunction $\psi(x)$ as a superposition of normal solutions to the Schrödinger equation $\phi_n(x)$, and solve for the superposition's parameters using Fourier transforms [11, p. 95].

$$\psi(x) = \sum_n c_n\phi_n(x)$$

The Fourier transform is useful for calculating the c_n as follows:

$$c_n = \int_{-\infty}^{\infty} \phi_n^*(x)\psi(x)\,dx.$$

If we can represent any wavefunction in our Hilbert space with a weighted sum (linear combination) of our $\phi_n(x)$, then the set $\{\phi_0(x), \phi_1(x), \ldots\}$ is said to be complete [11, p. 95]. Any $Operator$'s action on any wavefunction can be evaluated if we know how the $Operator$ operates on each function in the basis set. This is called linearity and highlights for us that quantum mechanics is a linear theory.

What are conjugate variables?
Momentum and position are called Fourier transform duals, or conjugate variables. Conjugate variables become useful for conjugate coding. Conjugate coding's major application is quantum key distribution, which is arguably the most practical quantum cryptographic scheme and has already penetrated industry.

x and p follow Heisenberg's[6] uncertainty principle: $\Delta x \Delta p \geq \frac{\hbar}{2}$. It is also important that they do not commute, meaning $[x, p] = xp - px \neq 0$. Instead, $[x, p] = i\hbar$ which is known as the canonical commutation relation [11, p. 43].

[6] Heisenberg played a leading role in Nazi Germany's nuclear research program. The failure of this program is sometimes attributed in part to incompetence on Heisenberg's part, and sometimes to his intentional efforts to sabotage progress [19].

$$xp(\psi) = x(-i\hbar\frac{\partial\psi}{\partial x}) = -i\hbar x\frac{\partial\psi}{\partial x}$$
$$px(\psi) = -i\hbar\frac{\partial}{\partial x}(x\psi) = -i\hbar x\frac{\partial\psi}{\partial x} - i\hbar\psi(x)$$

$$[x, p]\psi(x) = i\hbar\psi(x)$$
$$\therefore [x, p] = i\hbar$$

Example: Our First Commutator with Sympy

```
from sympy.physics.quantum import qapply

psi = Wavefunction(exp(I * (k * x − o * t)), x)
x_psi = Wavefunction(x * exp(I * (k * x − o * t)), x)

p = −I * hbar * DifferentialOperator(Derivative(f(x), x), f(x))

first_term = x * qapply(p * psi)
second_term = qapply(p * x_psi) # performs chain rule

commutator = first_term − second_term
```

We have now encountered commutation relations which are a derivative of group theory. To fully appreciate this we will need a basic understanding of group theory.

2.4 Elementary Group Theory and Commutators

For a set of elements to form a group under multiplication modulo-N, the following must be true [20, p. 193]. The set must contain the multiplicative identity element I, the modulo-N multiplicative inverses of all members of the group, and the modulo-N products of all pairs of its members. Additive groups can also be defined using a similar set of criteria. Instead of a multiplicative identity (i.e. 1), inverse (i.e. $\div$) and product (i.e. $\times$), we require the additive identity (i.e. 0), inverse (i.e. $-$) and sum (i.e. +). A subset of a group G that also meets the criteria for being a group is called a subgroup of G. The number of elements in a group is called the order of the group.

A ring is a group with both additive and multiplicative operations defined on its elements. It inherits the requirements for additive groups. The ring's multiplicative operation must be associative, and its multiplicative operation must be left and right distributive with respect to addition. It must also be Abelian under addition, meaning the additive operation is associative and commutative. A non-Abelian group's operation is not commutative. Non-Abelian statistics play a large role in quantum physics, but more on this later.

What are commutators?
This allows us to define a general *Commutator*, which is $[x, y] = x^{-1}y^{-1}xy$ in the case of a group and $[x, y] = xy - yx$ in the case of a ring.

The *Commutator* class from Sympy is general enough to handle group of discretized operators (matrices) via Sympy's *MatrixOperator* class, etc., and group of differential operators via Sympy's *DifferentialOperator*.

Example: Built-in Commutators with Sympy

```
from sympy.physics.quantum import Commutator

print(Commutator(x, p)) # represents the commutator symbolically
print(Commutator(x, p).doit()). # expands the commutator
```

```
-hbar*I*[x,DifferentialOperator(Derivative(f(x), x),f(x))]
-hbar*I*(x*DifferentialOperator(Derivative(f(x), x),f(x)) - DifferentialOperator(Derivative(f(x), x),f(x))*x)
```

We can manipulate commutators according to a few rules.

$$\begin{aligned}
[A, B] &= -[B, A] \\
[A, B + C] &= [A, B] + [A, C] \\
[A, BC] &= [A, B]C + B[A, C]
\end{aligned}$$

Example: Commutation Relations with Sympy

```
A, B, C = symbols('A B C', commutative=False)

print(Eq(Commutator(A, B), -Commutator(B, A)))
print(Commutator(A, B+C).expand(commutator=True))
print(Commutator(A, B*C).expand(commutator=True))
```

```
True
[A,B] + [A,C]
B*[A,C] + [A,B]*C
```

An important set of commutation relations that will come in handy later involve the Pauli operators S_x, S_y and S_z. In an upcoming section, we will derive these operators from the quantum mechanical properties of particles.

$$\begin{aligned}
[S_x, S_y] &= 2i S_z \\
[S_y, S_z] &= 2i S_x \\
[S_z, S_x] &= 2i S_y
\end{aligned}$$

2.5 Generalized Conjugate Variables and the Measurement Postulate

How do conjugate variables work?
Conjugate pairs in general follow the generalized uncertainty principle where σ_A, σ_B are standard deviations corresponding to observables A and B [11, p. 111].

$$\sigma_A^2 \sigma_B^2 \geq \left(\frac{1}{2i} \langle [A, B] \rangle \right)^2$$

Whenever you hear that something is generalized, if you are someone with software engineering experience you should think of abstraction. In order to prepare ourselves in advance for future topics involving Fourier transform duals besides x and p we will familiarize ourselves with the *Operator* class provided by the Sympy quantum package which defines this abstraction.

This class captures the behaviors of observables. We have made use of expectations of general observables like $\langle A \rangle$ and $\langle B \rangle$. The details of how an observable and its expectation value relate to a physical system have not yet been provided however. We will need to understand the measurement postulate of quantum mechanics to fully appreciate the generalized uncertainty principle.

We know from statistics that an expected value or average outcome $\langle j \rangle$ is a sum over outcomes multiplied with their probabilities of occurring $\sum_j j P(j)$. We have also seen that operators act on wavefunctions in quantum mechanics. An observation of an outcome in quantum mechanics is called a measurement and a measurement "collapses" the wavefunction. This is why a measurement operator in quantum mechanics can interchangeably be called an observable. The wavefunction collapses to an eigenstate of the projective measurement operator. This measurement operator M has a spectral decomposition $\sum_m m P_m$ for outcomes (eigenvalues) m and projectors P_m onto the eigenstates. Given outcome m occurred, the state of a quantum system after measurement is $\frac{P_m \psi}{\sqrt{P(m)}}$ where $P(m) = \psi^\dagger P_m \psi$. We can combine these points to derive an expression for the expectation of a quantum observable $\langle B \rangle$.

$$\begin{aligned}
\langle B \rangle &= \sum_b b\ P(b) \\
&= \sum_b b\ \psi^\dagger P_b \psi \\
&= \psi^\dagger \sum_b b P_b \psi \\
&= \psi^\dagger B \psi
\end{aligned}$$

Example: Expected Values with Sympy

```
from sympy import Indexed, Sum, IndexedBase
from sympy.physics.quantum import Operator, Dagger

Indexed.is_commutative = False

B = Operator('B')
i = Symbol('i') # our sum index
P = Indexed('P', i) # the probabilities of the b
b = Indexed('b', i) # the eigenvalues of B

exp = Sum(b * P, (i, 0, 2)) # say there are two eigenvalues, the two possible outcomes
print(exp.doit())

Pb = IndexedBase('Pb') # the projectors
exp = Sum(b * Dagger(psi) * Pb[i] * psi, (i, 0, 2))
print(exp)
exp = Dagger(psi) * Sum(b * Pb[i] * psi, (i, 0, 2))
print(exp)

Eq(exp, Dagger(psi) * B * psi)
```

```
b[0]*P[0] + b[1]*P[1]
Wavefunction(exp(-I*(-3*conjugate(t) + conjugate(x))), x)*(b[0]*Pb[0]*Wavefunction(exp(I*(-3*t + x)), x) +
    b[1]*Pb[1]*Wavefunction(exp(I*(-3*t + x)), x))
```

Now we are ready to reason about the generalized uncertainty principle. The variance of an operator A is $\sigma_A = \langle A^2 \rangle - \langle A \rangle^2 = \langle (A - \langle A \rangle)^2 \rangle$. The uncertainty of A is $\sqrt{\sigma_A} = \Delta A$. Let $\bar{A} = A - \langle A \rangle$. Then, $\sigma_A = \psi^\dagger \bar{A}\bar{A}\psi$. If for some operator B we have $\sigma_B = \psi^\dagger \bar{B}\bar{B}\psi$ then by the Schwartz inequality we have that $\sigma_A \sigma_B \geq |\psi^\dagger \bar{A}\bar{B}\psi|^2$. Consider that $\psi^\dagger \bar{A}\bar{B}\psi$ is an imaginary number z and the conjugate transpose generalizes the complex conjugate for operators.

$$\begin{aligned} Re(\psi^\dagger \bar{A}\bar{B}\psi) &= \frac{z + z^*}{2} \\ Im(\psi^\dagger \bar{A}\bar{B}\psi) &= \frac{z - z^*}{2i} \end{aligned}$$

$$\begin{aligned} Re(z) &= \frac{\psi^\dagger \bar{A}\bar{B}\psi + \psi^\dagger \bar{B}\bar{A}\psi}{2} \\ Im(z) &= \frac{\psi^\dagger \bar{A}\bar{B}\psi - \psi^\dagger \bar{B}\bar{A}\psi}{2i} \\ &= \frac{\langle [A, B] \rangle}{2i} \end{aligned}$$

$$\begin{aligned} |z|^2 = | \langle \bar{A}\psi | \bar{B}\psi \rangle |^2 = |Re(z)|^2 + |Im(z)|^2 &\geq |Im(z)|^2 \\ \therefore \sigma_A \sigma_B &\geq |Im(z)|^2 \end{aligned}$$

Then this gives us the generalized uncertainty principle.

Quantum measurement.
The action of measurement on a quantum system can be described by a set of measurement operators, not just one M. Such a set of positive semi-definite operators $\{M_j\}$ is called a Positive Operator Valued Measure (POVM).

A POVM is informationally complete if the $\{M_j\}$ span the space of Hermitian operators. A Hermitian operator A is one that satisfies the following, where $\int d\tau$ is integration over all coordinates.

$$\int \psi^* A \psi d\tau = \int A^* \psi^* \psi d\tau$$

The adjoint $\dagger$ of an operator is defined exactly such that:

$$\int \psi_2^* A \psi_1 d\tau = \int ((A^\dagger)\psi_2)^* A \psi_1 d\tau.$$

Therefore the definition of a Hermitian says that it is equal to its adjoint. The expected value of a Hermitian is also real.

$$\begin{aligned} &\langle A\psi|\psi\rangle \\ &= \langle\psi| A |\psi\rangle^* \\ &= \langle\psi| A |\psi\rangle \because \langle A\rangle^* = \langle A\rangle \end{aligned}$$

2.6 Linear Algebra

Connection to discretized linear algebra.
We have already seen that a Hilbert space is suitable for describing quantum states. In fact, any finite dimensional space is a Hilbert space if it is equipped with an inner product.

Our formulation using the square root of the probability was informed by experimental evidence from experiments like Planck's study of blackbody radiation, Young's double-slit experiment with photons and much more. So, we know that our inner products will always sum to one as per $\int_{-\infty}^{\infty} \psi^*(x,t)\psi(x,t)dx = 1$. What's more is that quantum mechanics is a linear theory. Linearity means that if we have an *Operator* A applying to states ϕ_i, then $A \sum_i a_i \phi_i = \sum_i a_i (A\phi_i)$. It is more precise to say that quantum mechanics is a linear theory with nonlinear phenomena since the collapse of the state is nonlinear; however, the properties of quantum mechanics are still highly suggestive of linear algebra overall.

If we define our inner product to be the vector inner product and represent our state by a vector, this is also highly useful for computer simulations. Say our state was represented by a bitstring with k bits of precision. If we knew this were a normalized basis state, then we could represent one qubit per bit and would have a state space with a size 2^k. 300 such bits would give us a state space so large that to represent all its bitstrings at once, all the degrees of freedom of every particle in the observable universe would be needed, encoding one bit per degree of freedom [21].

$$\begin{bmatrix} \alpha_0 \\ \vdots \\ \alpha_k \end{bmatrix}$$

When we begin to work with linear algebra, we find there are many options in terms of software packages that we can use for quantum information. Whereas before we were quite narrowly limited to just a few choices and used Sympy (Qutip would be another choice), now there are oodles of options. In the next section we will see how to use Sympy's linear algebra features to implement our first quantum algorithm, but after this we will explore Pennylane, QisKit and Numpy which allow us to work with vectors and matrices.

Say we performed matrix multiplication on this state to represent some kind of evolution. Then updating each α would take 2^k time and k^2 is the complexity of multiplication. Say s was the number of matrices we applied. We would have an overall complexity of $O(sk^2 2^k)$ [22], and we could limit this complexity to something feasible on our classical computer by limiting the size / precision of the vector representation.

Now that we are working with state vectors, we will denote the state $|\psi\rangle$. In general, for a state that evolves in time we can find the (unitary) matrix $U(t_2 - t_1)$ that we would use by exponentiating the Hamiltonian like $\exp(-\frac{iH(t_2-t_1)}{\hbar})$ where $i\hbar\frac{\partial|\psi\rangle}{\partial t} = H|\psi\rangle$. This unitary is called a propagator of the state. Then we get our time dependence in terms of a matrix $Operator$ that evolves the state $|\psi(t)\rangle = U(t_2 - t_1)|\psi\rangle$. A valid unitary matrix has the following properties: $UU^\dagger = U^\dagger U = I$

The equation giving the action of an operator on a quantum state looks like an eigenvalue equation when our operators are unitary matrices. Indeed, this is how we understand the action of a unitary operator that evolves the state in time. It acts on the state, scaling it by introducing a real coefficient in the case that the state is an energy eigenstate of the Hamiltonian. These real coefficients are the outcomes of our measurements.

Recall that the action of measurement on a quantum system can be described by a set of measurement operators, which is complete if it spans the space of Hermitian operators. When using linear algebra, Hermitians are by definition matrices with real eigenvalues. In quantum mechanics, observables are represented by Hermitian operators for this reason: the expected values of our measurement outcomes must be real. What would an imaginary outcome even mean? An important property of Hermitians is that they are equal to their own adjoints. A Hermitian matrix is equal to its complex conjugate transpose, since the adjoint of a matrix is its complex conjugate transpose. A Hermitian operator is called positive semi-definite iff its eigenvalues are not negative. A Hermitian unitary has real eigenvalues *and* the eigenvalues' absolutes values equal to 1. Not all unitaries are Hermitian.

We can represent our state vector in a particular basis using the following representation. Here the c_i are complex coefficients and the $|\phi_i\rangle$ are basis vectors. Our $c_i = \langle\phi_i|\psi_n\rangle$.

$$|\psi_n\rangle = \sum_i c_i |\phi_i\rangle$$

Notice how this is essentially equivalent to the equation we saw earlier.

$$\psi_n(x) = \sum_i c_i \phi_i(x).$$

A Fourier transform was useful for calculating the c_i as follows, which is like $c_i = \langle\phi_i|\psi_n\rangle$.

$$c_i = \int_{-\infty}^{\infty} \phi_i^*(x)\psi(x)\, dx$$

An important piece of the puzzle when switching to a vector representation is to realize that taking the Fourier series of a function is functionally the same as projecting a vector onto a basis.

When it comes to the vector inner product, we are talking about the dot product $\psi_n \cdot \psi_m = a_0 b_0 + a_1 b_1 + \cdots + a_k b_k$. This product gives us a complex scalar. In Bra-Ket notation we specify an inner product as $\langle\psi_n|\psi_m\rangle$. The Bra-Ket inner product of a vector with itself $\langle\psi_n|\psi_n\rangle$ is similar to the expression $\int_{-\infty}^{\infty} \psi^*(x,t)\psi(x,t)dx$ from earlier and both evaluate to a scalar.

$$\langle\psi_n|\psi_m\rangle = \begin{bmatrix} a_0^* & a_1^* & \dots a_k^* \end{bmatrix} \begin{bmatrix} b_0 \\ b_1 \\ \vdots \\ b_k \end{bmatrix}$$

$$|\psi_n\rangle = \sum_i a_i |\phi_i\rangle$$

$$|\psi_m\rangle = \sum_i b_i |\phi_i\rangle$$

We could specify $|\phi_0\rangle = |0\rangle = \begin{bmatrix}1\\0\end{bmatrix}$ and $|\phi_1\rangle = |1\rangle = \begin{bmatrix}0\\1\end{bmatrix}$, but that is only one example of a choice we could make. It is important however that these ϕ_n are orthonormal so that we get the property that $\langle\phi_n|\phi_m\rangle = \delta_{nm}$. Two vectors ϕ_n, ϕ_m are orthonormal if $\langle\phi_n|\phi_m\rangle = 0$.

We get the following properties from linear algebra for free: linearity ($\langle\psi_n|\sum_i b_i |\phi_i\rangle = \sum_i b_i \langle\psi_n|\phi_i\rangle$), complex conjugation ($\langle\psi_n|\psi_m\rangle^* = \langle\psi_m|\psi_n\rangle$) and completeness. Completeness refers to the ability of a set of basis vectors to span our entire Hilbert space.

$$\psi_n = \sum_i c_i |\phi_i\rangle = \sum_i \langle\phi_i|\psi_n\rangle |\phi_i\rangle = (\sum_i |\phi_i\rangle \langle\phi_i|) |\psi_n\rangle$$

$$\to \sum_i |\phi_i\rangle \langle\phi_i| = I$$

If the sum of a set of vectors is the identity then the entire space is spanned by those vectors since there is a 1 in every column and row.

Now that we are in the realm of discrete linear algebra, we can make use of any number of quantum computing software packages to implement quantum algorithms and are less restricted to sympy and symbolic computing.

We have shown some similarities in the wavefunction and the statevector representations of quantum states, such as the completeness relation and linearity found in both cases. It will however become much more concretely apparent in subsequent sections how linear algebra becomes a necessary part of working with quantum objects, this is only a preview. This preview is however enough to let us look at a famous example and application of quantum information processing called quantum key distribution.

2.7 Quantum Key Distribution

Quantum key distribution takes advantage of the property of conjugate variables, or Fourier transform duals, that says there is an uncertainty principle between them. This uncertainty principle means that the more accurately one variable is measured the less information about the other is learned.

In the first chapter we learned about polarization qubits which can have states that are vertical, horizontal, diagonal or circular, for example. When we measure such a qubit, we can measure in one of several orthonormal bases. One choice of basis could be the 0, $\frac{\pi}{2}$ basis. Another could be the circular basis. These bases are conjugate to each other meaning that measuring in one of these bases will yield a meaningful result (for our purposes in QKD) only when the polarization prior to measurement was one of the corresponding basis states. If the state was instead a basis state of one of the bases *not* used for measurement, the outcome of the measurement will be random.

Any orthonormal bases can be used to perform quantum key distribution in theory and there exist for example protocols such as the six state protocol which use six states to encode quantum keys [23, 24], but we will first consider BB84, named after its inventors Bennett and Brassard,[7] which is generally the most discussed QKD protocol [26].

BB84 involves two parties, often named Alice and Bob, who wish to share a secret key by the end. Critically, they want to avoid the scenario in which their key is intercepted in transport by a third party, Eve. The protocol makes use of the randomness of measurement outcomes to achieve this.

The first step in the protocol is for Alice, the sender of the key, to encode a bit of the key in one of the two orthonormal bases that will be used for the protocol. The specific basis used for this bit is chosen randomly. The encoding is simple: a 0 is encoded as one basis vector and a 1 is encoded as the other. Alice then transmits the photon to Bob. This is repeated until all the bits of the secret key have been encoded and transmitted. Alice keeps track of what basis she used for each bit.

In the event that there is a third party, Eve, who performs a man-in-the-middle attack, the photons will be measured en-route. There is however no way for the attacker to know the basis in which the message was encoded for any particular bit.

You may be thinking, hold up! I've heard of state tomography (which we will get to in a later section in detail)! Couldn't Eve use that to figure out the exact state? However, in quantum state tomography it is necessary to measure the state multiple times in different bases. Since each bit of the key is not repeated in BB84, there is no opportunity for this kind of attack. The attacker must measure the bit in a random basis, and they only get one guess as to the right one. They can then spoof the measured photon by transmitting one to Bob with the state that they observed, but there is no guarantee that the spoof will match what Alice originally sent since there is a chance the measurement was in the wrong basis and the outcome was random.

Bob actually does not know the bases used by Alice either, but chooses random bases and measures the incoming photons, recording the time, basis used and the outcome for each. Only after all the photons are sent, Alice communicates publicly with Bob all the bases that she used and Bob communicates what bases he measured in. They discard the bits that don't match and the remaining bitstring is their shared secret.

There should be no difference between the secrets that Alice and Bob share, but there will be deviations if Eve measured any of the bits in the wrong basis, since measuring in a given basis yields outcomes in that basis. Alice and Bob compare a subset of their bitstrings in order to check for man-in-the-middle attacks.

In practice, BB84 looks like the following, for example.

- Alice randomly generates a string of n bits. $x = x_{n-1}x_{n-2}\dots x_0$.
- Alice prepares one of the following qubits for each of the $0 \leq i \leq n$, choosing at random whether to encode in the $\{|0\rangle, |1\rangle\}$ Z basis or the $\{|+\rangle, |-\rangle\}$ X basis. She records her choices of basis in bitstring y_n.

[7] Gilles Brassard was inducted into the Order of Canada in 2013 [25].

If $x_i = 0$ and using Z basis: $|\psi_i\rangle = |0\rangle$.
If $x_i = 1$ and using Z basis: $|\psi_i\rangle = |1\rangle$.
If $x_i = 0$ and using X basis: $|\psi_i\rangle = H\,|0\rangle$.
If $x_i = 1$ and using X basis: $|\psi_i\rangle = H\,|1\rangle$.

- Alice sends each of the prepared qubits to Bob via a public quantum channel.
- Bob measures each qubit, recording his basis choices in string y'_n.
- The choices of bases y_n, y'_n are made public and every jth bit where $y_j \neq y'_j$ is discarded by both Alice and Bob.
- An eavesdropper is ruled out by comparing some subset of the remaining bits.

To work out a specific example, we will make use of the notion of projective measurement discussed earlier.

Say we wanted to determine what measurement outcomes of a measurement in a specific basis corresponded to what states given in another basis. There may be randomness involved, but what does it look like exactly? Say, the measurement basis in question was the X basis. The Hadamard operator maps the Z basis' $|0\rangle$ and $|1\rangle$ states to the X's eigenstates $|+\rangle$ and $|-\rangle$ by way of the following transformation. Say the measured state was the eigenstate $|0\rangle$ of the Z basis.

$$H = \frac{1}{\sqrt{2}}\begin{bmatrix}1 & 1\\ 1 & -1\end{bmatrix}$$

We can expand the X gate into its spectral decomposition to determine the projectors. In the expression for the projectors $P = \sum_{i=1}^{k} |i\rangle\langle i|$, the $|i\rangle$ are the observable's eigenvectors. The eigenvalues of the X matrix are ± 1 since $X^2 = I$.

$$\begin{aligned} X &= \begin{bmatrix}0 & 1\\ 1 & 0\end{bmatrix} \\ &= |0\rangle\langle 1| + |1\rangle\langle 0| \\ &= |+\rangle\langle +| - |-\rangle\langle -| \end{aligned}$$

$$\begin{aligned} |+\rangle &= \frac{1}{\sqrt{2}}(|0\rangle + |1\rangle) \\ |-\rangle &= \frac{1}{\sqrt{2}}(|0\rangle - |1\rangle) \end{aligned}$$

Then the probability of a +1 outcome would be $\langle 0|\, P_{+1}\, |0\rangle$. Note that for basis vectors $|i\rangle, |j\rangle$ of an orthonormal basis, $\langle i|j\rangle = \delta_{ij}$.

$$
\begin{aligned}
&\langle 0|\, P_{+1}\, |0\rangle \\
&= \langle 0|+\rangle \, \langle +|0\rangle \\
&= |\, \langle 0|+\rangle \,|^2 \\
&= |\frac{1}{\sqrt{2}}(\langle 0|0\rangle + \langle 0|1\rangle)|^2 \\
&= \frac{1}{2}
\end{aligned}
$$

Therefore we are just as likely to measure a +1 outcome as a -1 outcome when we measure in the wrong basis.

Example: Alice Sends a 7-bit Message to Bob

For example, Alice might transmit $x = 0100100$ as $|+-00-00\rangle$. Bob might measure in bases XZZZZXZ. The measurement of the first qubit is in the right basis so a +1 outcome is guaranteed. However, the 2nd, 5th and 6th qubit measurements will have random outcomes for Bob. These will be discarded. Say Eve tried to measure the third qubit in transit and used the wrong basis. Then, the outcome might be 1 in which case Alice and Bob could compare this bit to detect the eavesdropper.

```
from sympy.physics.quantum.qubit import Qubit, measure_partial_oneshot
from sympy.physics.quantum.gate import HadamardGate
from random import getrandbits

bases = ['Z', 'X']

# Alice's variables
a_bases = []
a_secret = []
bits = []

# Bob's variables
b_bases = []
b_secret = []
outcomes = []

# Alice generates her bits to transmit
bits = '{0:07b}'.format(getrandbits(7))

for i in range(len(bits)):

    # Alice preps a qubit
    qubit = Qubit(bits[i])

    # Alice selects an encoding basis
    basis = bases[int(getrandbits(1))]
    a_bases.append(basis)

    # Alice encodes her bit
    if basis == 'X':
        qubit = qapply(HadamardGate(0) * qubit)

    # Alice sends the qubit to Bob

    # Bob chooses a measurement basis
    m_basis = bases[int(getrandbits(1))]
    b_bases.append(m_basis)
```

```
    # Bob measures the bit
    if m_basis == 'X':
        qapply(HadamardGate(0) * qubit)
    outcome = measure_partial_oneshot(qubit, [0])
    outcomes.append(outcome)

# Alice and Bob compare bases
to_remove = []
while i < len(a_bases):
    if b_bases[i] == a_bases[i]:
        i += 1
    else:
        to_remove.append(i)

# Alice throws away the bits corresponding with mismatched bases
for j in range(len(bits)):
    if j not in to_remove:
        a_secret.append(bits[j])
del bits

# Bob throws away the bits corresponding with mismatched bases
for k in range(len(outcomes)):
    if k not in to_remove:
        b_secret.append(outcomes[k])
del outcomes

# Alice and Bob have a shared secret a_secret == b_secret
```

2.8 Quantum Computing Foundations

A quick review of some of the concepts we have seen in the previous section will follow, which we will build upon until we reach our basic setup for quantum computing. Consider that the expectation of a quantum observable is $|\psi|^2 = \langle\psi|\psi\rangle$. Then consider that a measurement of some observable property of a quantum particle tells us only, for example, its position and not its trajectory. We therefore cannot know from a measurement whether particles are interchanged. We can say identically that their interchange must not change the expectation value of the measurement. This gives us the following condition [11, p. 204].

$$\psi(\xi_1, \xi_2, \ldots, \xi_i, \ldots, \xi_j, \ldots, \xi_N) = \pm\psi(\xi_1, \xi_2, \ldots, \xi_j, \ldots, \xi_i, \ldots, \xi_N)$$

Since the Hamiltonian[8] is invariant with particle interchange, its eigenvalues are degenerate. Electrons are fermions and therefore exhibit an interchange asymmetry meaning the wavefunction $\psi_a(\xi_1, \xi_2)$ of a system of two identical electrons is antisymmetric.

[8] The Hamiltonian is named after Sir William Rowan Hamilton, who was appointed professor of astronomy before finishing his undergraduate degree in 1827 and was knighted in Ireland in 1835 [27].

Table 2.2 Quantum numbers

Quantum number	Definition
n	$\in \mathsf{Z}$
l	$\in \mathsf{Z}, < n$
m	$\in [-l, \ldots, l]$
s	$\in [\frac{1}{2}, -\frac{1}{2}]$

$$\psi_a(\xi_1, \xi_2) = \frac{1}{\sqrt{2}}(\psi(\xi_1, \xi_2) - \psi(\xi_2, \xi_1))$$

This wavefunction expresses that two identical fermions cannot occupy the same state since the terms $\psi(\xi_1, \xi_2)$ and $\psi(\xi_2, \xi_1)$ cancel out when they are the same. Therefore, these terms must differ by a quantum number such as spin. This property of non-integer-spin particles is taken as an axiom in non-relativistic quantum mechanics.

A further implication of the Pauli exclusion principle is that the orbitals that may constitute the valence shells of atoms are actually degenerate. They may only contain more than two electrons because they in actuality contain more than one discrete energy level. Each of these slightly detuned energy levels may be filled by at most two identical fermions of differing state ($\frac{1}{2}$ and $-\frac{1}{2}$). Therefore, when we count the electrons in atomic orbitals by the following convention ((Table 2.2)).

Quantum dots (the author's speciality) measured in hundredths of microns (for reference, coffee might be ground to 525 microns) can approximate atoms with one electron like the hydrogen atom. At this scale the wavelength nature of the electron comes into play. To consider the wavelike properties of a particle we can start with the familiar classical notions of momentum and kinetic energy.

$$p = mv$$

$$KE = \frac{1}{2}mv^2 = \frac{p^2}{2m}$$

Then we can choose to write an equation in terms of energy instead of mass by mass-energy equivalence. The frequency of a particle multiplied by the Planck constant h gives its energy. We can also consider that velocity is the derivative of position to write momentum as a differential operator.

$$\hbar = \frac{h}{2\pi}$$
$$p^2 = -\hbar^2 \frac{\partial^2}{\partial x^2} = -\hbar^2 \nabla^2$$
$$p = -i\hbar \frac{\partial}{\partial x}$$
$$\nabla^2 = \frac{\partial^2}{\partial x^2}$$

This operator will operate on a quantum state of the form $\psi(x, t) = \exp(i(kx - \omega t))$ which encodes frequency and position. Schrödinger took similar steps to express the kinetic energy as an operator in the stationary state Schrödinger equation for a non-relativistic particle in one dimension.

$$KE = -\frac{\hbar^2}{2m} \frac{\partial^2}{\partial x^2}$$
$$-\frac{\hbar^2}{2m} \frac{\partial^2}{\partial x^2} \psi(x, t) + U(x, t)\psi(x, t) = i\hbar \frac{\partial}{\partial t} \psi(x, t)$$

$\frac{\partial}{\partial t}\psi(x, t)$ gives us the system's energy as Planck's constant times the frequency. The left side of the equation splits the energy up into kinetic (KE) and potential energy $U(x, t)$ components.

$$\frac{\partial}{\partial t} \psi(x, t) = -i\omega\psi(x, t)$$
$$i\hbar \frac{\partial}{\partial t} \psi(x, t) = \hbar\omega\psi(x, t)$$
$$= \frac{h\omega}{2\pi} \psi(x, t)$$

The time-dependent Schrödinger equation can be summarized as the following eigenvalue problem of which the stationary state Schrödinger equation for a non-relativistic particle is a special case.

$$i\hbar \frac{\partial}{\partial t} |\psi(t)\rangle = E |\psi(t)\rangle$$

We can replace E by the operator H to emphasize that this is an eigenvalue problem where the energy will be an eigenvalue associated with a particular eigenstate after the operation is performed on ψ. This is then called a Hamiltonian H.

$$i\hbar \frac{\partial}{\partial t} |\psi(t)\rangle = H |\psi(t)\rangle$$

When we model a quantum dot we can think of it as a particle in a box. We'll consider the idealized model of a quantum harmonic oscillator in a parabolic potential well. (Note, however, that the energy spacing for a square potential well is given by $\frac{n^2\pi^2\hbar^2}{2mL^2}$ rather than $(n - 1/2)\hbar\omega$, where L is the length of the well). For every allowed energy of such an idealized particle there is at least one solution to the stationary state Schrödinger equation.

$$U(x) = \frac{1}{2}m\omega^2 x^2$$

$$-\frac{\hbar^2}{2m}\frac{\partial^2}{\partial x^2}\psi(x) + \frac{1}{2}m\omega^2 x^2\psi(x) = E\psi(x)$$

This is the one-particle case of the general many-body Hamiltonian H_N.

$$H = -\sum_{i=1}^{N}\frac{\hbar^2}{sm_i}\nabla_i^2 + U(x_1, x_2, \ldots, x_N, t)$$

The exact eigenstates for a general many-body Hamiltonian are impossible to obtain directly. However, we can reason about the operators that make up H and how they relate to one another, i.e. by commutation.

$$xp - px = [x, p]$$

$$xp(\psi) = x(-i\hbar\frac{\partial}{\partial x}\psi)$$

$$px(\psi) = -i\hbar\frac{\partial}{\partial x}x\psi = -i\hbar x\frac{\partial}{\partial x}\psi - i\hbar\psi \rightarrow [x, p] = i\hbar$$

The operators x and p give us a particle's position and momentum but we can think about other observable properties of a particle as well. We can consider three dimensional angular momentum $L = x \times p$ using what we've seen before [5, p. 201].

$$[L_x, L_y] = [yp_z - zp_y, zp_x - xp_z] = [yp_z, zp_x] - [zp_y, xp_z] - [yp_z, zp_z] + [zp_y, xp_z]$$

$$[x_i, p_j] = i\hbar\delta_{ij}$$

$$[yp_z, zp_x] = yp_z zp_x - zp_x yp_z = yp_x(p_z z - zp_z) = yp_x[p_z, z] = -yp_x i\hbar$$

$$[zp_y, xp_z] = -xp_y i\hbar$$

$$[zp_y, zp_x] = zp_y zp_x - zp_x zp_y = 0$$

$$[yp_z, xp_z] = 0$$

$$\therefore\ [L_x, L_y] = i\hbar(xp_y - yp_x) = i\hbar L_z$$

In general for any of the three dimensional angular momentum operators we have the following. ε_{ijk} is -1 is i, j, k in cyclic order, otherwise it is 1.

$$[L_i, L_j] = i\hbar\varepsilon_{ijk}L_k$$

We can also represent a general angular momentum operator L by a row-column matrix with entries $L_{mm'} = \langle l, m'| L |l, m\rangle$ in terms of quantum numbers l, m and m'. For a specific example, $L_z |l, m\rangle = \mu_m |l, m\rangle$. To help us solve these entries we introduce $L_+ = L_x + iL_y$.

$$\begin{aligned}
[L_z, L_+] &= [L_z, L_x] + i[L_z, L_y] = \hbar(L_y + iL_x) = \hbar L_+ \\
L_z L_+ |l, m\rangle &= (L_+ L_z + \hbar L_+) |l, m\rangle \\
L_z L_+ &= (\mu_m + \hbar) L_+ |l, m\rangle
\end{aligned}$$

$L_+ |l, m\rangle$ is an eigenfunction of L_z with eigenvalue $(\mu_m + \hbar)$. Successive applications of L_+ continue to raise the eigenvalue by $\hbar$, finding additional solutions. We call this a "raising operator". Then we can solve for the maximum μ_m by setting $L_+ |l, m_{max}\rangle = 0$ [5, pp. 195–196]. Let $L^2 = L_x^2 + L_y^2 + L_z^2$ and $L^2 |l, m_{max}\rangle = \lambda |l, m_{max}\rangle$.

$$\begin{aligned}
L_- L_+ |l, m_{max}\rangle &= (L^2 - L_z^2 - \hbar L_z) |l, m_{max}\rangle \\
&= (\lambda - \mu_{m_{max}}^2 - \hbar\mu_{m_{max}}) |l, m_{max}\rangle = 0 \\
\lambda - \mu_{m_{max}}^2 - \hbar\mu_{m_{max}} &= 0 \\
\lambda = \mu_{m_{max}}^2 - \hbar\mu_{m_{max}}
\end{aligned}$$

Similarly,

$$\lambda = \mu_{m_{min}}^2 - \hbar\mu_{m_{min}}$$

$$\therefore \; \mu_{m_{max}} = \mu_{m_{min}}.$$

This gives us the following picture: $\mu_m \in [-l\hbar, -l\hbar + \hbar, -l\hbar + 2\hbar, \ldots, l\hbar]$. For a general particle, the matrix representation of L_z is therefore the following.

$$L_z = \begin{bmatrix} -l\hbar & 0 & \ldots & 0 \\ 0 & (-l+1)\hbar & & \\ \vdots & & \ddots & \vdots \\ 0 & \ldots & & l\hbar \end{bmatrix}$$

We also like to consider spin s, which for spin-1/2 particles is $s \in [-\frac{1}{2}, \frac{1}{2}]$. Spin, or intrinsic angular momentum, is similar to what we have explored if we set $l = -\frac{1}{2}$. The degeneracy of the eigenvalue problem becomes 2.

$$S_z = \frac{\hbar}{2}\begin{bmatrix} 1 & 0 \\ 0 & -1 \end{bmatrix}$$

$$S_z \left| l, m_s \right\rangle = \mu_{m_s} \left| l, m_s \right\rangle = \pm\frac{\hbar}{2} \left| l, m_s \right\rangle$$

The eigenvectors are obvious to see as they relate to each eigenvalue. The state $|l, m_s\rangle$ in this case can be represented by a vector that may take either of the following values, or complex linear combinations thereof, called energy eigenstates. These form the computational basis for quantum computing.

$$|1\rangle = \begin{bmatrix} 0 \\ 1 \end{bmatrix} \quad |0\rangle = \begin{bmatrix} 1 \\ 0 \end{bmatrix}$$

Additional observable properties can be solved for such as the angular momentum (or spin) in the x dimension. Let $S_- = S_x - iS_y$.

$$\begin{aligned} S_{x_{m_s m_s'}} &= \frac{1}{2} \left\langle l, m_s' \right| S_+ + S_- \left| l, m_s \right\rangle \\ &= \frac{\hbar}{2} \left\langle l, m_s' | l, m_s + 1 \right\rangle + \frac{\hbar}{2} \left\langle l, m_s' | l, m_s - 1 \right\rangle \\ &= \frac{\hbar}{2}\delta_{m_s', m_s+1} + \frac{\hbar}{2}\delta_{m_s', m_s-1} \\ \therefore\ S_x &= \frac{\hbar}{2}\begin{bmatrix} 0 & 1 \\ 1 & 0 \end{bmatrix} \end{aligned}$$

Spin is a magnetic property. For an electron in a magnetic field, the classical equation $U = -\boldsymbol{\mu} \cdot \boldsymbol{B}$ becomes $U = (\frac{\mu_B}{\hbar})\boldsymbol{S} \cdot \boldsymbol{B}$. Here U is the perturbation due to the magnetic field and $\boldsymbol{\mu}$ is the magnetic moment. μ_B is the Bohr magneton and $\boldsymbol{S}$ is spin angular momentum. This is called the Zeeman effect [5, p. 336]. The Bohr magneton is approximately the intrinsic electron magnetic moment caused by the electron's spin. This moment results in the electron g-factor, which relates the spin angular momentum to the magnetic moment of the particle.

If we are not dealing with electrons or other spin-1/2 particles we can alternatively solve for the properties of a spin-3/2 particle, say. For a spin-3/2 particle our L_z looks like the following.

$$L_z = \frac{\hbar}{2}\begin{bmatrix} 3 & 0 & 0 & 0 \\ 0 & 1 & 0 & 0 \\ 0 & 0 & -1 & 0 \\ 0 & 0 & 0 & -3 \end{bmatrix}$$

Our basis states are the following.

$$|0\rangle = \begin{bmatrix} 1 \\ 0 \\ 0 \\ 0 \end{bmatrix} \quad |1\rangle = \begin{bmatrix} 0 \\ 1 \\ 0 \\ 0 \end{bmatrix} \qquad |2\rangle = \begin{bmatrix} 0 \\ 0 \\ 1 \\ 0 \end{bmatrix} \quad |3\rangle = \begin{bmatrix} 0 \\ 0 \\ 0 \\ 1 \end{bmatrix}$$

The algebras of the groups of operators corresponding to each type of spinful particle are separate. The matrices themselves make up members of the new groups. The respective groups are also defined with separate zero and identity members.

In order to support commutation of the type discussed earlier, the groups must be rings. That means they are defined by an additive operation, multiplicative operation, their members (and their additive inverses), an additive identity, a multiplicative identity. Ring members added to the additive identity must not change just as ring members multiplied with the multiplicative identity must not change. In addition, the ring multiplicative operation must be associative and be left and right distributive with respect to addition. These are the defining properties of a ring.

The multiplicative operation natural to the matrices derived earlier is matrix multiplication. The additive operation is element-wise matrix addition with its inverse subtraction. The additive and multiplicative identities for an electronic spin qubit system are then necessarily the following.

$$zero = \begin{bmatrix} 0 & 0 \\ 0 & 0 \end{bmatrix} \quad I = \begin{bmatrix} 1 & 0 \\ 0 & 1 \end{bmatrix}$$

Similarly for a spin-3/2 system,

$$zero = \begin{bmatrix} 0 & 0 & 0 & 0 \\ 0 & 0 & 0 & 0 \\ 0 & 0 & 0 & 0 \\ 0 & 0 & 0 & 0 \end{bmatrix} \quad I = \begin{bmatrix} 1 & 0 & 0 & 0 \\ 0 & 1 & 0 & 0 \\ 0 & 0 & 1 & 0 \\ 0 & 0 & 0 & 1 \end{bmatrix}.$$

A minimal setup for quantum computing with this kind of qubit could be achieved in Python with the following code where only a simple group theory class is used as a basis. This class may be found in the appendices.

Example: Quantum Computing with Python

```
import numpy as np
from group import add, mul, Ring, additive_inverse, Commutator

hbar = 1.054571817 * 10 ** -34 # J * s

# spin-1/2 matrices (electron spin)
Sx = 0.5 * hbar * np.matrix([
    [0., 1.],
    [1., 0.]
])

Sz = 0.5 * hbar * np.matrix([
```

```
    [1., 0.],
    [0., −1.]
])

Sy = − complex(0, 1) * Sx * Sz

zero2 = np.matrix([
    [0., 0.],
    [0., 0.]
])

id2 = np.matrix([
    [1., 0.],
    [0., 1.]
])

ring2 = Ring(
    additive_operation=add,
    multiplicative_operation=mul,
    members=[Sx, Sy, Sz],
    additive_identity=zero2,
    multiplicative_identity=id2,
    additive_inverse_function=additive_inverse
)

S2 = Sx ** 2 + Sy ** 2 + Sz ** 2

print(Commutator(ring2).commute(Sx, Sy))
print(Commutator(ring2).commute(Sy, Sz))
print(Commutator(ring2).commute(Sx, Sz))

raising = Sx + complex(0, 1) * Sy
lowering = Sx − complex(0, 1) * Sy

# computational basis vectors of quantum computing
print(np.linalg.eig(Sz))
```

We can use this code to check the Commutators $[S_x, S_y]$, $[S_y, S_z]$, $[S_x, S_z]$ which should obey $[S_i, S_j] = i\hbar\varepsilon_{ijk}S_k$. From this starting point we can also confirm the Pauli algebra. For the even permutations of jkl:

$$S_j S_k = -S_k S_j = i S_l$$

> **How does quantum info arise from physics?**
> These physical spin operators S_x, S_y, S_z are used to do computing as they are identical to the theoretical Pauli operators (ignoring a factor of $\hbar/2$) X, Y and Z which can be used to construct universal sets of "gates".

A universal gateset must be able to create superposition, create entanglement and have real as well as imaginary gates among other things in order to fully realize quantum computing.

In gate model quantum computing, we use gates such as X, Y and Z to create quantum circuits that look like successive applications of these gates to qubits, or spin-1/2 systems.

Because computing can be done with Fermions of spin-3/2 just as well as with qubits, consider that we could lay a foundation for computing with spin-3/2 particles in a similar way.

Example: Spin-3/2 Quantum Computing with Python

```
# spin−3/2 matrices
Jx = 0.5 * hbar * np.matrix([
    [0, np.sqrt(3), 0, 0],
    [np.sqrt(3), 0, 2, 0],
    [0, 2, 0, np.sqrt(3)],
    [0, 0, np.sqrt(3), 0]
])

Jy = 0.5 * hbar * np.matrix([
    [0, complex(0, np.sqrt(3)), 0, 0],
    [complex(0, np.sqrt(3)), 0, −2j, 0],
    [0, 2j, 0, complex(0, −np.sqrt(3))],
    [0, 0, complex(0, np.sqrt(3)), 0]
])

Jz = 0.5 * hbar * np.matrix([
    [3, 0, 0, 0],
    [0, 1, 0, 0],
    [0, 0, −1, 0],
    [0, 0, 0, −3]
])

zero4 = np.matrix([
    [0., 0., 0., 0.],
    [0., 0., 0., 0],
    [0., 0., 0., 0.],
    [0., 0., 0., 0.]
])

id4 = np.matrix([
    [1., 0., 0., 0.],
    [0., 1., 0., 0],
    [0., 0., 1., 0.],
    [0., 0., 0., 1.]
])

ring4 = Ring(add, mul, [Jx, Jy, Jz], zero4, id4, additive_inverse)

print(Commutator(ring4).commute(Jx, Jy))
print(Commutator(ring4).commute(Jy, Jz))
print(Commutator(ring4).commute(Jx, Jz))

J2 = Jx ** 2 + Jy ** 2 + Jz ** 2

raising4 = Jx + complex(0, 1) * Jy
lowering4 = Jx − complex(0, 1) * Jy

# spin states
print(np.linalg.eig(Jz))
```

This is however less popular just like analog computing is less popular than digital computing. Computing with bits is easier than analog computing. However, we do see examples where spin-3/2 systems are used to encode qubits, such as with hole spin qubits. We can treat the two $m = \pm\frac{3}{2}$ states as $|0\rangle$ and $|1\rangle$. In this case we would call our qubit as isospin.

While the Pauli group is an excellent starting point for qubit manipulation, it consists only of so-called single-qubit gates (the Pauli matrices). There is therefore no way to represent the interaction of one spin with another in its algebra. This interaction is necessary for computation since qubits must be networked to meet the criteria for computing.

So, we most often make use of the Clifford group instead of the Pauli group for quantum computing. The Clifford group's members include each of the Pauli matrices as well as the CNOT gate. All these gates are unitary, meaning they take vectors on the Bloch sphere to vectors on the Bloch sphere. For those with a bit of a math background, the Clifford group normalizes the Pauli group. It is generated by CNOT, H and phase gates. The number of Cliffords is upper bounded by the number of permutations of Paulis, permutations on 4^n. With one additional non-Clifford gate like T, we have a universal gateset for quantum computing.

$$CNOT = \begin{bmatrix} 1\,0\,0\,0 \\ 0\,1\,0\,0 \\ 0\,0\,0\,1 \\ 0\,0\,1\,0 \end{bmatrix}$$

When the Pauli matrices are applied via matrix multiplication to a given quantum state it is necessary to restate them in the correct dimensionality so that the entries acting on a qubit not targeted by the operation have no effect on the untargeted qubit, much like the upper left quadrant of the CNOT itself.

$$S_{x_0} = S_x \otimes I = \begin{bmatrix} 0\,0\,1\,0 \\ 0\,0\,0\,1 \\ 1\,0\,0\,0 \\ 0\,1\,0\,0 \end{bmatrix}$$

$$S_{x_1} = I \otimes S_x = \begin{bmatrix} 0\,1\,0\,0 \\ 1\,0\,0\,0 \\ 0\,0\,0\,1 \\ 0\,0\,1\,0 \end{bmatrix},$$

etc.

This can be done in Python with the following code snippet.

```
def correct_gate_dimensionality(gate, target, num_qubits):
    "@\textcolor{mauve}{""}@
    Correct a gate's dimensionality so it may be applied to a circuit.
    :param target: The target qubit.
    :param gate: The gate whose dimensionality to correct.
    :return: The corrected gate.
    """
    return reduce(
        lambda state, d_qubit: np.kron(state, d_qubit),
        [(np.eye(2) if q != target else gate) for q in range(num_qubits)]
    )
```

The multiple N-qubit state is a vector with a length appropriate to the group and is a tensor product of individual qubit states.

$$(\alpha_0 |0\rangle + \beta_0 |1\rangle) \otimes (\alpha_1 |0\rangle + \beta_1 |1\rangle) \otimes \cdots \otimes (\alpha_N |0\rangle + \beta_N |1\rangle)$$

A tensor product of two vectors or matrices is defined as the block matrix (or vector) of the second matrix (or vector) multiplied by each element of the first, i.e. for 2×2 matrices:

$$\begin{bmatrix} a_{00} & a_{01} \\ a_{10} & a_{11} \end{bmatrix} \otimes \begin{bmatrix} b_{00} & b_{01} \\ b_{10} & b_{11} \end{bmatrix} = \begin{bmatrix} a_{00} \begin{bmatrix} b_{00} & b_{01} \\ b_{10} & b_{11} \end{bmatrix} & a_{01} \begin{bmatrix} b_{00} & b_{01} \\ b_{10} & b_{11} \end{bmatrix} \\ a_{10} \begin{bmatrix} b_{00} & b_{01} \\ b_{10} & b_{11} \end{bmatrix} & a_{11} \begin{bmatrix} b_{00} & b_{01} \\ b_{10} & b_{11} \end{bmatrix} \end{bmatrix}.$$

More generally, for two states defined over orthonormal bases $\{|e_i\rangle\}$ and $\{|f_j\rangle\}$,

$$\left(\sum_i a_i |e_i\rangle\right) \otimes \left(\sum_j b_j |f_j\rangle\right) = \sum_{ij} a_i b_j |e_i\rangle \otimes |f_j\rangle$$

and the tensor product has an orthonormal basis $\{|e_i\rangle \otimes |f_j\rangle\}$.

There are a number of ways to engineer CNOT gates that each have their own physical interpretations depending on the mechanism for qubit interaction. However, it is very desirable to support the conditional (controlled) application of a gate such as the X gate in quantum computing architectures since this is a basic networked switching function.

2.9 Quantum Algorithms

All gate model quantum computing software packages include gate libraries. So, while we can implement them ourselves (as we just did for the sake of doing it ourselves once), we can usually skip this step. Let's consider a few examples of how we might implement a simple circuit in Pennylane, Cirq and QisKit by making use of their gate libraries.

The simple circuit that we choose to present first is the superdense coding circuit, for its illustrativeness of quantum superposition and entanglement. Then we look at order finding, because it is the basis of Shor's algorithm and the threat of quantum computing to modern cryptographic systems. There are other useful quantum algorithms such as Grover's search, the basis of many quantum algorithms that might have advantage over classical approaches due to the search's use of quantum parallelism; and phase estimation, the basis of quantum algorithms that are useful for chemical simulation and drug discovery via solving protein folding [28]. This is not a long list and it excludes some algorithms such as Deutsch-Jozsa, Bernstein-Vazirani and

Simons algorithms which are more like toy algorithms that illustrate something about quantum computing rather than provide practical utility. Indeed, there are really only a few useful quantum algorithms and all the fancy applications either borrow from or build upon them. Even still, our goal in this section is not to disseminate everything about every program that can be run on a quantum computer. We try instead to expose the reader to a few good examples and invite the interested reader to further study with a book such as [29] or [20] which make these algorithms their focus. We will also include some more on algorithms, such as the Quantum Fourier Transform, in the appendices. However, these topics are really quantum computer science and less quantum computer engineering and therefore semi-adjacent to our focus in this book.

The general setup for gate model quantum computing involves an input register $|x\rangle$, output register $|y\rangle$ and unitary operator U.

$$U\,|x\rangle\,|y\rangle = |x\rangle\,|f(x) \oplus y\rangle\,;\;\; U : x \rightarrow f(x)$$

The input register is left unchanged and the output register contains the output of the computation at the end of the algorithm.

2.9.1 *Superdense Coding*

We'll begin with a simple quantum circuit called the quantum superdense coding circuit [20, p. 146]. The circuit in question looks like the following (Fig. 2.2).

This circuit achieves something that a classical computer can't by using quantum entanglement as a resource. By creating a Bell state, and sending half of it to a receiving party, the sender of a message sets the stage for "superdense" communication. After encoding a two-bit message by applying Paulis (equal to spin operators S_x and S_z in many quantum computing implementations) to the qubit he/she still has, the sender then transmits the final half of the entangled two-qubit state to the receiver. Then the receiver performs a measurement that allows him/her to learn the two-bit message. The neat part is that the two halves of the entangled state can be as far apart as you like and the entanglement will still be just as valid. Also, at a later time, only one quantum bit (qubit) needs to be operated on and transmitted in order for the receiver to learn two bits of information. Usually, we name this sender Alice and the receiver Bob.

The superdense coding circuit is a good introduction to quantum algorithms because it is simple, and it makes use of both superposition and entanglement. The

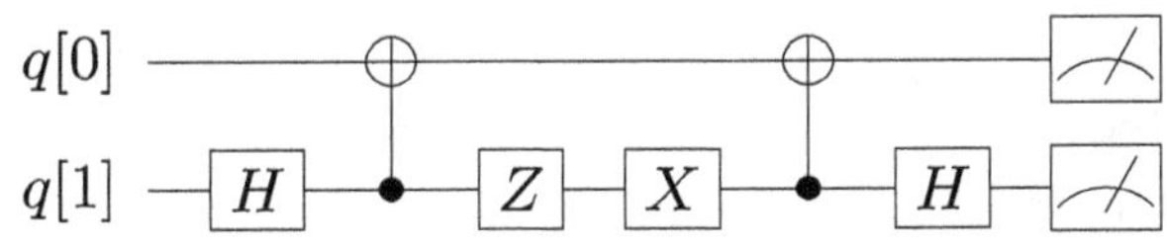

Fig. 2.2 Superdense coding circuit

first step of the algorithm is in fact to put the quantum state (such as the intrinsic angular momentum or spin of a particle) of a qubit into a superposition. This is done with a Hadamard gate, which we have seen previously.

$$H = \frac{1}{\sqrt{2}} \begin{bmatrix} 1 & 1 \\ 1 & -1 \end{bmatrix}$$

When applied to a basis state vector by simple matrix multiplication, you can easily see how this operator creates one of the superpositions (which we have also seen before).

$$H\,|0\rangle = \frac{1}{\sqrt{2}} \begin{bmatrix} 1 & 1 \\ 1 & -1 \end{bmatrix} \begin{bmatrix} 1 \\ 0 \end{bmatrix} = \frac{1}{\sqrt{2}} \begin{bmatrix} 1 \\ 1 \end{bmatrix} = \frac{1}{\sqrt{2}}(|0\rangle + |1\rangle)$$

$$H\,|1\rangle = \frac{1}{\sqrt{2}} \begin{bmatrix} 1 & 1 \\ 1 & -1 \end{bmatrix} \begin{bmatrix} 0 \\ 1 \end{bmatrix} = \frac{1}{\sqrt{2}} \begin{bmatrix} 1 \\ -1 \end{bmatrix} = \frac{1}{\sqrt{2}}(|0\rangle - |1\rangle)$$

The probability of finding this property in the $|0\rangle$ state is equal to the probability of finding it in the $|1\rangle$ state. These are only two of infinitely many superpositions of the two basis states.

The next step in the algorithm is to finish creating entanglement between the two particles. The four two-qubit computational basis states are $|00\rangle, |01\rangle, |10\rangle, |11\rangle$. The entanglement procedure used, which includes the creation of superposition we already covered, will create a Bell state. The specific Bell state will depend on the initial computational basis state of the particles. The final step to the entanglement creation procedure is the application of a CNOT with the superposition state as the control. This simply flips the state of the target bit when the control is in the $|1\rangle$ state and does nothing when it is in the $|0\rangle$ state. Since quantum mechanics is a linear theory and any state an be expressed in the computational basis, as a superposition of the basis states in the most complex case, we can always figure out the effect of the CNOT on any state by simply looking at each ket and flipping the appropriate bits in each one. The outcome of the entangling procedure is the following where $a \in \{0, 1\}$ and $b \in \{0, 1\}$ are the labels in the initial basis state $|ab\rangle$.

$$\frac{1}{\sqrt{2}}(|0b\rangle + (-1)^a\,|1\bar{b}\rangle)$$

Consider the case where we begin with a $|00\rangle$ state. The particles are then entangled (creating the $\frac{1}{\sqrt{2}}(|00\rangle + |11\rangle)$ state) and one of the two particles is shared with Bob. Alice keeps the other one. Then when Alice performs her encoding operations X and/or Z, we can again check how these operations affect the overall state by checking how each Pauli operates on each of the basis states that make up the superposition. The $X = \begin{bmatrix} 0 & 1 \\ 1 & 0 \end{bmatrix}$ operator flips the target bit while the $Z = \begin{bmatrix} 1 & 0 \\ 0 & -1 \end{bmatrix}$ operator flips the sign of the term containing the ket if the target bit is in the $|1\rangle$ state ((Table 2.3)).

Table 2.3 Superdense message encodings

Message	Encoding ops	State after ops
00	I	$\frac{1}{\sqrt{2}}(\|00\rangle + \|11\rangle)$
01	Z	$\frac{1}{\sqrt{2}}(\|00\rangle - \|11\rangle)$
10	X	$\frac{1}{\sqrt{2}}(\|01\rangle + \|10\rangle)$
11	−iY=XZ	$\frac{1}{\sqrt{2}}(\|01\rangle - \|10\rangle)$

The last step in the superdense coding algorithm is to send Bob the second qubit so that he can perform a Bell measurement. The Bell measurement involves unentangling the two particles by applying a CNOT followed by a Hadamard.[9] You can check that flipping the first bit whenever the second bit is 1 in the "state after ops" column of the above table always yields an overall state where the first bit is the same in all the kets. This means that the state can once again be expressed as a tensor product of the first particle's state with the second. It can be factored out. The definition of entanglement is when we can't do this. Finally, the Hadamard collapses the superposition of the second qubit to a basis state. This is also easy to check. The final state exactly reads as the encoded message.

Below are the promised Python implementations. Note how each package slightly differs. In QisKit the gates are methods on the circuit object, while in Pennylane they are defined under the top-level package and belong to a circuit by way of being called in a function decorated by the *@qml.qnode* decorator. In Cirq they must be appended to the circuit as though the circuit is a list of gates.

Example: Superdense Coding with Pennylane

```
import pennylane as qml

MSG = '01' # change to send another message

# Setup Two-Qubit Device
NUM_QUBITS = 2
dev = qml.device("default.qubit", wires=NUM_QUBITS)

@qml.qnode(dev)
def circuit():
    # Create Bell State
    qml.Hadamard(wires=1)
    qml.CNOT(wires=[1, 0])

    # Alice sends the first qubit to Bob

    # Encoding
    if msg[0] == '1':
        qml.PauliX(wires=1)
    if msg[1] == '1':
        qml.PauliZ(wires=1)

    # Alice sends the second qubit
```

[9] The Hadamard gate is named after Jacques Hadamard, who worked on radar for the Allies during WWII and who wrote on the psychology of invention in his book, *The Psychology of Invention in the Mathematical Field* [30].

```
    # Bell Measurement
    qml.CNOT(wires=[1, 0])
    qml.Hadamard(wires=1)

    return qml.expval(qml.Z(0)), qml.expval(qml.Z(1))
```

Example: Superdense Coding with Cirq

```
import cirq

MSG = '01' # change to send another message

# Setup Two-Qubit Device
circuit = cirq.Circuit()
q0, q1 = cirq.LineQubit.range(2)

# Create Bell State
circuit.append([cirq.H(q1), cirq.CNOT(q1, q0)])

# Alice sends the first qubit to Bob

# Encoding
if msg[0] == '1':
    circuit.append(cirq.X(q1))
if msg[1] == '1':
    circuit.append(cirq.Z(q1))

# Alice sends the second qubit

# Bell Measurement
circuit.append([cirq.CNOT(q1, q0), cirq.H(q1)])

circuit.append(cirq.measure(q1, q0, key="output"))
```

Example: Superdense Coding with Qiskit

```
from qiskit import QuantumCircuit

MSG = '01' # change to send another message

# Setup Two-Qubit Device
qc = QuantumCircuit(2)

# Create Bell State
qc.h(1)
qc.cx(1, 0)

# Alice sends the first qubit to Bob

# Encoding
if msg[0] == '1':
    qc.x(1)
if msg[1] == '1':
    qc.z(1)

# Alice sends the second qubit

# Bell Measurement
qc.cx(1, 0)
qc.h(1)

qc.measure_all()
```

2.9.2 Order Finding

Order finding [20, p. 63] is an application of quantum computers with significant implications due to the reliance of RSA encryption on the hardness of the problem of factorization. Factorization is a problem related to order finding in a way we will see shortly. Classically, the algorithm we will study scales with $O(\exp(n^{1/3}))$ complexity which is very poor. This is why RSA encryption works. The scaling of a quantum Shor's algorithm is however $O(n^2)$. Shor's algorithm combines quantum order finding with a classical step to crack RSA encryption. A quantum computer was first reported to experimentally "hack" encryption algorithms that underlie AES encryption in 2024 [31].

Before we begin Shor's algorithm, it is important to understand how RSA encryption works exactly. The protocol involves key generation, encryption and decryption. Key generation involves calculating a public key from two prime numbers p and q. First, the receiver of a message (Bob) computes $p \cdot q = N$ and $(p-1) \cdot (q-1) = M$. Next Bob chooses an integer c such that $gcd(M, c) = 1$. d is computed next so that $c \cdot d = 1(mod\ M)$. Finally, Bob announces N and c, the public key. The key generation protocol is complete. To encrypt a message a, Alice may now compute $b = a^c(mod\ N)$. This b may then be sent to Bob. Bob may decrypt the message by applying his private key, calculating $b^d(mod\ N) = a$.

We begin Shor's algorithm with the goal of factoring a public key N. We can do this easily if N is even (has a factor of 2) or if N is a prime power. In the case that N is not even or a prime power, however, quantum computing may be useful. The first step in the algorithm is to choose an integer $l < N$. It is easy to efficiently calculate the greatest common denominators of two integers using the Euclidean algorithm, and we do this to estimate a factor $N/gcd(l, N)$ of N.

If this yields the trivial factor $gcd(l, N) = 1$ then we know that l, N are co-prime so that $l^r = 1(mod\ N)$ for some r. We call this r the multiplicative order of l. We use the quantum computer to perform order finding which tells us the value of r.

The quantum order finding routine begins with a state of $|0\ldots0\rangle$ in both input and output registers. A Hadamard is applied to each input qubit and then a function $f(x) = l^x(modN)$ so that our input and output registers become $\sum_{x=0}^{2^n-1} |x\rangle\,|f(x)\rangle$. Note how we can label the states by $|0\rangle, |1\rangle, |2\rangle, \ldots, |2^n-1\rangle$ to denote $|0\ldots00\rangle$, $|0\ldots01\rangle, |0\ldots10\rangle, \ldots, |1\ldots11\rangle$. The size of our second register is $\lceil \log_2 N \rceil$ and the size of the first one can be chosen to be large enough that we represent r with the desired accuracy.

The period of $f(x)$ is r so that $f(x+r) = f(x) = l^x l^r = l^x$ iff $l^r = 1$. We measure the output register obtaining outcome $f(x_{out})$ which corresponds to an input $x_{in} = x_{out} + kr;\ k \in \mathbf{Z}$. The state becomes $\sum_{k=0}^{m-1} |x_{in} + kr\rangle\,|f(x_{out})\rangle$. Next we apply a Quantum Fourier Transform (QFT).

A QFT performs a Discrete Fourier Transform (DFT) on the complex probability amplitudes of the quantum state, converting a function to the frequency domain. The DFT function is the following.

$$y_k = \frac{1}{\sqrt{n}} \sum_{j=0}^{n-1} x_j \exp(2\pi ijk/n)$$

In the case of the Quantum Fourier Transform, we have a unitary operator U_{QFT}. This operator is unitary because products of unitaries are unitary. For an explicit decomposition of this U_{QFT} to gates and controlled gates please see appendix B. There we also include Pennylane code to implement a QFT.

$$U_{QFT} = \frac{1}{\sqrt{2^n}} \sum_{y=0}^{2^n-1} \exp(\frac{2\pi ixy}{2^n}) |y\rangle$$

We apply the QFT to our state in the order finding problem.

$$\begin{aligned} U_{QFT} &= \sum_{k=0}^{m-1} \frac{1}{\sqrt{2^n}} \sum_{y=0}^{2^n-1} \exp(\frac{2\pi iy(x_{out}+kr)}{2^n}) |y\rangle \\ &= \frac{1}{\sqrt{2^n}} \sum_{y=0}^{2^n-1} \exp(\frac{2\pi iyx_{out}}{2^n}) \sum_{k=0}^{2^n-1} \exp(\frac{2\pi iykr}{2^n}) |y\rangle \end{aligned}$$

We measure the input register and the most likely outcomes are multiples of r. This is because $(\exp(\frac{2\pi iyx_{out}}{2^n}))^2 = 1$ and $|\sum_{k=0}^{m-1} \exp(\frac{2\pi iy_0kr}{2^n})|^2$ approaches 1 when $k = \frac{2^n}{r}$ and y_0 is a multiple of r.

We can implement Shor's algorithm in Pennylane, which is an exercise left to the reader in their codebook. Pennylane provides two functions for us, one called *get_matrix_l_mod_N* that takes N, l as parameters and returns the unitary matrix $U_{N,l}$ so that $U_{N,l}^x = |f_{N,l}(x)\rangle$. The other, *get_period*, computes r using this $U_{N,l}$.

To see how we can implement something like these provided functions ourselves, we can refer to [32] which provides for us examples of quantum circuits that implement such functions as required to implement the order finding algorithm using CNOTs and Toffolis, which are like CNOTs but have two controls. These circuits are essentially constructed straightforwardly from our functions' $(f_{N,l}(x))$ truth tables.

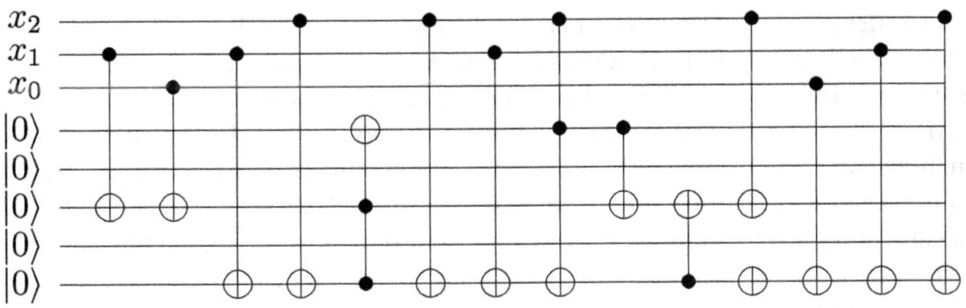

Fig. 2.3 $f_{21,4}(x)$ circuit

Example: $4^x\,(mod\ 21)$ **Implementation** For example, for the case where $N = 21$, we have an input register of three qubits and $l = 4$, and the truth table is (Fig. 2.3; Tables 2.4, 2.5):

```
@qml.qnode(dev)
def f_21_4():
    qml.CNOT([1, 5])
    qml.CNOT([2, 5])
    qml.CNOT([1, 7])
    qml.CNOT([0, 7])
    qml.Toffoli([7, 5, 3])
    qml.CNOT([0, 7])
    qml.CNOT([1, 7])
    qml.Toffoli([0, 3, 7])
    qml.CNOT([3, 5])
    qml.CNOT([7, 5])
    qml.PauliX(7)
    qml.CNOT([0, 5])
    qml.CNOT([2, 7])
    qml.CNOT([1, 7])
    qml.CNOT([0, 7])
```

Table 2.4 $f_{21,4}(x)$

x	0	1	2	3	4	5	6	7
$4^x\,(mod\ 21)$	1	4	16	1	4	16	1	4

Table 2.5 $f_{21,4}(x)$ truth table

x_2	x_1	x_0	y_4	y_3	y_2	y_1	y_0
0	0	0	0	0	0	0	1
0	0	1	0	0	1	0	0
0	1	0	1	0	0	0	0
0	1	1	0	0	0	0	1
1	0	0	0	0	1	0	0
1	0	1	1	0	0	0	0
1	1	0	0	0	0	0	1
1	1	1	0	0	1	0	0

A similar circuit would exist for each selection of $|x|$, N, l. We could build CNOT circuits that implement these sorts of functions programmatically using the implementation provided in the appendices that uses 5 output qubits.

An approach is first tried which relies on the ability to solve linear systems of equations with binary variables. Less brute-force constructions also exist, such as that presented in [33]. This approach, based on the Cuccaro ripple carry adder [34], is a valid approach to building any modular multiplication circuit, but the automated linear system solver may solve for more T-gate-count efficient circuits at the cost of additional runtime in theory. We default to the Cuccaro adder-based approach when the linear system solver fails.

Let's consider the linear system solver approach. Gaussian elimination and LU-decomposition can be used to solve for the circuit that implements a reversible classical function using CNOT gates, but more efficient techniques also exist [35].

Bijective functions are those such that the inputs can be recovered from the outputs and all output values are achievable. Reversible circuits implement bijective functions. A quantum circuit implements a unitary function and is, therefore, reversible. Linear reversible classical circuits can be generated by CNOT gates without needing any other types of gates.

Synthesizing arbitrary linear reversible circuits on n wires using CNOTs is equivalent to row reducing an n x n binary matrix. Gaussian elimination and LU-decomposition take $O(n^2)$ ops to do this. The authors' proposed approach in [35] does it in $O(n^2/\log(n))$. The CNOT gate corresponds to multiplication by an elementary matrix and row operations can be performed by the multiplication of an elementary matrix. To be reversible, an operator's matrix should be such that all non-trivial sums of the rows are nonzero; i.e. the first row can be anything but the 0 row, the second can be anything but the 0 row or the first row, etc., to be bijective. This means the operator's matrix is non-singular. Any non-singular matrix can be reduced to the identity using row operations, so the operator matrix is equal to a product of elementary matrices. Therefore, any linear reversible function can be synthesized using CNOTs. After saying all this, the authors of [35] prove that $O(n^2/\log(n))$ is a lower bound. The sketch of the proof is as follows. They argue that the number of different CNOTs that can be applied on n wires is $n(n-1)$. The number of unique CNOT circuits with no more than d gates, therefore, must be no more than $(n^2 - n + 1)^d$. The number of circuits with no more than d CNOT gates must be greater than the number of unique linear reversible functions on n wires. This gives us an inequality. Taking the log of the left and right sides gives $d \geq O(n^2/\log(n))$. This inequality suggests that there should be a way to implement things efficiently. Gaussian elimination uses row operations to place ones on the diagonal and to eliminate remaining ones and takes $O(n^2)$ row ops since there are n^2 matrix entries. Instead, the [35] propose to group entries together and eliminate repeating patterns in these subrows, reducing the number of needed row ops and, therefore, the number of required gates. The remaining rows are handled by Gaussian elimination. The length of the synthesized circuit is the number of row ops. They show that the worst case for their algorithm is indeed $O(n^2/\log(n))$. They also show some simulation results demonstrating that their result is better than straight Gaussian elimination.

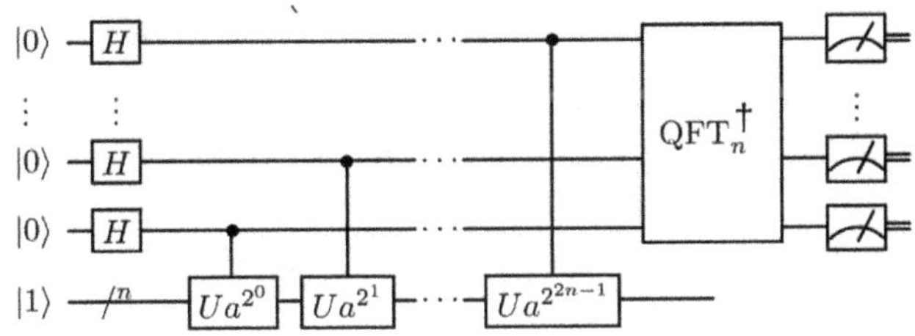

Fig. 2.4 Shor circuit

Now let's consider the approach based on the Cuccaro adder. The quantum circuit used in Shor's algorithm reduces to the following, involving several controlled constant modular multiplications Ul^{2^i} where $Ul^{2^i} : |b\rangle \to |ba^{2^i} \bmod N\rangle$ (Fig. 2.4).

Each modular multiplication involves a number of shifted additions of b: $a_0^{2^i} b + a_1^{2^i} 2b + a_2^{2^i} 2^2 b + \cdots + a_{n-1}^{2^i} 2^{n-1} b$. Instead of shifting b by a different amount to facilitate each individual addition in the long sum, the shifting can be done by adding the register containing b to the most significant $n - j$ bits of the output register, where j is the subscript on the $a_j^{2^i}$ and the amount of the shift. The building of the Cuccaro adder is implemented in the appendix recursively, evaluating one addition in the sum $(((a_0^{2^i} b + a_1^{2^i} 2b) + a_2^{2^i} 2^2 b) + \cdots + a_{n-1}^{2^i} 2^{n-1} b)$ per function call, where the scope of the evaluation in each stack frame is denoted by the brackets. The value b was persisted in an ancilla register. To make the modular multiplication in-place, a register swap and an inverse out-if-place multiplication operation were used.

$$|b\rangle\,|0\rangle \xrightarrow{Ua^{2^i}} |b\rangle\,|a^{2^i} \cdot b \bmod N\rangle \xrightarrow{\text{swap}} |a^{2^i} \cdot b \bmod N\rangle\,|b\rangle$$
$$\xrightarrow{(U(a^{2^i})^{-1})^\dagger} |a^{2^i} \cdot b \bmod N\rangle\,|0\rangle$$

Here, $(U(a^{2^i})^{-1})^\dagger : |x\rangle\,|(a^{2^i})^{-1} \cdot x \bmod N\rangle \to |x\rangle\,|0\rangle$.

References

1. Ghose, S.: Quantum Computing at Wilfrid Laurier University. Undergraduate Course (2017)
2. Ghose, S.: Quantum Mechanics 1 at Wilfrid Laurier University. Undergraduate Course (2017)
3. Haranas, I.: Quantum Mechanics 2 at Wilfrid Laurier University. Undergraduate Course (2018)
4. Salfi, J.: Quantum Information and Computation at University of British Columbia. Graduate Course (2023)
5. Sakurai, J.J., Napolitano, J.: Modern Quantum Mechanics, 2nd edn. Dorling Kindersley, New Delhi (2014)
6. Serrano, M.A., Pérez-Castillo, R., Piattini, M. (eds.): Quantum Software Engineering. Springer International Publishing, Cham (2022). https://doi.org/10.1007/978-3-031-05324-5. https://link.springer.com/10.1007/978-3-031-05324-5

7. Wong, H.Y.: Introduction to Quantum Computing: From a Layperson to a Programmer in 30 Steps. Springer International Publishing, Cham (2024). https://doi.org/10.1007/978-3-031-36985-8. https://link.springer.com/10.1007/978-3-031-36985-8
8. Cline, B.L.: Louis de Broglie. Biography, Atomic Theory, Discovery, & Facts, Britannica. https://www.britannica.com/biography/Louis-de-Broglie
9. de Broglie, L.: Recherches sur la Théorie des Quanta. Masson, Paris (1924)
10. Quantum numbers from h to ne. https://www.chemteam.info/Electrons/QuantumNumbers-H-to-Ne.html
11. Griffiths, D.J.: Introduction to Quantum Mechanics, 2nd edn. Pearson Prentice Hall, Upper Saddle River, NJ (2005)
12. Hawking, S.: The Dreams that Stuff is Made of: The Most Astounding Papers on Quantum Physics and How They Shook the Scientific World. Running Press (2011)
13. Hosch, W.L.: Ludwig Boltzmann. Statistical mechanics, Thermodynamics, Kinetic theory, Britannica (2025). https://www.britannica.com/biography/Ludwig-Boltzmann
14. Sharp, K., Matschinsky, F.: Translation of ludwig boltzmann's paper "on the relationship between the second fundamental theorem of the mechanical theory of heat and probability calculations regarding the conditions for thermal equilibrium" sitzungberichte der kaiserlichen akademie der wissenschaften. mathematisch-naturwissen classe. abt. ii, lxxvi 1877, pp 373–435 (wien. ber. 1877, 76:373–435). reprinted in wiss. abhandlungen, vol. ii, reprint 42, p. 164–223, barth, leipzig, 1909. Entropy **17**(4), 1971–2009 (2015). https://doi.org/10.3390/e17041971. https://www.mdpi.com/1099-4300/17/4/1971
15. Planck, M.: The Theory of Heat Radiation. P. Blakiston's Son & Co., Pennsylvania (1914)
16. Stuewer, R.H.: Max Planck. Biography, Discoveries, & Quantum Theory, Britannica (2025). https://www.britannica.com/biography/Max-Planck
17. Einstein, A.: On the electrodynamics of moving bodies. Annalen Phys. **17**, 891–921 (1905). https://doi.org/10.1002/andp.200590006
18. Struik, D.J.: Joseph Fourier. Biography & Facts, Britannica (2025). https://www.britannica.com/biography/Joseph-Baron-Fourier
19. Beyler, R.: Werner Heisenberg-Quantum Physicist, Nobel Prize, Nazi Party, Britannica (2025). https://www.britannica.com/biography/Werner-Heisenberg
20. Mermin, N.D.: Quantum Computer Science: An Introduction. Cambridge University Press (2007). Google-Books-ID: q2S9APxFdUQC
21. Bardin, J.C., Sank, D., Naaman, O., Jeffrey, E.: Quantum computing: an introduction for microwave engineers. IEEE Microwave Mag. **21**(8), 24–44 (2020). https://doi.org/10.1109/MMM.2020.2993475. https://ieeexplore.ieee.org/document/9134701/?arnumber=9134701
22. Yard, J.: Quantum Information Processing at University of Waterloo. Graduate Course (2018)
23. Bruß, D.: Optimal eavesdropping in quantum cryptography with six states. Phys. Rev. Lett. **81**(14), 3018–3021 (1998). https://doi.org/10.1103/physrevlett.81.3018. http://dx.doi.org/10.1103/PhysRevLett.81.3018
24. Renner, R.: Security of Quantum Key Distribution (2006). https://arxiv.org/abs/quant-ph/0512258
25. Gregersen, E.: Gilles Brassard. Biography, BB84, Quantum Teleportation, & Facts | Britannica. https://www.britannica.com/biography/Gilles-Brassard
26. Bennett, C.H., Brassard, G.: Quantum Cryptography: Public Key Distribution and Coin Tossing (1984)
27. Wilkins, D.: Sir William Rowan Hamilton. Irish Mathematician, Astronomer & Physicist, Britannica (2025). https://www.britannica.com/biography/William-Rowan-Hamilton
28. Mustafa, H., Morapakula, S.N., Jain, P., Ganguly, S.: Variational Quantum Algorithms for Chemical Simulation and Drug Discovery (2022). https://arxiv.org/abs/2211.07854
29. Silva, V.: Quantum Computing by Practice: Python Programming in the Cloud with Qiskit and IBM-Q. Apress, Berkeley (2024). https://doi.org/10.1007/978-1-4842-9991-3. https://link.springer.com/10.1007/978-1-4842-9991-3
30. Hadamard, J.-S.: French Mathematician. Britannica. https://www.britannica.com/biography/Jacques-Salomon-Hadamard

31. Swayne, M.: Chinese Scientists Report Using Quantum Computer to Hack Military-Grade Encryption (2024). https://thequantuminsider.com/2024/10/11/chinese-scientists-report-using-quantum-computer-to-hack-military-grade-encryption/
32. Gamel, O., James, D.F.V.: Simplified Factoring Algorithms for Validating Small-Scale Quantum Information Processing Technologies (2013). https://arxiv.org/abs/1310.6446
33. Beauregard, S.: Circuit for Shor's Algorithm Using 2n+3 Qubits (2003). https://doi.org/10.48550/arXiv.quant-ph/0205095. http://arxiv.org/abs/quant-ph/0205095. ArXiv:quant-ph/0205095
34. Cuccaro, S.A., Draper, T.G., Kutin, S.A., Moulton, D.P.: A New Quantum Ripple-carry Addition Circuit (2004). https://doi.org/10.48550/arXiv.quant-ph/0410184. http://arxiv.org/abs/quant-ph/0410184. ArXiv:quant-ph/0410184
35. Patel, K.N., Markov, I.L., Hayes, J.P.: Efficient Synthesis of Linear Reversible Circuits (2003). https://doi.org/10.48550/arXiv.quant-ph/0302002. http://arxiv.org/abs/quant-ph/0302002. ArXiv:quant-ph/0302002

Chapter 3
Quantum Computing in the Lab

Abstract As the title suggests, the third chapter focuses on the realities of how a qubit behaves in a laboratory setting in the real world (Fig. 3.1). The many types of non-idealities that one encounters are described and modeled mathematically. A tutorial is provided for countering decoherence and errors with dynamical decoupling, and error correction techniques are also covered.

You could find the information synthesized into this chapter across multiple papers that will be cited throughout as well as the overall framework in textbooks including [1, 2] and (very briefly) in [3]. You might also find some of the broad topics covered in an upper-level quantum information course at any university, such as [4–6]. We provide (we hope) a good, concise and unique summary as well as very practical examples which you would not find in an information theory focused text or course.

3.1 Learning Objectives

In this third chapter, we will look at how the essentials we have learned thus far translate into a more realistic setting for quantum computing. This realistic setting involves things like decoherence and noise which come from interactions with the environment. We will aim to understand these and how to address them. Some specific learning objectives include:

- Understand how to model decoherence and noise.
- See how these models apply specifically to superconducting qubits.
- See how these models apply specifically to spin qubits.
- Know how to transform between the rotating and lab frames.
- Understand error correction and how to code it.

© The Author(s), under exclusive license to Springer Nature Switzerland AG 2026

M. S. Edwards, *An Introduction to Quantum Computing for Computer Engineers*,
https://doi.org/10.1007/978-3-032-03650-6_3

Fig. 3.1 Learning objective for this chapter: realities of quantum computing

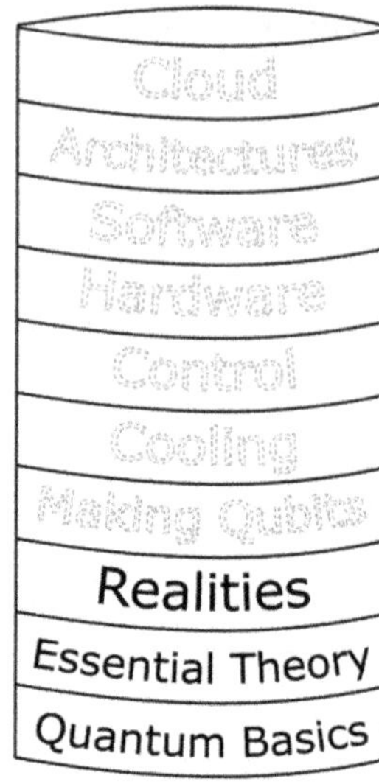

3.2 Recap

In summary of what we have seen thus far, we can arrive at the specific computational basis vectors typically used for quantum computing with electrons or hole spins by considering the Pauli[1] spin operators **S** of a particle such as an electron [2, p. 165].

$$S_x = \frac{\hbar}{2}\begin{bmatrix} 0 & 1 \\ 1 & 0 \end{bmatrix} \; S_z = \frac{\hbar}{2}\begin{bmatrix} 1 & 0 \\ 0 & -1 \end{bmatrix} \tag{3.1}$$

$$S_y = -i S_x S_z$$

We take our computational basis vectors $|l, m_s\rangle \in \{|0\rangle, |1\rangle\}$ to be defined by the eigenvalue equation $S_z |l, m_s\rangle = \pm\frac{\hbar}{2} |l, m_s\rangle$.

$$|1\rangle = \begin{bmatrix} 0 \\ 1 \end{bmatrix} \quad |0\rangle = \begin{bmatrix} 1 \\ 0 \end{bmatrix}$$

Together with matrix multiplication, a few other matrices and the CNOT gate, the Pauli matrices form the Clifford algebra. The Clifford algebra is the mathematical backdrop of quantum computing in most cases, though there are alternative universal gatesets [8, 9].

[1] Wolfgang Pauli published his first academic paper, on relativity, at the young age of 18 [7].

In terms of the basis vectors $\{|0\rangle, |1\rangle\}$ we can write a quantum (qubit) state as $|\psi\rangle$.

$$|\psi\rangle = \alpha|0\rangle + \beta|1\rangle \ ; \alpha, \beta \in \mathrm{C}$$

Since α and β are complex, they are normalized by taking their squares. Physically, the squares of these coefficients are the probabilities of observing either basis state via the a measurement in the appropriate basis.

$$p(\psi = 0) = |\alpha|^2$$

$$p(\psi = 1) = |\beta|^2$$

$$|\alpha|^2 + |\beta|^2 = 1$$

$$M_0 = \begin{bmatrix} 1 & 0 \\ 0 & 0 \end{bmatrix} \quad M_1 = \begin{bmatrix} 0 & 0 \\ 0 & 1 \end{bmatrix}$$

It is a simple process to normalize a quantum state. The normalization requirement restricts our state vector to a sphere called the Bloch sphere. This makes it natural to express the state in polar coordinates.

$$\begin{aligned}
|\psi_{\text{normalized}}\rangle &= \frac{1}{\sqrt{|\alpha|^2 + |\beta|^2}} |\psi\rangle \\
|\alpha|^2 &= \cos^2\left(\frac{\theta}{2}\right) \\
|\beta|^2 &= \left(\sin\left(\frac{\theta}{2}\right)\exp(i\phi)\right)^2 \\
|\psi\rangle &= \cos\left(\frac{\theta}{2}\right)|0\rangle + \sin\left(\frac{\theta}{2}\right)\exp(i\phi)|1\rangle
\end{aligned}$$

The set of matrices that take a state on the Bloch sphere to another state on the Bloch sphere are the same that rotate a vector on a unit sphere. These are the unitary transformations U with the following form and properties. U rotates a state vector by γ about (n_x, n_y, n_z) (Fig. 3.2).

$$\begin{aligned}
U^\dagger U &= I \\
U^\dagger &= (U^*)^T \\
U &= \exp(i\delta)[\cos\left(\frac{\gamma}{2}\right) I + i\sin\left(\frac{\gamma}{2}\right)(n_x S_x + n_y S_y + n_z S_z)]
\end{aligned}$$

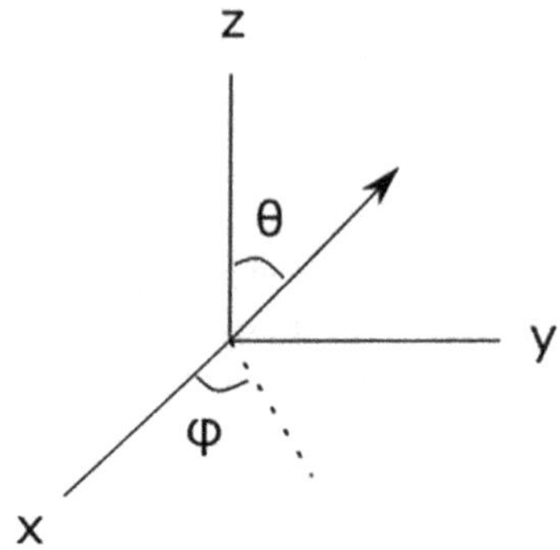

Fig. 3.2 Bloch sphere

Any state vector within one of these spheres can also be represented by a density matrix ϱ, which is a 2×2 square matrix. The density matrix corresponding to a pure quantum state is given by the following [1, p. 99]:

$$\varrho = \begin{bmatrix} |\alpha|^2 & \alpha\beta^* \\ \alpha^*\beta & |\beta|^2 \end{bmatrix}.$$

In the Schrödinger picture, any system dynamics will be modeled by changes to the state. Changes to the state in time will be modeled by the applications of unitary operators [1, p. 99].

$$\varrho(t) = U_t \varrho(0) U_t^\dagger$$

3.3 Decoherence and Noise

What we have covered thus far is an idealized description. In reality, the spin state of a given particle tends to decohere which is not conducive to quantum computing. It can even experience non-unitary evolution and is in general no longer represented by a vector on the Bloch sphere but instead within it. In the worst case, the state decoheres to the maximally mixed state at the center of the Bloch sphere. A state with non-unitary evolution can be best represented by a density matrix [1, p. 99]. A density matrix is only pure (on the Bloch sphere) if it has rank 1 and most generally has the form below for pure states ψ_s.

$$\varrho = \sum_s p_s |\psi_s\rangle \langle\psi_s|$$

For a given $|\psi_s\rangle$ we have for some α, β:

$$|\psi_s\rangle = \alpha\,|0\rangle + \beta\,|1\rangle$$
$$|\psi_s\rangle\,\langle\psi_s| = \begin{bmatrix}\alpha\\ \beta\end{bmatrix}\begin{bmatrix}\alpha^* & \beta^*\end{bmatrix}$$
$$= \begin{bmatrix}|\alpha|^2 & \alpha\beta^*\\ \alpha^*\beta & |\beta|^2\end{bmatrix}$$

The measurement of an outcome k would have the probability $\langle v_k|\varrho|v_k\rangle = |\langle v_k|\psi_s\rangle|^2$ for measurement basis state composed of orthonormal vectors $|v_1\rangle \ldots |v_d\rangle$. If k is observed, ϱ collapses to $\varrho = |v_k\rangle\,\langle v_k|$.

To apply a unitary to the state, we apply it to both sides of ϱ like $U\varrho U^\dagger$. Qubit density matrices have the following form, where the Bloch vector $\begin{bmatrix}r_x & r_y & r_z\end{bmatrix}$ gives a point on or within the Bloch sphere so that $|r| \leq 1$ and $r \in \mathsf{R}^3$.

$$\varrho_r = \frac{1}{2}(I + r_x X + r_y Y + r_z Z)$$

If we recall that a given qubit state can be provided in polar form like $|\psi\rangle = \cos(\theta/2)\,|0\rangle + \exp(i\phi)\sin(\theta/2)\,|1\rangle$ where $0 \leq \theta \leq \pi$ and $0 \leq \phi \leq 2\pi$, then we can also give a Bloch vector as $r = \begin{bmatrix}\sin\theta\cos\theta & \sin\theta\sin\phi & \cos\theta\end{bmatrix}$.

How do we model decoherence and noise?
The density matrix, which can represent decoherence in the form of mixed states, can be expressed as

$$\varrho = \frac{1}{2}\begin{bmatrix}1+\cos\theta & \exp(i\phi)\sin\theta\\ \exp(i\phi)\sin\theta & 1-\cos\theta\end{bmatrix}$$

The action of a measurement of the state can be represented, as stated earlier, by a POVM $\{M_j\}$ which is now a set of matrices defined such that $\sum_j M_j = I$ [1, p. 102]. The probability associated with a single outcome j is $p(j) = Tr\varrho M_j$, and we can also generalize to the probability of outcomes like $p(j \in J) = Tr\varrho\sum_{j\in J} M_j$. For a concrete example, consider the following M_0 and M_1 which correspond to a computational basis measurement.

$$M_0 = \begin{bmatrix}1 & 0\\ 0 & 0\end{bmatrix}$$
$$M_1 = \begin{bmatrix}0 & 0\\ 0 & 1\end{bmatrix}$$

The trace, or more generally the partial trace, of a matrix is defined as follows [1, p. 105]. $Tr_B\varrho$ is the unique linear operator such that $Tr_B(P_A \otimes P_B) = P_A$. That is why we call this action "tracing out B" from $P_A \otimes P_B$. We can similarly trace out A.

In general, if $\varrho_{AB} = \begin{bmatrix} G_{1,1} & \dots & G_{1,a} \\ \vdots & \ddots & \vdots \\ G_{a,1} & \dots & G_{a,a} \end{bmatrix}$

then, $Tr_B(\varrho_{AB}) = \begin{bmatrix} Tr\ G_{1,1} & \dots & Tr\ G_{1,a} \\ \vdots & \ddots & \vdots \\ Tr\ G_{a,1} & \dots & Tr\ G_{a,a} \end{bmatrix}$

and $Tr_A(\varrho_{AB}) = G_{1,1} + \cdots + G_{a,a}$ (summing up the diagonal blocks).

A physically relevant description of the quantum state in the Bloch–Redfield density matrix ϱ_{BR} [10]. Here the longitudinal and transverse relaxations of the spin state are represented by $\exp(-\Gamma_1 t)$ and $\exp(-\Gamma_2 t)$, respectively. Transverse refers to xy rotation. $\delta\omega$ gives an offset of the qubit frequency ω_q from the rotating frame frequency ω_d (we will discuss in a subsequent section). Relaxation times T_1, T_2 are given in terms of relaxation rates Γ_1, Γ_2 by $\frac{1}{\Gamma_1}$, $\frac{1}{\Gamma_2}$, respectively. T_1 gives the energy relaxation time and T_2 the dephasing time. Typical T_1 relaxation times for hole spin qubits in gated quantum dots range from 0.3 μs at $B_z = 1.5T$ to 60 μs at $B_z = 0.5T$ [11], for example. Notably, there hasn't been a lot of experimental analysis of T1 in p-type quantum dots in high magnetic fields [12].

$$\varrho_{BR} = \begin{bmatrix} 1 + (|\alpha|^2 - 1)\exp(-\Gamma_1 t) & \alpha\beta^* \exp(i\delta\omega t)\exp(-\Gamma_2 t) \\ \alpha^*\beta \exp(-i\delta\omega t)\exp(-\Gamma_2 t) & |\beta|^2 \exp(-\Gamma_1 t) \end{bmatrix}$$

The longitudinal relaxation time is related to the rates of erroneous spin flips $\Gamma_{1\uparrow}$ and $\Gamma_{1\downarrow}$. We have a relaxation to the ground state for transverse vectors similar to $\Gamma_{1\downarrow}$.

$$\Gamma_1 = \frac{1}{T_1} = \Gamma_{1\uparrow} + \Gamma_{1\downarrow}$$

For a typical qubit we will reach an equilibrium due to Boltzmann statistics. Typical experimental values are $\frac{\omega_q}{2\pi} = 5$ GHz and $T = 20$ mK. For hole spins in gated quantum dots, the $|\uparrow\rangle \rightarrow |\downarrow\rangle$ transition depends on B_z and can be measured using a spin-readout protocol [11] (Fig. 3.3).

$$\Gamma_{1\uparrow} = \exp(-\hbar\omega_q / k_B T)\Gamma_{1\downarrow}$$

Stochastic fluctuations of qubit frequency result in different precession frequencies. This is called "pure dephasing" Γ_ϕ.

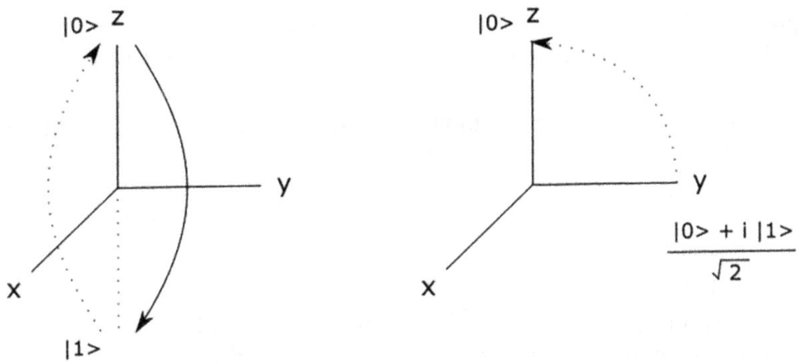

Fig. 3.3 Spin flips and relaxation

Fig. 3.4 Pure dephasing

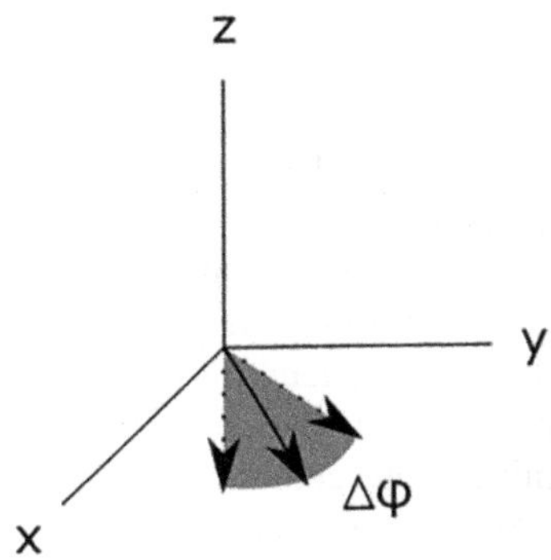

T_2 is given by a combination of the transverse relaxation and dephasing [13] (Fig. 3.4).

$$\frac{1}{T_2} = \frac{1}{2T_1} \frac{1}{T_\phi}$$

> **How do we model errors?**
> We can model errors as quantum channels. In general, a quantum channel is a completely positive, trace preserving linear map or "CPTP" map. A quantum channel $\mathsf{N}(p)$ can often be written in what is called Kraus form.

$$\mathsf{N}(p) = \sum_i N_i p N_i^\dagger$$

Here the N are matrices of the same size such that $\sum N_i^\dagger N_i = I$.

A quantum channel operates on complex matrices and takes Hermitians to Hermitians. Even if only part of a state is acted upon, N will take density matrices to density matrices.

Some types of noise that are common are bit flips and phase flips. A bit flip channel is described by the following channel [1, p. 429]:

$$\mathsf{N} : \varrho \to (1-p)\varrho + pX\varrho X.$$

Here X is the Pauli X matrix which maps the $|0\rangle$ state to the $|1\rangle$ state and vice versa. It is also often denoted σ_x. p is the probability of a bit flip. We apply X to the left of ϱ and the right of ϱ since the way to apply a unitary to a density matrix is $U\varrho U^{\dagger}$, but the Pauli matrices X, Y, Z are their own adjoints.

A phase flip channel is given by the following [1, p. 430]:

$$\mathsf{N} : \varrho \to (1-p)\varrho + pZ\varrho Z.$$

Here Z is the Pauli Z operator which changes the sign of a $|1\rangle$ state. It is also often denoted σ_z.

When we are dealing with decoherence, our state can decohere to the point that it is not a pure state any longer. This means it is not longer represented by a vector on the Bloch sphere. Instead, the state is best represented by a probabilistic mixture of pure states and a vector somewhere inside the Bloch sphere. We call the maximally mixed state, at the center of the Bloch sphere, a thermalized state. The channel which takes a state to the maximally mixed state is the following.

$$\mathsf{N}(\varrho) = \frac{1}{2}\begin{bmatrix} 1 & 0 \\ 0 & 1 \end{bmatrix}$$
$$\mathsf{N} : \varrho \to \frac{1}{4}(\varrho + X\varrho X + Y\varrho Y + Z\varrho Z)$$

There are also other channels which represent types of noise processes that are often encountered. Depolarizing channels are very general, and model Pauli noise. Any of the Paulis are erroneously applied with equal probability.

$$\varrho \to (1-p)\varrho + p/3 \sum_i \sigma_i \varrho \sigma_i$$

Dephasing channels tend to reduce the off-diagonal entries of the density matrix and contracts the sphere toward the Z axis.

$$\varrho \to D_0 \varrho D_0^{\dagger} + D_1 \varrho D_1^{\dagger}$$

$$D_0 = \begin{bmatrix} 1 & 0 \\ 0 & \sqrt{1-\gamma} \end{bmatrix}$$
$$D_1 = \begin{bmatrix} 1 & 0 \\ 0 & \sqrt{\gamma}. \end{bmatrix}$$

We can simulate quantum channels using superoperators. If $\mathsf{N}(\varrho) = \sum_{j=1}^{e} N_i \varrho N_j^\dagger$ is a quantum channel, then there exists a unitary U on a larger system that simulates or realizes this channel. This is called a Stinespring extension.

$$\mathsf{N}(\varrho) = Tr_E U(|0\rangle\langle 0| \otimes \varrho) U^\dagger$$

3.3.1 Decoherence in Superconducting Qubits

Let's drill down into a specific regime: superconducting qubits. We will get into the details of superconducting qubits in a later section, but for now suffice it to say that they are LC circuit in which the energy oscillates between the capacitor C and the nonlinear inductor J. A nonlinear inductor is used so that the energy levels of the qubit subspace are spaced differently than the other energy levels of the system. Qubit rotations are driven by mircowave signals and flux tuning.

Decoherence in superconducting qubits has multiple sources including noise on drivelines and thermal noise. For example, noise on the Z flux driveline results in qubit frequency fluctuations and noise on the other lines cause phase fluctuations [14]. A smaller capacitance of the LC circuit corresponds with more dephasing, so the more anharmonicity the more dephasing noise there is. This presents a trade-off between a well-defined and qubit subspace and one isolated from noise. Gate times on the order of 10ns are required to avoid dephasing noise.

We continue to model noise as follows [10]. The qubit is coupled to several fluctuators. Together these fluctuations generate Gaussian noise.[2] Here Chi generalizes pure dephasing gamma to include non-exponential (i.e. Gaussian) decay $\exp(-(t/T_{\phi,G})^2)$.

$$\varrho = \begin{bmatrix} 1 + (|\alpha|^2 - 1)\exp(-\Gamma_1 t) & \alpha\beta^* \exp(i\delta\omega t)\exp(-\frac{\Gamma_1}{2}t)\exp(-\chi_N(t)) \\ \alpha^*\beta\exp(-i\delta\omega t)\exp(-\frac{\Gamma_1}{2}t)\exp(-\chi_N(t)) & |\beta|^2\exp(-\Gamma_1 t) \end{bmatrix}$$

Noise power spectral density is given in terms of the autocorrelation function $c_\lambda(\mathsf{T}) = \langle \lambda(\mathsf{T})\lambda(0)\rangle$ for a noise source λ. This is connected to the longitudinal relaxation rate via the qubit Hamiltonian H_q. There is a similar relationship for transverse relaxation.

[2] Carl Friedrich Gauss, the namesake of Gaussian noise, famously was prodigious at calculating complex mathematical expressions entirely in his head [15].

$$S_\lambda(\omega) = \int_{-\infty}^{\infty} d\mathsf{T}\, \langle \lambda(\mathsf{T})\lambda(0)\rangle \exp(-i\omega \mathsf{T})$$

$$\Gamma_1 = \frac{1}{\hbar^2} |\langle 0| \frac{\partial H_q}{\partial \lambda} |1\rangle|^2 S_\lambda(\omega_q)$$

We can model charge/ohmic noise by $S_Q(\omega)$. The transmon is a type of superconducting charge qubit, meaning that its basis states are charge states representing the occupation of excess Cooper pairs on an island defined by a superconducting-insulating-superconducting junction. Charge noise arises from occupation fluctuations of these states and leads to fluctuations in the local electric field. The transmon is a type of superconducting qubit that is less sensitive to charge noise than its predecessor, the Cooper Pair Box (CPB). This is due to the fact that we operate the CPB and transmon in different regimes ($E_C >> E_J$ vs. $E_J >> E_C$, respectively). In transmons specifically, the anharmonicity reduces algebraically with a power law in E_J/E_C, but the charge dispersion reduces exponentially. Therefore, by operating the qubit at a large E_J/E_C we combat charge noise with minimal compromise in the anharmonicity [16].

$$S_Q(\omega) = \sum_n |\frac{\partial J}{\partial V_n}|^2\, SV_n(\omega)$$

Here J is an exchange coupling between two spins, δJ is exchange noise, $\frac{\partial J}{\partial V_n}$ gives the charge noise sensitivity and V_n are fluctuating gate voltages. Operating regions of low charge sensitivity are dubbed "sweet spots".

Magnetic flux noise's physical manifestation remains uncertain, but it can be modeled by $S_\phi(\omega)$. It is possible that it comes from paramagnetic center fluctuations, which might eventually be important to the performance of very clean devices. Here Φ_0 is the smallest amount of magnetic flux allowed by quantum mechanics.

$$S_\phi(\omega) = A_\phi^2 \left(\frac{2\pi \times 1\,\mathrm{Hz}}{\omega}\right)^{\gamma_\phi}$$

$$A_\phi^2 = (1\mu\Phi_0)^2/\mathrm{Hz}$$

$$\gamma_\phi \doteq 0.8 - 1.0$$

Strategies for noise mitigation [10] include substrate cleaning, substrate annealing, trenching, optical surface treatment and quasiparticle traps. Capacitor geometry can also be altered to reduce electric field density. Asymmetric junctions can be used to reduce flux noise. The so-called spin-echo technique disrupts the free evolution of the quantum state by a pi pulse to combat dephasing. A CPMG sequence of π pulses can be used to achieve dynamical decoupling of the state. The stochastic pumping of quasiparticles is another approach to combating noise sources since quasiparticles (unpaired electrons) can lead to both longitudinal relaxation and dephasing. Furthermore, we can reduce photon flux by efficient heat sinking and/or absorptive black material. We can also make use of additional cavity filters for thermalization. QPUs should also be carefully isolated from dangerous electrostatic discharges before being isolated from the rest of the environment. It is important to use grounding straps and anti-static mats when carefully handling chips to avoid this kind of electrostatic discharge, on the more practical side.

3.3.2 Decoherence in Spin Qubits

We can model a spin qubit's coupling to a noisy magnetic field using $H_{\text{noise}} = -\hbar(g\mu_B\delta B(t)/\hbar)\cdot S$ [17]. We can say that if the fluctuations of this noisy magnetic field are parallel to a large applied magnetic field causing spin precession, then $b(t) = |g\mu_B\delta B(t)/\hbar|$ is the fluctuation in the Larmor[3] frequency of the spin caused by the noise.

Hole spin qubits have large spin-orbit fields meaning that T_2, T_2^* are limited by charge noise [17]. Spin-orbit interaction (SOI) refers to the coupling of a particle's spin with a magnetic field as it travels through the magnetic field due to its orbit. A strong SOI means that the spin can be manipulated using electric fields without the need for a micromagnet implanted in the substrate, which is a common approach to creating spin-orbit effects when working with electrons. The energy separation between states is also larger for holes than electrons, so weaker magnetic fields can be used for Zeeman splitting. Qubit lifetimes are much longer for hole spin qubits in Germanium than in Gallium Arsenide, for example, since the dominant isotopes of Ge have no nuclear spin. Enhanced SOI however increases the effects of charge noise on the spin [11]. Therefore there is a trade-off between our ability to make use of Electric-Dipole Spin Resonance (EDSR) to drive our qubit with an electric field and our sensitivity to charge noise.

Besides charge noise, the main contributor to decoherence in spin qubits are phonons. Phonons are quanta of motional or vibrational energy in a crystal of oscillating atoms. Phonons introduce spin-flip elements to the matrix describing a spin qubit's evolution which are similar to those that enable electric driving. However, in the case of phonons, these elements contribute to qubits' relaxation. Fluctuations induced by both charge noise and phonons can also affect the g-factor of our qubit and cause dephasing [1, p. 344] (the g-factor gives the relationship between our Larmor frequency and the oriented magnetic field). EDSR strength is linear in

[3] Sir Joseph Larmor believed that all matter was composed of electric particles moving through the mysterious substance called ether which is no longer considered to exist [18].

SOI, but relaxation and dephasing rates are quadratic [11]. Therefore it is important to find SOI in a range that optimizes the trade-off between decoherence and fast EDSR. Phonon-electron and phonon-hole couplings can be modeled using Density Functional Theory (DFT) perturbation theory or KP perturbation theory, describing the positions of atoms in the material relative to ideal positions and then create a deformation potential which is seen by the electron or hole [19, 20].

Another contributor to decoherence in spin qubits is interaction with the nuclear bath [11]. The charge carrier moves through the electric and magnetic fields generated by a collection of nuclei. The electron or hole orbital and spin degrees of freedom are coupled to the nuclear spin. This "hyperfine" interaction has three terms: the Fermi contact interaction, dipolar coupling and orbital coupling. Solid-state holes interact less strongly in this way than electrons do [12].

3.4 Rotating Frames

A concept that is essential when we are working with qubits is the rotating frame. We will often convert between the "lab frame" and a "rotating frame" which precesses with the qubit at the Larmor frequency [2, p. 167].

An electron in a magnetic field has a Hamiltonian with the form below. Here e is the charge of the electron, 1.6×10^{-19} C and m_e is the effective mass of the electron. **B** is a magnetic field and **S** gives the qubit state.

$$H = \frac{|e|}{m_e}\mathbf{S} \cdot \mathbf{B}$$

We are often concerned with the case where the magnetic field is in the z direction.

$$\begin{aligned} H &= \frac{|e|}{m_e} B_z \cdot S_z \\ \omega_0 &= \frac{|e|}{m_e} B_z \end{aligned}$$

Say we had an initial state that was an eigenstate of S_x.

$$\begin{aligned} |x_+\rangle &= \frac{1}{\sqrt{2}}(|+\rangle + |-\rangle) \\ < \mathbf{s} >_{x_+} &= \frac{\hbar i}{2} \end{aligned}$$

The time evolution operator is given by $U(t) = \exp(\frac{-i\omega_0 S_z t}{\hbar})$. Then,

$$
\begin{aligned}
|\alpha(t)\rangle &= U(t)\,|x_+\rangle = \frac{1}{]\sqrt{2}}\left(\exp\left(\frac{-i\omega_0 t}{2}\right)|+\rangle + \exp\left(\frac{i\omega_0 t}{2}\right)|-\rangle\right) \\
< S_x >_\alpha &= \langle\alpha(t)|\, S_x\, |\alpha(t)\rangle \\
&= \langle\alpha(t)|\, \frac{\hbar}{2}(|+\rangle\langle-| + |-\rangle\langle+|)\, |\alpha(t)\rangle \\
&= \frac{\hbar}{2}\cos(\omega_0 t)
\end{aligned}
$$

Similarly, $< S_y >_\alpha = \frac{\hbar}{2}\sin(\omega_0 t)$ and $< S_z >_\alpha = 0$. Combining these, $< \mathbf{S} >_\alpha = \frac{\hbar}{2}(cos(\omega_0 t)i + \sin(\omega_0 t)j)$ where i, j are unit vectors in x, y coordinates.

One can see that the spin precesses around the z-axis in the $x - y$ plane. The frequency at which this precession takes place is called the "Larmor" frequency.

To help us present qubit evolution in a concise and understandable way, it is typical to make a change of coordinates to the so-called rotating frame which has no relative motion to this spin precession.

Consider a time-dependent Hamiltonian $H(t)$.

$$H(t) = H_0 + V(t)$$

Let $U(t_0, t)$ be the propagator of a state $|\psi(t)\rangle$ such that $|\psi(t)\rangle = U(t_0, t)\,|\psi(t_0)\rangle$. Then consider the propagator of the time-independent state $U_0(t) = U_0$.

$$U_0(t) = \exp(\frac{-i}{\hbar}\int_{t_0}^{t} dt'\, H_0(t'))$$

This $U_0(t)$ follows from the fact that discrete solutions of the time-dependent Schrödinger equation have the form $\psi(x, t) = \sum_n c_n \phi_n(x) \exp(\frac{-i}{\hbar} E_n t)$.

Consider time-independent forces.

$$U(x, t) = U(x)$$

Then the solutions to the time-dependent Schrödinger equation are separable such that:

$$
\begin{aligned}
&\psi(x, t) = \phi(x)\chi(t) \\
&\frac{-\hbar^2}{2m}\chi(t)\frac{\partial^2\phi(x)}{\partial x^2} + U(x)\phi(x)\chi(t) = i\hbar\phi(x)\frac{\partial\chi(t)}{\partial t} \\
&\frac{\hbar^2}{2m\phi(x)}\frac{\partial^2\phi(x)}{\partial x^2} + U(x) = i\hbar\frac{1}{\chi(t)}\frac{\partial\chi(t)}{\partial t}
\end{aligned}
$$

These can only be equal if $\frac{-\hbar^2}{2m}\frac{\partial^2\phi(x)}{\partial x^2} + U(x)\phi(x) = E\phi(x)$ and $i\hbar\frac{\partial\chi(t)}{\partial t} = E\chi(t)$ for some constant E.

$$\therefore \ \chi(t) = \chi(0)\exp(\frac{-iEt}{\hbar})$$

See that $i\hbar\partial_t U_0(t_0, t) = H_0(t)U_0(t_0, t)$. Let $|\phi(t)\rangle = R(t)\,|\psi(t)\rangle$ for some rotation operator $R(t)$. Then solve for the time dependence of $\phi(t)$.

$$\begin{aligned} i\hbar\partial_t\,|\phi(t)\rangle &= i\hbar\partial_t(R(t)\,|\psi(t)\rangle) \\ &= i\hbar\partial_t R(t)\,|\psi(t)\rangle + R(t)i\hbar\partial_t\,|\psi(t)\rangle \\ &= i\hbar\dot{R}(t)\,|\psi(t)\rangle + R(t)H(t)\,|\psi(t)\rangle \\ &= i\hbar\dot{R}(t)R^\dagger(t)\,|\phi(t)\rangle + R(t)H(t)R^\dagger(t)\,|\phi(t)\rangle \\ &= (i\hbar\dot{R}(t)R^\dagger(t) + R(t)H(t)R^\dagger(t))\,|\phi(t)\rangle \end{aligned}$$

$\therefore$ $|\phi(t)\rangle$ obeys Schrödinger's equation $i\hbar\partial_t\,|\phi(t)\rangle = H\,|\phi(t)\rangle$ with a Hamiltonian $H'(t)$.

What is a rotating frame?

$$H'(t) = i\hbar\dot{R}(t)R^\dagger(t) + R(t)H(t)R^\dagger(t)$$

This is called a rotating frame. For a specific rotating frame consider the example $R(t) = U_0(t_0, t)$ which gives the so-called interaction picture.

Spin precession is a key concept required for us to understand strategies to combat noise such as dynamical decoupling.

3.5 Tutorial: Combating Decoherence

There are various strategies for combating decoherence in qubits, which vary from simplistic to more complex schemes. A common technique is called refocusing the state, and we will see it used in both Hahn echo and the dynamical decoupling sequences CP and CPMG named after Carr, Purcell, Meiboom and Gill. Carr and Purcell optimized the pulse sequence used, while Meiboom and Gill improved upon the approach by using better control electronics [21].

Each of the schemes we will address stem from the discoveries of Erwin Hahn,[4] who was the first to see echoes in nuclear magnetic resonance (NMR) in 1950 [23]. The original experiment performed by Hahn involved two $\pi/2$ pulses, with a time

[4] The uses of nuclear magnetic resonance include magnetic resonance imaging (MRI). As such, Erwin Hahn now has an institute for magnetic resonance imaging named after him [22].

delay between their applications, which had the effect of refocusing spins which experienced decoherence during the time delay.

A primary source of dephasing is local field inhomogeneity, which may differ between qubits at different sites in a molecular liquid, for example. In NMR quantum computing experiments, unlike superconducting experiments, we usually work with a molecular bath involving many qubits which we program at once, and so we are more likely to think about groups of spins rather than singular qubits. However, the techniques apply equally to both settings.

Consider a group of spins that have different precession frequencies due to pure dephasing caused by local field inhomogeneities. The cross-product of the local magnetic field with the spinful particle's magnetic dipole moment gives a torque. The angular momentum is calculated from this moment by dividing by the gyromagnetic ratio, which is dependent on the particle's charge and mass.

The higher the field, the faster the precession and vice versa. Depending on the local field polarity, the spin will either rotate clockwise or counter-clockwise in the xy plane of the Bloch sphere.

In order to study dephasing on its own, we prepare a state in the xy plane by the application of a $\pi/2$ pulse to the $|0\rangle$ state. Then, imagine that this shared initial state begins to dephase differently for n individual qubits. Some gain phase faster and some slower. The cumulative signal that we would measure if we did a readout of our group of qubits loses strength as a result, as their states no longer agree.

Hahn assumed a rectangular spectrum. For every spin accumulating phase clockwise on the Bloch sphere there is a spin losing phase (rotating counterclockwise in xy in the rotating frame) at the same rate so that we have isochromatic moment pairs. As these pairs of spins dephase, they maintain symmetry about the initial non-dephased state.

What Hahn noticed in 1950 was that when he applied a second $\pi/2$ pulse at time τ, at a time equal to 2τ a signal with an intensity less than that of the original signal appeared. This "echo" of the original is also called a refocusing. Prior to time 2τ, all the state vectors interfere destructively, but at time 2τ they once again interfere constructively as a result of the dephasing continuing to maintain symmetry about the initial state, until they refocus on the other side of the sphere.

In the way that Hahn performed this in initial experiment more than one echo resulted. If the second pulse is applied at the time that the first moment pair reaches the x axis, then the most intense echo occurs at time 2τ. This most intense pulse represents the special case where the moment pair vectors were on the x axis at the time of the second $\pi/2$ pulse. This would be the moment pair with the fastest phase accumulation. The rest of the pairs precess in planes parallel to xy and refocus at later times due to their slower phase accumulations, creating echoes of smaller magnitude.

Carr and Purcell proposed in 1954 that instead of a second $\pi/2$ pulse, a π pulse could be applied in order to maximize the echo signal at 2τ. The π pulse rotates moment pairs about the x axis until they are reflected in the xy plane about the x axis. The exact phase that has been accumulated as a result of dephasing for each moment pair is then the same as the amount that must be accumulated before the vectors recombine to form the reflection across the x axis of the initial non-dephased state. All the moment pairs will recombine at the same time as a result, maximizing the echo signal.

Repeating the wait of the τ followed by another π pulse and final wait of time τ can yield another refocusing which can be compared to the previous measurement. One will find a decaying signal intensity this way. Repetitions of this wait, π pulse, wait, measure sequence can therefore be used to measure the T2 relaxation time and to refocus the state over longer periods of time [21]. The envelope of the subsequent echoes shows the decay of the polarization in the xy plane.

These "spin echoes" can be applied to otherwise idling qubits in order to reduce the effects of dephasing. See below how we might insert spin echoes to a circuit in Cirq in order to fight decoherence during a computation [24].

Example: Dynamical Decoupling in Cirq

```
from itertools import map
import numpy as np
import cirq

# generate several pi pulses
pi_pulses = [
    cirq.PhasedXPowGate(phase_exponent=p, exponent=1.0) for p in (−0.5, 0.0, 0.5, 1.0)
]

# choose three random pi pulses
echo_count = 3
state = np.random.RandomState(1)

echo = [
    random_state.choice(pi_pulses) for _ in range(echo_count)
]

# apply the pulses to the qubit
map(lambda pulse: [pulse.on(qubit[0]), cirq.inverse(pulse.on(qubit[0]))])

# insert the pulses into the circuit
circuit.insert(4, echo)

circuit = cirq.stratified_circuit(
    circuit,
    categories=[lambda op : len(op.qubits) == 1, lambda op : len(op.qubits) == 2]
)
```

Here the *cirq.PhasedXPowGate(phase_exponent=p, exponent=t)* creates an operator with the following matrix [25].

$$\begin{bmatrix} \exp(i\pi\, t/2)\cos(\pi t/2) & -i\exp(i\pi(t/2-p))\sin(\pi t/2) \\ -i\exp(i\pi(t/2+p))\sin(\pi t/2) & \exp(i\pi\, t/2)\cos(\pi t/2) \end{bmatrix}$$

Compare this to the traditional Rx gate to see the differences.

$$\begin{bmatrix} \cos(t/2) & -i\sin(t/2) \\ -i\sin(t/2) & \cos(t/2) \end{bmatrix}$$

The *cirq.PhasedXPowGate* gate differs from this in that it has an additional phase shift on the off-diagonals. t is also multiplies π in *cirq.PhasedXPowGate*.

Based on our parameters, we generate the following unitaries, which are all related to the Pauli X gate and are all rotations by π.

$$\begin{bmatrix} 0 & i \\ -i & 0 \end{bmatrix} \begin{bmatrix} 0 & 1 \\ 1 & 0 \end{bmatrix} \begin{bmatrix} 0 & -i \\ i & 0 \end{bmatrix} \begin{bmatrix} 0 & -1 \\ -1 & 0 \end{bmatrix}$$

The cirq.PhasedXPowGate is equal to $Z^{-p}X^{t}Z^{p}$. A rotation of the amount given by t is caused in the xy plane, and p specifies the axis in the xy plane about which this rotation occurs. A π pulse lasts twice as long as a $\pi/2$ pulse and rotates the state vector twice the amount as a result.

After generating these rotations, we randomly choose three of them and insert them into the circuit. Finally, we call *cirq.stratified_circuit* to align our inserted spin echo operations with any other single-qubit gates in the circuit. The idea here is that while qubit zero is idling during time steps where other qubits have operations, we will instead apply a refocusing pulse to combat dephasing.

However, in general there are many pulse level considerations that need to be taken into account when scheduling dynamical decoupling sequences, i.e. what is the state I am refocusing? The example above skips over some of these considerations which you as a pulse sequencer would have to think about. In reality you might need to schedule refocusing pulses more carefully so that you know that the timing of the pulses is correct, i.e. at exactly τ and 2τ.

Compare the Cirq code above to the following snippet of Qiskit code which creates a three qubit Greenberger–Horne–Zelinger[5] state.

Example: Dynamical Decoupling in Qiskit

```
import numpy as np
from qiskit.circuit import QuantumCircuit
from qiskit.circuit.library import XGate
from qiskit.transpiler import PassManager, InstructionDurations
from qiskit.transpiler.passes import ALAPScheduleAnalysis, PadDynamicalDecoupling

# define circuit
circ = QuantumCircuit(4)
circ.h(0)
circ.cx(0, 1)
circ.cx(1, 2)
circ.measure_all()

# provide pulse-level gate details
durations = InstructionDurations(
    [("h", 0, 50), ("cx", [0, 1], 700), ("reset", None, 10),
    ("cx", [1, 2], 200), ("x", None, 50), ("measure", None, 1000)]
)

# generate several pi pulses
```

[5] This simple entangled three-particle state became known as the "GHZ" state, after its namesakes. In the early nineties, these authors along with Abner Shimony made an effort to show Bell's points we discussed briefly before using three-particle GHZ states [26], and they published a paper in 1990 [27].

```
dd_sequence = [XGate(), XGate()]

# insert the pulses into the circuit
pm = PassManager([ALAPScheduleAnalysis(durations),
                  PadDynamicalDecoupling(durations, dd_sequence)])
circ_dd = pm.run(circ)
```

This code snippet applies a sequence XX that multiplies to the identity but refocuses the state when applied at the right time. You only need to provide the gate times associated with each gate implementation in your quantum computer and the *qiskit.transpiler.passes.PadDynamicalDecoupling* call fills in the idle time with XX sequences. This is similar to what we did with *cirq.stratified_circuit* in Cirq. In this case, the padding of the circuit with our decoupling sequences is a transpiler step. This relates to the Qiskit framework which features a concept of transpilers which map circuits to circuits by some transformations provided by the transpiler pass implementations themselves. A user can implement a transpiler pass, or use built-in passes like this one.

There is another simple approach to programming dynamical decoupling pulses provided in Qiskit which is worth mentioning if you are just interested in getting the job done. This approach only requires that you import a package and write a few lines of python, but the 'qiskit_research' package will take the pulse level considerations into account for you.

```
from qiskit_research.utils.convenience import add_dynamical_decoupling

new_circuit = add_dynamical_decoupling(
    circuit,
    backend,
    'XY4',
    add_pulse_cals=True
)
```

Here we specify a particular dynamical decoupling sequence named "XY4" which is XYXY, one of several refocusing sequence options which is useful because it equals the identity but due to the successive pi rotations refocuses the state. Qiskit will insert this sequence where appropriate so that instead of idling, qubits will refocus their states so that dephasing during what would be idle time is mitigated. It will fill idling while qubits would wait for single-qubit gates on other qubits, but more importantly while they would wait for two qubit gates which are generally around ten times longer.

3.6 Quantum Error Correction

We have seen a relatively simple approach to combating dephasing noise by using the noise processes themselves which are generally held to be fixed during an experiment to our advantage. But what about the variety of other noise channels we discussed earlier? Say, a spin flip occurs at a time different from 2τ and ruins our dynamical decoupling sequence. What do we do then?

A random spin flip is a logical error since it looks like the effect of a logical operation (Pauli X) on our qubit. How can we distinguish between errors, logical and otherwise, and correct them systematically? This is where quantum error correction comes in.

Let's first think about how we model errors. Say we have a generalized "message" a that we wish to be preserved through a channel. Then consider a message b that is either the preserved message or a transformation of the original message to an erroneous one.

We can consider the values which b may take on and the values which a may take. Let's denote values that a may take on as a_i for the ith possible value, and similarly for b_j. There should be a probability of finding message b_j after the channel action on the message a_i, $p(b_j|a_j)$. This introductory setup is valid whether our message is a quantum state or a literal message in a bottle.

One approach to combating errors is to simply duplicate the message in hopes that an error does not occur on all of our copies. This is called a repetition code, since it encodes the original message into a larger one by simple repetition. If p_f is the probability that each bit flips then the probability that there is a single bit flip in an n-bit message is $(1 - p_f)^{n-1} p_f$. The probability of two or more bits flipping is $3p_f^2(1 - p_f) + p_f^3$. The repetition code works when $3p_f^2(1 - p_f) + p_f^3 < p_f$.

What is a quantum error correction code?
In general, there are diverse ways one can encode a message into a larger space to introduce tolerance. In this more general case, our Error Correction Code (ECC) is composed of "codewords" which may be used to encode unique pieces of information, like unique quantum states.

When we are considering messages which are binary strings, such as those which label the basis vectors belonging to an orthonormal basis of quantum states, then we concern ourselves with the Hamming distances between the codewords of our code.

$$d(c_i \oplus c_j)$$

The exclusive or operation results in a string with a number of 1s equal to the number of differing bits in the two codewords. Finally, to get our Hamming distance, we count these 1s. The Hamming distance is therefore a measure of how different our codewords are. Our code C itself is characterized by a minimum distance $d(C)$.

$$d(C) = \min_{c_i, c_j \in C\,; i \neq j} d(c_i \oplus c_j)$$

A simple way to "decode" a message encoded into a noisy codeword c_k which has experienced the action of some error channel is to find the codeword c_l with the minimum Hamming distance to the noisy codeword. This kind of decoder is successful if $d(c_k \oplus c_l) < \lfloor \frac{d-1}{2} \rfloor$.

The rate of a code is the ratio between the number of bits in the message a and the number of bits in the codewords c.

$$R_k(C) = \frac{|a|}{|c|}$$

A family of codes is characterized by the number of bits $|c|$ in the codewords. There is a theorem called the Gilbert–Varshamov theorem which bounds the rate of a family of codes $C_{|c|}$ as the size $|a|$ of the message increases.

$$\lim_{|a| \to \infty} R_k(C_{|c|}) \leq 1 - h(p)$$

Here $h(p)$ is the binary entropy function and p is the probability that a binary variable takes a certain value. $h(p)$ gives the amount of information that can be encoded into a variable based on how much freedom there is in the values it can take. It is maximized for a bit, then, when the probability of its taking a given value is $\frac{1}{2}$.

$$h(p) = -p \log_2 p - (1-p) \log_2(1-p)$$

A similar concept to binary entropy is Shannon[6] entropy [1, p. 572]. Given any random variable X which can take values in the set $\chi = \{x_i, x_{i+1}, \dots\}$ with probabilities $p(x_i), p(x_{i+1}), \dots$, the Shannon entropy is $H(X) = -\sum_{x \in \chi} p(x) \log_2 p(x)$. $0 \leq H(X) \leq \log_2(|X|)|X|^{-1}$. It gives the amount of information needed to describe the state of the variable. Another related quantity is Shannon information $I(X, Y)$ which gives us a measure of how surprising a particular outcome is [1, p. 553].

$$\begin{aligned} H(X, Y) &= -\sum_{i,j} p(x_i, y_j) \log_2 p(x_i, y_j) \\ &= H(X) + H(X|Y) \\ I(X, Y) &= H(X) + H(Y) - H(X, Y) \\ &= H(Y) - H(Y|X) \end{aligned}$$

Error correction starts to resemble something useful as we begin to consider how it can be represented in terms of linear algebra. A so-called linear ECC is given by a binary generation matrix $G \in \{0, 1\}^{|a| \times |c|}$. The top $|a|$ rows of G are denoted $\mathcal{I}_{|a|}$ and the bottom $|c| - |a|$ denoted G'.

[6] Claude Shannon, an electrical engineer, is often referred to as the father of information theory, and his Masters thesis on using Boolean algebra to describe electric circuits is considered one of the 20th century's most important theses [28].

The minimum distance of a code generated by a binary generation matrix is given by the Hamming weight minimized over all the linear combinations of the columns of G.

Decoding a linear ECC involves a parity check matrix H where $HG = 0$, $H \in \{0, 1\}^{(|c|-|a|)\times|c|}$. The first $|a|$ columns of H are denoted H' and the rest are $\mathcal{I}_{|c|-|a|}$. $H' = G'$.

To encode a message a into a codeword c, we apply the matrix G s.t. $c = Ga$. Next the channel action on the codeword yields c'. $c' = c \oplus e$. We finally decode the noisy codeword into a syndrome s where $s = Hc' = H(c \oplus e) = He \because Hc = HGa = 0$.

This error syndrome gives us a way to infer the error s.t. $E_s = \{e|He = s\}$. Then we can also infer a correction $c' + \bar{e}$.

Let's think about how we could use the same framework to correct errors on quantum states. We need a way to decode error syndromes. The ability to measure Hermitian unitaries comes in handy here.

$M_2 H_2 cU_{2,1} H_2$ measures the observable U on the state $|\psi\rangle$. $|\psi\rangle$ may be a multi-qubit state and U a multi-qubit unitary. This is called a "Hadamard test" (Fig. 3.5).

In most cases, quantum computers are not equipped with physical measurement apparatuses besides the single-qubit Z measurement. Therefore, it is very helpful to be able to measure arbitrary unitaries by using this trick. It only requires that we are able to implement the controlled unitary and the single-qubit Hadamards.

Let's take a look at how this works for a useful two-qubit unitary: $Z \otimes Z$. After the first time step (vertical column of gates) in the circuit we have:

$$\frac{1}{\sqrt{2}}(|0_1\rangle + |1\rangle_1)\,|\psi_{23}\rangle$$

and after the second time step:

$$\frac{1}{\sqrt{2}}(|0_1\rangle\,|\psi_{23}\rangle + |1_1\rangle\,Z_2 Z_3\,|\psi_{23}\rangle)$$

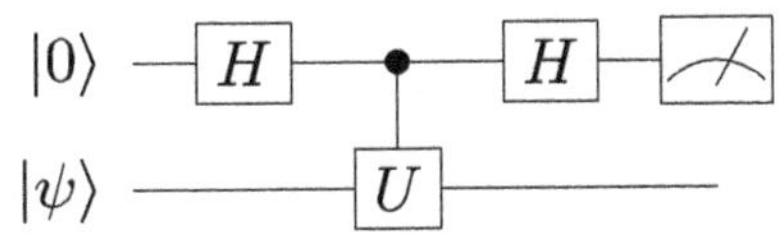

Fig. 3.5 Hadamard test circuit

and after the third:

$$\frac{1}{\sqrt{2}}(\frac{1}{\sqrt{2}}(|0_1\rangle + |1_1\rangle)\,|\psi_{23}\rangle + \frac{1}{\sqrt{2}}(|0_1\rangle - |1_1\rangle)Z_2Z_3\,|\psi_{23}\rangle)$$
$$= \frac{1}{2}(|0_1\rangle\,(Z_2Z_3\,|\psi_{23}\rangle + |\psi_{23}\rangle) + |1_1\rangle\,(|\psi_{23}\rangle - Z_2Z_3\,|\psi_{23}\rangle))$$

The Hadamard test allows us to map the eigenvalues of the unitary U onto our single-qubit Z measurement outcomes. The unitary $Z \otimes Z$ (or Z_2Z_3) has the following matrix form:

$$\begin{bmatrix} 1 & 0 & 0 & 0 \\ 0 & -1 & 0 & 0 \\ 0 & 0 & -1 & 0 \\ 0 & 0 & 0 & 1 \end{bmatrix}.$$

This matrix has eigenvalues $+1$ and -1. The eigenvectors $(1, 0, 0, 0)^T$ and $(0, 0, 0, 1)^T$ correspond to the eigenvalue $+1$ and the others $((0, 1, 0, 0)^T$ and $(0, 0, 1, 0)^T)$ to the eigenvalue -1. Remember what these eigenvectors mean to us. They are the eigenstates $|00\rangle$, $|11\rangle$, $|01\rangle$ and $|10\rangle$ in the order that they appeared in the previous sentence.

Now that we know these eigenvector and eigenvalue pairings, let's set $|\psi_{23}\rangle = |00\rangle$ to see how the expected outcome from measuring the unitary $(+1)$ maps to the outcome of the single-qubit Z measurement. Recall that the Pauli Z matrix $\begin{bmatrix} 1 & 0 \\ 0 & -1 \end{bmatrix}$ $= |0\rangle\,\langle 0| - |1\rangle\,\langle 1|$.

Example: Hadamard Test Measurement

$$\frac{1}{2}(|0_1\rangle\,((|0\rangle\,\langle 0| - |1\rangle\,\langle 1|)_2(|0\rangle\,\langle 0| - |1\rangle\,\langle 1|)_3\,|0_20_3\rangle + |0_20_3\rangle)$$
$$+ |1_1\rangle\,(|0_20_3\rangle - (|0\rangle\,\langle 0| - |1\rangle\,\langle 1|)_2(|0\rangle\,\langle 0| - |1\rangle\,\langle 1|)_3\,|0_20_3\rangle))$$
$$= \frac{1}{2}(|0_1\rangle\,((|0\rangle\,\langle 0| - |1\rangle\,\langle 1|)_2\,|0_20_3\rangle$$
$$+ |0_20_3\rangle) + |1_1\rangle\,(|0_20_3\rangle - (|0\rangle\,\langle 0| - |1\rangle\,\langle 1|)_2\,|0_20_3\rangle))$$
$$= \frac{1}{2}(|0_1\rangle\,(|0_20_3\rangle + |0_20_3\rangle) + |1_1\rangle\,(|0_20_3\rangle - |0_20_3\rangle))$$
$$= |0_10_20_3\rangle$$

The measurement outcome of the first qubit will be $+1$ since $+1$ is the eigenvalue corresponding with the eigenstate $|0\rangle$ of Z with $|\,\langle 0|0\rangle\,|^2 = 100\%$ likelihood. We can see that an eigenstate $|\psi_{23}\rangle = |01\rangle$ similarly maps to the appropriate eigenvalue (-1).

Table 3.1 Bit flip error detection circuit

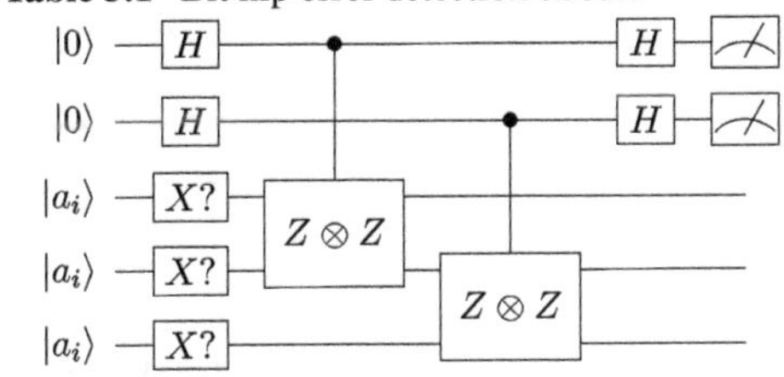

Table 3.2 Single bit flip error corrections

Z_1Z_2	Z_2Z_3	Error	Correction
$+1$	$+1$	No flip	$I_1I_2I_3$
$+1$	-1	Qubit 3 flipped	$I_1I_2X_3$
-1	$+1$	Qubit 1 flipped	$X_1I_2I_3$
-1	-1	Qubit 2 flipped	$I_1X_2I_3$

$$\begin{aligned}
&\frac{1}{2}(|0_1\rangle\,((|0\rangle\langle 0| - |1\rangle\langle 1|)_2(|0\rangle\langle 0| - |1\rangle\langle 1|)_3\,|0_2 1_3\rangle + |0_2 1_3\rangle) \\
&\quad + |1_1\rangle\,(|0_2 1_3\rangle - (|0\rangle\langle 0| - |1\rangle\langle 1|)_2(|0\rangle\langle 0| - |1\rangle\langle 1|)_3\,|0_2 1_3\rangle)) \\
&= \frac{1}{2}(|0_1\rangle\,(-(|0\rangle\langle 0| - |1\rangle\langle 1|)_2\,|0_2 1_3\rangle + |0_2 1_3\rangle) \\
&\quad + |1_1\rangle\,(|0_2 0_3\rangle\,(|0\rangle\langle 0| - |1\rangle\langle 1|)_2\,|0_2 1_3\rangle)) \\
&= \frac{1}{2}(|0_1\rangle\,(|0_2 1_3\rangle - |0_2 1_3\rangle) + |1_1\rangle\,(|0_2 1_3\rangle + |0_2 1_3\rangle)) \\
&= |1_1 0_2 1_3\rangle
\end{aligned}$$

The measurement outcome of the first qubit will be -1 since -1 is the eigenvalue corresponding with the eigenstate $|1\rangle$ of Z with $|\langle 1|1\rangle|^2 = 100\%$ likelihood. We can check the cases for eigenstates $|\psi_{23}\rangle = |11\rangle, |10\rangle$ similarly. If ψ is not an eigenstate then the measurement outcome is probabilistic, which we can also check. This is left as an exercise to the reader.

Consider how we could employ this tool to detect a bit flip error in a quantum channel. The following circuit helps us to detect one bit flip when we have a 3-qubit repetition code. These Z_1Z_2 and Z_2Z_3 are called syndrome operators. The measurement outcomes themselves are the error syndromes. We can infer where the erroneous bit flip has occurred based on Table 3.1.

We can also trivially implement a single phase flip correcting code that corrects for one erroneous phase flip (erroneous Pauli Z operation) by straightforwardly replacing all the Xs in this scheme with Zs and Zs with Xs (Table 3.2).

A 7-qubit code which can correct a phase flip and a bit flip called the Steane code [9, p. 121], introduced by Steane in 1996 [29], is the next logical scheme for us to tackle.

Table 3.3 Steane code syndrome operators

$X_0X_4X_5X_6$	$Z_0Z_4Z_5Z_6$
$X_1X_3X_5X_6$	$Z_1Z_3Z_5Z_6$
$X_2X_3X_4X_6$	$Z_2Z_3Z_4Z_6$

Instead of two syndrome operators, the Steane code uses six. These six syndrome operators are mutually commuting, square to the identity and have eigenvalues ± 1. The Steane code has an error threshold around 10^{-2} to 10^{-5} (Table 3.3).

The Steane codewords are the states "stabilized" by these operators. A state that is stabilized by an operator is an eigenvector of that operator with an eigenvalue of $+1$. These codewords encode a logical qubit state $|0_L\rangle$ or $|1_L\rangle$.

$$|0_L\rangle = 2^{-3/2}(I + X_0X_4X_5X_6)(I + X_1X_3X_5X_6)(I + X_2X_3X_4X_6)\,|0000000\rangle$$
$$|1_L\rangle = 2^{-3/2}(I + X_0X_4X_5X_6)(I + X_1X_3X_5X_6)(I + X_2X_3X_4X_6)X^{\otimes 7}\,|0000000\rangle$$

In general, a stabilizer code is an Abelian subgroup of the Pauli group that defines our codespace. To get the stabilizer group, you take a set of commuting Paulis. The $+1$ simultaneous eigenspace gives us the codespace.

Measuring in the Steane code circuit projects the state $|\psi\rangle$ into either the subspace with eigenvalue $+1$ or -1. The -1 subspace is the "error subspace". We have six ancilla qubits (measured qubits) in the Steane code. An ancilla qubit is a qubit that is added to any algorithm which does not carry the state we are primarily concerned with, in this case $|\psi\rangle$.

The Steane code is an example of how we can use a more interesting code with codewords that are not simple repetitions to encode a correctable qubit state. This original correctable qubit state is called our logical qubit. It is the logical qubit's state that we are interested in; it is our "message". There are yet more interesting approaches to encoding a qubit into a higher dimensional state. One important family of codes are the surface codes, originally introduced by Kitaev[7] [30]. The simplest of the surface codes is the Toric code.

A toric code is defined over a number of qubits arranged in an $L \times L$ square lattice on the surface of a torus such that we have periodic boundary conditions when we walk the graph in each dimension. The qubits are on the edges of the lattice rather than the nodes. The minimum code distance is the minimum size of a non-trivial logical operator which is L.

[7] Kitaev is also responsible for the Solovay-Kitaev algorithm [30] which is a fundamental result in quantum simulation that allows gatesets like Clifford+T to approximate arbitrary unitaries with error ϵ. This algorithm is used widely but is no longer the most optimal, and is beginning to be replaced for example by the Ross-Selinger algorithm [31] and others [32]. None of these yet achieve the lower bound [33].

The toric code is defined in terms of stabilizer operators which are products of Pauli operators. These stabilizers each involve four Paulis, arranged either around a node or a plaquette. The stabilizers consisting of Paulis around a node are dubbed star operators and the ones consisting of Pauli around a plaquette are called plaquette operators.

$$A_s = \prod_{i \in \text{star}(s)} X_i$$
$$B_p = \prod_{j \in \text{boundary}(p)} Z_j$$

We get the following Hamiltonian and commutation relations. The codewords of the code are the ground states of this Hamiltonian.

$$H = -J_{\text{charge}} \sum_s A_s - J_{\text{vortex}} \sum_p B_p + \text{perturbations}$$
$$[A_s, A_{s'}] = 0$$
$$[B_p, B_{p'}] = 0$$
$$[A_s, B_p] = 0$$

These commutations hold even in the case that the operators apply to overlapping edges.

We say that a state $|\psi\rangle \in \mathcal{L}$ the eigenbasis iff $\forall s, p: \ A_s |\psi\rangle = |\psi\rangle$ and $B_p |\psi\rangle = |\psi\rangle$, i.e. $dim(\mathcal{L}) = 4$, meaning that the code encodes two logical qubits.

The number of spins in the lattice is 2^{2k^2} from some k. The ground state space's size is constrained by the condition that the star and plaquette operators equal the identity. In our Hamiltonian there are $k^2 - 1$ independent star operators and the same number of independent plaquette operators. Then the ground state degeneracy is given by $2^{2k^2-(k^2-1)-(k^2-1)} = 2^2 = 4$. The ground states correspond to windings around the torus.

The toric code ground state is $\prod_p (\frac{HB_p}{2}) |00\rangle = \sum_c |c\rangle$ where c are all the possible loop configurations. One can perform degenerate perturbation theory to find how robust the ground states are to perturbations. Non-local operators are needed to distinguish the ground states, and therefore higher order perturbations are required to lift degeneracy.

The toric code assumes that any error can be decomposed into Pauli operators. A Pauli error E applies to a codeword $|\psi\rangle$ of the code defined by stabilizer operators $s_1, \ldots, s_l$.

$$s_j E = (-1)^{\mu_j} E s_j$$
$$\mu \in \{0, 1\}, j \leq l$$

We determine the syndrome $(\mu_1, \ldots, \mu_l)$ from whether the s_j and E commute or anti-commute. Since the syndrome is "imprinted" on $|\psi\rangle$, it can be measured by applying the stabilizer operators.

We would apply controlled operations to measure the stabilizer on an ancilla qubit a and find the error syndrome. For example, the following circuit measures the $ZZZZ$ stabilizer (Fig. 3.6).

The measurement scheme for an XXXX stabilizer involves Hadamards on either side of CNOTs to map the X eigenstates (Fig. 3.7).

However, in practice, we cannot do this "Hadamard test" style measurement with the control in the opposite direction (from the measured qubit) because we don't want to change state of logical qubit with the controlled U.

Consider an example where the order of a Z error and a star operator overlapping on an edge matters.

Example: Phase Difference Based on Order of Operations

$$\begin{aligned} ZX\,|0\rangle &= (|0\rangle\langle 0| - |1\rangle\langle 1|)(|0\rangle\langle 1| + |1\rangle\langle 0|)\,|0\rangle \\ &= (|0\rangle\langle 0| - |1\rangle\langle 1|)\,|1\rangle \\ &= -\,|1\rangle \end{aligned}$$

$$\begin{aligned} XZ\,|0\rangle &= (|0\rangle\langle 1| + |1\rangle\langle 0|)(|0\rangle\langle 0| - |1\rangle\langle 1|)\,|0\rangle \\ &= (|0\rangle\langle 1| + |1\rangle\langle 0|)\,|0\rangle \\ &= |1\rangle \end{aligned}$$

Depending on the order of the operations on the overlapping edge, the qubit's state there differs by a phase.

Also, if we have a product of Paulis that braid a closed loop around a qubit in the $|1\rangle$ state, the ground state picks up a -1.

There are three types of Pauli errors that we consider in the toric code. The first is when the error acts trivially on the code s.t. $E = \pm \prod_j s_j^{\gamma_j}$; $\gamma_j \in \{0, 1\}$. The second case is when the error is detectable $\because \mu \neq 0$. The final case is when E is a non-trivial logical operator $E \neq \pm \prod_j s_j^{\gamma_j}$, $\mu = 0$. This third case causes logical errors.

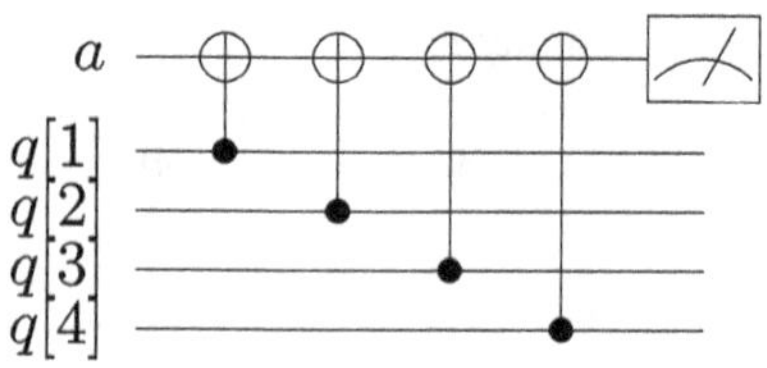

Fig. 3.6 ZZZZ measurement circuit

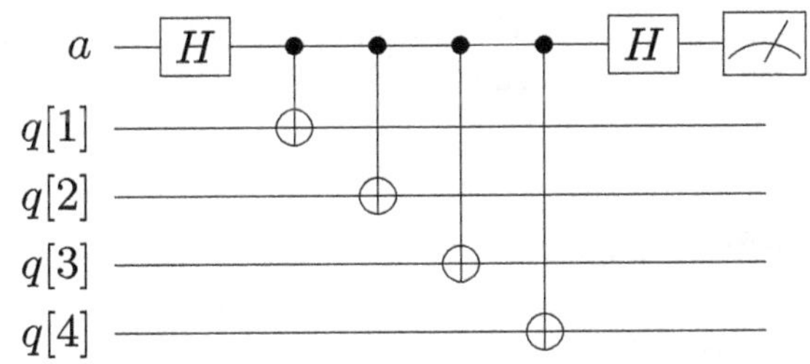

Fig. 3.7 XXXX measurement circuit

Let's consider Z type errors $E = \prod_{j \in \zeta} Z_j$. This is a trivial error if the qubits ζ form the mod 2 boundary of some region. The error is detectable if $\mu(\zeta) \neq 0$. The error is non-trivial if $\mu = 0$ and E is not a product of stabilizers, i.e. a line going around the Torus.

Logical operators preserve the code and commute with the stabilizer operators. The logical Pauli operators are formed by products of physical Paulis along lines on the torus. X_1 is the product of Xs along a horizontal line, X_2 is the product of Xs along a vertical line, Z_1 the product of Zs along a vertical line and Z_2 the product of Zs on a horizontal line. We get the Pauli algebra.

$$\begin{aligned} X_1 Z_1 &= -Z_1 X_1 \\ Z_2 X_2 &= -X_2 Z_2 \\ X_2 Z_2 &= -Z_2 X_1 \\ &\text{etc.} \end{aligned}$$

We can correct z type errors in the following way. We call nonzero syndrome bits from z-type errors "charges" and nonzero syndrome bits from x-type errors "vortices". After measuring the error syndrome, the syndrome must be sent to a classical computer which will infer the error that occurred by connecting the charges by lines in the lattice. The inferred error operator is applied to cancel the actual error. Inference can be error-prone and when an incorrect inference occurs, it can result in further logical errors. An optimal inference algorithm is maximum likelihood decoding in which errors with a particular syndrome are broken into four topological classes (i.e. vertical, horizontal, trivial, diagonal). If we know the probability of a particular type of error occurring on our device, we simply choose infer that this most likely error occurred.

The toric code is only the simplest of the surface codes. Notably, Google Quantum AI presented [34] two surface codes operating below a critical threshold in August 2024. This threshold called the surface code threshold, is a threshold of the physical error rate which determines whether the logical error rate will be suppressed exponentially in the number of physical qubits used to encode the logical qubit. Google's distance-7 and distance-5 codes were the first to perform below this threshold and beyond break-even in terms of the lifetime of the encoded state. Then, in a ground-breaking follow-up in November 2024 [34], Google demonstrated below-threshold error correction on a 105 qubit processor.

References

1. Nielsen, M.A., Chuang, I.L.: Quantum Computation and Quantum Information, 10th Anniversary. Cambridge University Press, Cambridge (2010)
2. Sakurai, J.J., Napolitano, J.: Modern Quantum Mechanics, 2nd edn. Dorling Kindersley, New Delhi (2014)
3. Wong, H.Y.: Introduction to Quantum Computing: From a Layperson to a Programmer in 30 Steps. Springer International Publishing, Cham (2024). https://doi.org/10.1007/978-3-031-36985-8
4. Ghose, S.: Quantum Computing at Wilfrid Laurier University. Undergraduate Course (2017)
5. Salfi, J.: Quantum Information and Computation at University of British Columbia. Graduate course (2023)
6. Yard, J.: Quantum information processing at university of waterloo. Graduate course (2018)
7. Pauli, W.: Nobel Prize-Winning Physicist, Britannica (2025). https://www.britannica.com/biography/Wolfgang-Pauli
8. Barenco, A., Bennett, C.H., Cleve, R., DiVincenzo, D.P., Margolus, N., Shor, P., Sleator, T., Smolin, J.A., Weinfurter, H.: Elementary gates for quantum computation. Phys. Rev. A **52**(5), 3457–3467 (1995). https://doi.org/10.1103/physreva.52.3457
9. Mermin, N.D.: Quantum Computer Science: An Introduction. Cambridge University Press (2007). Google-Books-ID: q2S9APxFdUQC
10. Krantz, P., Kjaergaard, M., Yan, F., Orlando, T.P., Gustavsson, S., Oliver, W.D.: A quantum engineer's guide to superconducting qubits. Appl. Phys. Rev. **6**(2), 021318 (2019). https://doi.org/10.1063/1.5089550
11. Fang, Y., Philippopoulos, P., Culcer, D., Coish, W.A., Chesi, S.: Recent advances in hole-spin qubits. Mater. Quant. Technol. **3**(1), 012003 (2023)
12. Bogan, A., Studenikin, S., Korkusinski, M., Gaudreau, L., Zawadzki, P., Sachrajda, A., Tracy, L., Reno, J., Hargett, T.: Single hole spin relaxation probed by fast single-shot latched charge sensing. Commun. Phys. **2**(1) (2019). https://doi.org/10.1038/s42005-019-0113-0
13. Chakraborty, S.: Cryogenic CMOS for Low Power Quantum Computing Applications: Roadmap, Present Status, Challenges and Opportunities (2023)
14. Bardin, J.C., Sank, D., Naaman, O., Jeffrey, E.: Quantum computing: an introduction for microwave engineers. IEEE Microwave Mag. **21**(8), 24–44 (2020)
15. Gray, J.J.: Carl Friedrich Gauss. Biography, Discoveries, & Facts | Britannica (2025). https://www.britannica.com/biography/Carl-Friedrich-Gauss
16. Koch, J., Yu, T.M., Gambetta, J., Houck, A.A., Schuster, D.I., Majer, J., Blais, A., Devoret, M.H., Girvin, S.M., Schoelkopf, R.J.: Charge insensitive qubit design derived from the Cooper pair box. Phys. Rev. A **76**(4), 042319 (2007).https://doi.org/10.1103/PhysRevA.76.042319
17. Burkard, G., Ladd, T.D., Pan, A., Nichol, J.M., Petta, J.R.: Semiconductor spin qubits. Rev. Mod. Phys. **95**(2), 025003 (2023). https://doi.org/10.1103/RevModPhys.95.025003
18. Larmor, S.J.: Quantum Theory, Electromagnetism, Mathematics, Britannica (2025). https://www.britannica.com/biography/Joseph-Larmor
19. Baroni, S., De Gironcoli, S., Dal Corso, A., Giannozzi, P.: Phonons and related crystal properties from density-functional perturbation theory. Rev. Mod. Phys. **73**(2), 515–562 (2001)
20. Kane, E.O.: Band structure of indium antimonide. J. Phys. Chem. Solids **1**(4), 249–261 (1957)
21. McIntosh, P.L.: CPMG, pp. 386–386. Springer, Berlin (2013). https://doi.org/10.1007/978-3-642-16712-6_320
22. Erwin, l.: Hahn Institute Homepage. https://hahn-institute.de/en

23. Hahn, E.L.: Spin echoes. Phys. Rev. **80**, 580–594 (1950). https://doi.org/10.1103/PhysRev.80.580
24. La Rose, R.: Circuit Optimization, Gate Alignment, and Spin Echoes. Cirq, Google Quantum AI (2021). https://quantumai.google/cirq/tutorials/google/spin_echoes
25. cirq.PhasedXPowGate. Cirq (2024). https://quantumai.google/reference/python/cirq/PhasedXPowGate
26. Horne, M.: American Institute of Physics Oral History: Michael Horne. In: Jaeger, G., Simon, D., Sergienko, A.V., Greenberger, D., Zeilinger, A. (eds.) Quantum Arrangements: Contributions in Honor of Michael Horne, pp. 11–34. Springer International Publishing, Cham (2021). https://doi.org/10.1007/978-3-030-77367-0_2
27. Greenberger, D.M., Horne, M.A., Shimony, A., Zeilinger, A.: Bell's theorem without inequalities. Am. J. Phys. **58**(12), 1131–1143 (1990)
28. Markowsky, G.: Claude Shannon. Father of Information Theory, American Engineer, Britannica (2025). https://www.britannica.com/biography/Claude-Shannon
29. Steane, A.: Multiple-particle interference and quantum error correction. Proc. R. Soc. Lond. Ser. A: Math. Phys. Eng. Sci. **452**(1954), 2551–2577 (1996). https://doi.org/10.1098/rspa.1996.0136
30. Kitaev, A.Y.: Quantum computations: algorithms and error correction. Russ. Math. Surv. **52**(6), 1191 (1997)
31. Ross, N.J., Selinger, P.: Optimal Ancilla-Free Clifford+T Approximation of z-Rotations (2014). https://arxiv.org/abs/1403.2975v3
32. Kliuchnikov, V., Maslov, D., Mosca, M.: Practical approximation of single-qubit unitaries by single-qubit quantum clifford and T circuits. IEEE Trans. Comput. **65**(1), 161–172 (2016)
33. Beverland, M., Campbell, E., Howard, M., Kliuchnikov, V.: Lower bounds on the non-Clifford resources for quantum computations. Quant. Sci. Technol. **5**(3), 035009 (2020). https://doi.org/10.1088/2058-9565/ab8963 [quant-ph]
34. Acharya, R., Aghababaie-Beni, L., Aleiner, I., Andersen, T.I., Ansmann, M., Arute, F., Arya, K., Asfaw, A., Astrakhantsev, N., Atalaya, J., Babbush, R., Bacon, D., Ballard, B., Bardin, J.C., Bausch, J., Bengtsson, A., Bilmes, A., Blackwell, S., Boixo, S., Bortoli, G., Bourassa, A., Bovaird, J., Brill, L., Broughton, M., Browne, D.A., Buchea, B., Buckley, B.B., Buell, D.A., Burger, T., Burkett, B., Bushnell, N., Cabrera, A., Campero, J., Chang, H.S., Chen, Y., Chen, Z., Chiaro, B., Chik, D., Chou, C., Claes, J., Cleland, A.Y., Cogan, J., Collins, R., Conner, P., Courtney, W., Crook, A.L., Curtin, B., Das, S., Davies, A., De Lorenzo, L., Debroy, D.M., Demura, S., Devoret, M., Di Paolo, A., Donohoe, P., Drozdov, I., Dunsworth, A., Earle, C., Edlich, T., Eickbusch, A., Elbag, A.M., Elzouka, M., Erickson, C., Faoro, L., Farhi, E., Ferreira, V.S., Burgos, L.F., Forati, E., Fowler, A.G., Foxen, B., Ganjam, S., Garcia, G., Gasca, R., Genois, ., Giang, W., Gidney, C., Gilboa, D., Gosula, R., Dau, A.G., Graumann, D., Greene, A., Gross, J.A., Habegger, S., Hall, J., Hamilton, M.C., Hansen, M., Harrigan, M.P., Harrington, S.D., Heras, F.J.H., Heslin, S., Heu, P., Higgott, O., Hill, G., Hilton, J., Holland, G., Hong, S., Huang, H.Y., Huff, A., Huggins, W.J., Ioffe, L.B., Isakov, S.V., Iveland, J., Jeffrey, E., Jiang, Z., Jones, C., Jordan, S., Joshi, C., Juhas, P., Kafri, D., Kang, H., Karamlou, A.H., Kechedzhi, K., Kelly, J., Khaire, T., Khattar, T., Khezri, M., Kim, S., Klimov, P.V., Klots, A.R., Kobrin, B., Kohli, P., Korotkov, A.N., Kostritsa, F., Kothari, R., Kozlovskii, B., Kreikebaum, J.M., Kurilovich, V.D., Lacroix, N., Landhuis, D., Lange-Dei, T., Langley, B.W., Laptev, P., Lau, K.M., Guevel, L.L., Ledford, J., Lee, K., Lensky, Y.D., Leon, S., Lester, B.J., Li, W.Y., Li, Y., Lill, A.T., Liu, W., Livingston, W.P., Locharla, A., Lucero, E., Lundahl, D., Lunt, A., Madhuk, S., Malone, F.D., Maloney, A., Mandrá, S., Martin, L.S., Martin, S., Martin, O., Maxfield, C., McClean, J.R., McEwen, M., Meeks, S., Megrant, A., Mi, X., Miao, K.C., Mieszala, A., Molavi, R., Molina, S., Montazeri, S., Morvan, A., Movassagh, R., Mruczkiewicz, W., Naaman, O., Neeley, M., Neill, C., Nersisyan, A., Neven, H., Newman, M., Ng, J.H., Nguyen, A., Nguyen, M., Ni, C.H., O'Brien, T.E., Oliver, W.D., Opremcak, A., Ottosson, K., Petukhov, A., Pizzuto, A., Platt, J., Potter, R., Pritchard, O., Pryadko, L.P., Quintana, C., Ramachandran, G., Reagor, M.J., Rhodes, D.M., Roberts, G., Rosenberg, E., Rosenfeld, E., Roushan, P., Rubin, N.C., Saei, N., Sank, D., Sankaragomathi, K., Satzinger, K.J., Schurkus, H.F., Schuster, C., Senior, A.W.,

Shearn, M.J., Shorter, A., Shutty, N., Shvarts, V., Singh, S., Sivak, V., Skruzny, J., Small, S., Smelyanskiy, V., Smith, W.C., Somma, R.D., Springer, S., Sterling, G., Strain, D., Suchard, J., Szasz, A., Sztein, A., Thor, D., Torres, A., Torunbalci, M.M., Vaishnav, A., Vargas, J., Vdovichev, S., Vidal, G., Villalonga, B., Heidweiller, C.V., Waltman, S., Wang, S.X., Ware, B., Weber, K., White, T., Wong, K., Woo, B.W.K., Xing, C., Yao, Z.J., Yeh, P., Ying, B., Yoo, J., Yosri, N., Young, G., Zalcman, A., Zhang, Y., Zhu, N., Zobrist, N.: Quantum Error Correction Below the Surface Code Threshold (2024). https://doi.org/10.48550/arXiv.2408.13687 [quant-ph]

Chapter 4
Building a Qubit

Abstract This chapter delves into the details of superconducting, photonic, annealing, nuclear magnetic resonance (NMR), ion trap and spin qubit quantum computers. As bits are to classical computers, qubits are the building blocks of quantum computers. Any quantum computer engineer must know them inside-out. At the end of the chapter, a tutorial on how to build your own small semiconductor spin qubit chip is provided.

4.1 Learning Objectives

In this chapter we finally get into how to build qubits! We have learned enough foundational theory and about the constraints of the environment (Fig. 4.1). Now we can really get into the engineering. Learning objectives for this chapter include:

- Understand the different types of qubits and their physical implementations.
- Learn how to design a semiconductor spin qubit chip.
- Gain practical skills with the CAD software K-Layout.

4.2 The Types of Qubits

In this chapter we will look at different ways we can represent a qubit physically. You might find similar information in textbooks such as [1, 2]. Certain specialized graduate courses offered for example at the Institute for Quantum Computing (IQC) [3], University of British Columbia [4] and recently the University of Victoria [5], for example, also cover certain aspects of this topic. In particular, [3] offered at IQC covers superconducting, optical, nuclear magnetic resonance and ion trap-based quantum computing. Having read and sampled multiple texts and courses, we present a condensed summary of these different physical implementations, including a few things that none of these other sources provide. The "Bible of quantum computing" [2] only covers harmonic oscillator, optical, ion trap and nuclear magnetic resonance

© The Author(s), under exclusive license to Springer Nature Switzerland AG 2026
M. S. Edwards, *An Introduction to Quantum Computing for Computer Engineers*,
https://doi.org/10.1007/978-3-032-03650-6_4

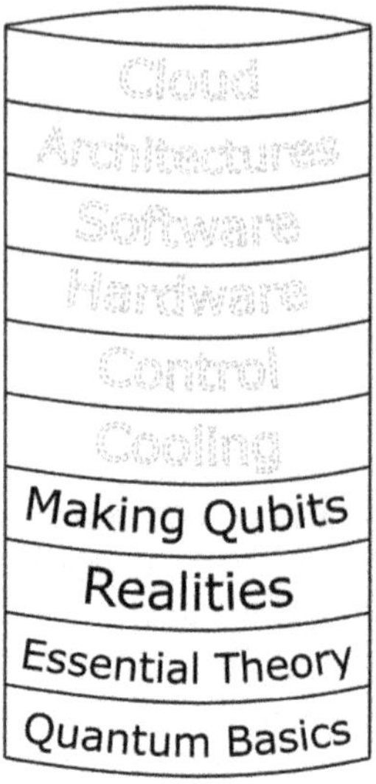

Fig. 4.1 Learning objective for this chapter: making qubits

quantum computers at any length, leaving only a few sentences for spin qubits, which is our specialty. We add to this a more thorough presentation of the basics of spin qubits, as well as a tutorial for an essential part of designing a realistic spin qubit yourself. There are few resources that cover this, though the recent course offering from TU Delft [6] is an excellent one.

When we cover superconducting qubits, optical qubits, flux qubits, NMR QCs, spin qubits and ion traps here we focus on different aspects than other sources, at times leaning more for example into the quantum mechanics. Across the board, we aim to combine and synthesize the best (most useful) information from all these sources into a single, compact reference that prefers to cite either original seminal papers that presented these implementations for the first time (we have gone to these original sources to improve our own understanding of the topics) or authoritative sources. Hence, to give our "intermediate" sources—such as courses taken in the past—due recognition, we acknowledge them now. To go more in-depth on these physical QC implementations we recommend reading textbooks that make them their whole focus and enrolling in great courses. Even better, go to the original sources too.

A requirement for a useful quantum computer is for it to be based on a well-defined two-level system, a qubit [7]. Not every qubit is created equal, however. Genuine spin-$\frac{1}{2}$ particles include electrons, leptons and quarks [8, p. 394]. Photons are technically spin-1, but the 0 state is forbidden. Subspaces in larger systems such as the first two energy levels in an anharmonic oscillator [9], or a hybridized state [10], for example, can also be used to encode a qubit. The key is that the two levels come from the energy eigenstates of a Hermitian operator.

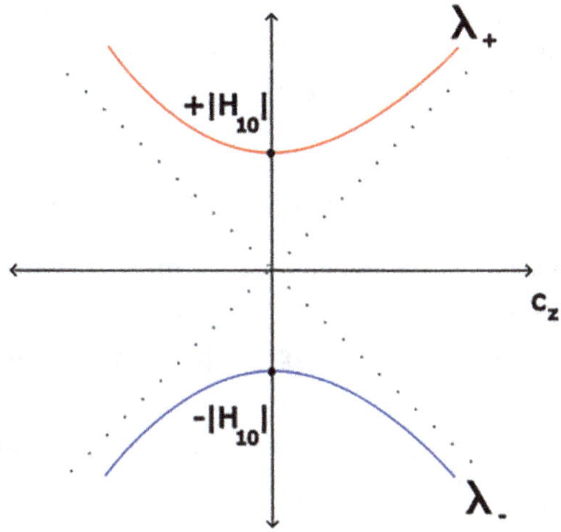

Fig. 4.2 Two-level system hyperbolic function

A Hermitian Hamiltonian has real diagonal entries, and its off-diagonal elements are complex conjugates. What's more, in the case of a qubit, we can represent the unitary evolution of the state by a Hermitian Hamiltonian which is a sum of weighted Paulis [11] with real coefficients $b \in R,\ c \in R^3$.

$$H = bI + \mathbf{c} \cdot \sigma = \begin{bmatrix} b + c_z & c_x - ic_y \\ c_x + ic_y & b - c_z \end{bmatrix}$$

We can substitute in $c_\pm = c_x \pm ic_y$ to simplify the off-diagonals. Then in order to solve for the eigenvalues (or energy levels) corresponding to our Hamiltonian, we would subtract λI and take the determinant.

$$\det\left(\begin{bmatrix} b + c_z - \lambda & c_- \\ c_+ & b - c_z - \lambda \end{bmatrix}\right) = 0$$
$$(b + c_z - \lambda)(b - c_z - \lambda) - c_- c_+ = 0$$
$$\lambda^2 - 2b\lambda = (c_x^2 + c_y^2 + c_z^2) - b^2$$

We complete the square by noticing that $\lambda^2 - 2b\lambda = (\lambda - b)^2 - b^2 \ldots$

$$(\lambda - b)^2 = c_x^2 + c_y^2 + c_z^2$$
$$\lambda_\pm = \pm\sqrt{c_x^2 + c_y^2 + c_z^2} + b$$

The energy level splitting comes from the energy difference in the two $\lambda_\pm$. This gives us a model for general two-level systems, where our splitting is a hyperbolic function (a graph of λ_+ and λ_- on the same axis) in c_z (Fig. 4.2).

This is what a two-level system looks like in general, and we call this hyperbolic shape an anti-crossing, but there are many types of qubits that are made at research institutions and in R & D focused companies. The candidates include qubits that are encoded in electron spins, hole spins, nucleus spins (in nuclear magnetic resonance setups), the electronic states of atoms (in ion traps), the state of a quantized circuit (in superconducting approaches), the polarizations of photons (in optical quantum computing), radiation damage centers in host materials like Silicon, and topological quantum states in condensed matter.

what is the landscape of quantum computing like?

The comparison has been made between quantum computing today and classical computing in the 1940s [12], when there was a zoo of interesting routes for realizing the dream of useful computers.

In the beginning researchers and engineers built ten-state computers, trinary computers and even analog computers. In the history of computing there have also been different semiconductor technologies. Did you know that the first transistor was made of Germanium, and not the Silicon which dominates semiconductor manufacturing today? Well, Germanium[1] may yet make a comeback as it has received renewed interest in the field of semiconductor spin qubits.

It is expected that one or two dominant qubit technologies will arise and lead to useful quantum computers. Arguably, no such technology has done so yet. A few have been used to achieve some impressive milestones called "quantum supremacy" and "quantum advantage" however. These terms are more or less interchangeable, so we will stick with the preferred term "quantum advantage" from now on. These technologies might arguably be the most interesting to study, so let's start with the ones involved in these milestones.

Just to cover our bases, we will also discuss NMR and spin qubits, which have not been involved in big claims like the advantage claims that have been made but have their own arguments for usefulness. For example, spin qubits are based on semiconductor technology, and this lends them to manufacturability using existing fabrication facilities at an industrial scale.

Quantum advantage has been demonstrated in superconducting [14], optical [15–17] and annealing-based [18] quantum computers. These claims of quantum advantage were spread out across the years 2019–2024. The first such claim came from Google, using their Sycamore quantum processor, a 2D array of 53 transmon qubits with tunable nearest-neighbor coupling.

---

[1] Germanium was discovered in 1886 by German chemist Clemens Winkler [13].

4.2.1 Superconducting Qubits

Sycamore is a 53 qubit computer based on superconducting circuits called transmons. To understand transmons [9, 19], we will first begin with a simple LC circuit. In such an LC circuit, "kinetic energy" is associated with the electrical energy in the capacitor and "potential energy" is stored in the magnetic field of the inductor. Energy in an LC circuit oscillates between these so-called potential and kinetic energies. The Lagrangian for this system looks like $L = T - U$, where T is our kinetic energy (capacitance energy) and U is our potential energy (inductance energy). Remember that the energy stored in a capacitor is given by $\frac{CV^2}{2}$ where C is the capacitance and V is the voltage. The potential energy U is $\frac{LI^2}{2}$ where L is inductance and I is current. Then we can express the Lagrangian in terms of flux.

$$\Phi(t) = \int V(t')\mathrm{d}t'$$

$$V = L\frac{\mathrm{d}I}{\mathrm{d}t}$$

$$I = C\frac{\mathrm{d}V}{\mathrm{d}t}$$

Using integration by parts we can solve for the energy $E(t) = \int_{-\infty}^{t} V(t')I(t')ft'$ of each circuit component.

$$\begin{aligned} T &= \frac{CV^2}{2} \\ &= \frac{C\dot{\Phi}^2}{2} \end{aligned}$$

The relationship $V = -\dot{\Phi}$ comes from Faraday's law of induction. We can alternatively derive it from the 2nd Josephson relation when the circuit includes a *nonlinear* inductor called a Josephson Junction. $\frac{\mathrm{d}\delta}{\mathrm{d}t} = \frac{2e}{\hbar}V$ for this kind of junction, where $\delta = \theta_1 - \theta_2 = -2\pi\frac{\Phi}{\Phi_0}$ is the difference of the Ginzburg–Landau order parameters across the junction.

$$V = \frac{\hbar}{2|e|}\dot{\delta} = \frac{\Phi_0}{2\pi}\left(-2\pi\frac{\dot{\Phi}}{\Phi_0}\right) = -\dot{\Phi}$$

We identify the energy in the inductor with the potential energy. This allows us to formulate a Lagrangian[2] and later a Hamiltonian.

$$
\begin{aligned}
U &= \frac{LI^2}{2} \\
&= \frac{(\Phi - \Phi_x)^2}{2L}
\end{aligned}
$$

Here Φ_x denotes where we would include an external flux, or flux due to an external field. For example, such a field may originate from a coil nearby to the device. Using these "kinetic" and "potential" energies we can write a Lagrangian. Here we assume going forward though that the external flux is zero.

$$
\mathsf{L} = \frac{C\dot{\Phi}^2}{2} - \frac{\Phi^2}{2L}
$$

We can also express the "kinetic" energy as $\frac{Q^2}{2C}$ since $C\dot{\Phi} = -CV = -Q$. Note that scientists and engineers will sometimes refer to the magnetic and kinetic contributions to an inductance as the "kinetic" inductance, while referring to the magnetic energy outside the wire as "geometric" inductance, but this is not what we are referring to here.

The energy of said system is given by the Hamiltonian H in terms of the Lagrangian, the conjugate momentum $p_{\text{conj.}}$ and the first derivative of the inductance $\dot{\Phi}$, using a Legendre transformation. Lagrangian mechanics come from the principle of least action. A Lagrangian is like a Hamiltonian, involving kinetic and potential energy, except that they are subtracted instead of added and the derivations for the two differ. A Legendre transformation converts functions $L(q, \dot{q})$ of a quantity and its derivative to functions $H(q, p_{\text{conj}})$ of the quantity and its conjugate such that $H(q, p_{\text{conj}}) = p_{\text{conj}}\dot{q} - L(q, \dot{q})$ [21]. In order to describe this as a quantum system, we need to choose one quantity, such as charge or inductance, as a coordinate. The conjugate variable can then be solved for. Let's choose inductance to be our coordinate. For now we leave the conjugate quantity as p_{conj}. Knowing that a Lagrangian and a Hamiltonian are the Legendre transforms of each other, we have the following Legendre transformation between them.

$$
H = p_{\text{conj.}}\dot{\Phi} - \mathsf{L} = \frac{Q^2}{2C} + \frac{\Phi^2}{2L}
$$

One immediately sees that the sign of the kinetic energy is positive despite the fact that the Lagrangian is subtracted from $p_{\text{conj}}\dot{\Phi}$. The potential term is exactly as we would suspect. The only explanation for this is that $p_{\text{conj}}\dot{\Phi}$ is twice the kinetic

[2] The Lagrangian is named after Joseph-Louis Lagrange, whom Frederick the Great called the greatest mathematician in Europe [20].

term and positive. Therefore, twice $\frac{Q^2}{2C} = \frac{Q^2}{C} = \frac{(C\dot{\Phi})^2}{C} = C\dot{\Phi}^2$ is $p_{\text{conj}}\dot{\Phi}$. p_{conj} must be $C\dot{\Phi} = -Q$.

The charge and flux coordinates Φ, Q become quantum operators which do not commute.

$$[\Phi, Q] = i\hbar$$

This implies for us the operator value of $p_{\text{conj}} = \frac{\hbar}{i}\frac{\mathrm{d}}{\mathrm{d}\Phi}$.

$$\begin{aligned}
[\Phi, p_{\text{conj}}]\psi(\Phi) &= \Phi p_{\text{conj}}\psi(\Phi) - p_{\text{conj}}\Phi\psi(\Phi) \\
&= \Phi\frac{\hbar}{i}\frac{\mathrm{d}}{\mathrm{d}\Phi}\psi(\Phi) - \frac{\hbar}{i}\frac{\mathrm{d}}{\mathrm{d}\Phi}(\Phi\psi(\Phi)) \\
= \Phi\frac{\hbar}{i}\frac{d}{d\Phi}\psi(\Phi) - \frac{\hbar}{i}\psi(\Phi) &- \frac{\hbar}{i}\Phi\frac{\mathrm{d}}{\mathrm{d}\Phi}\psi(\Phi) \\
&= -\frac{\hbar}{i}\psi(\Phi) \\
&= i\hbar\psi(\Phi)
\end{aligned}$$

This is the same relationship we saw between conjugate variables x and p in Chap. 2, where x was taken as a coordinate and p was given as a differential operator.

Also similar to how we saw raising and lowering operators in the second chapter, we have raising and lowering operators here which help us to construct our quantum harmonic oscillator's levels.

$$a = \sqrt{\frac{1}{2\hbar z_c}}(\Phi + iz_cQ)$$

$$a^\dagger = \sqrt{\frac{1}{2\hbar z_c}}(\Phi - iz_cQ)$$

Our wave function can be written in terms of flux or charge as $\psi(\Phi, t)$ or $\psi(Q, t)$. We have a probability density functions for each of these variables, such as $|\psi(\Phi, t)|^2$ for the flux, which gives us the probability of measuring a flux between Φ_a and Φ_b through the inductor at time t $\int_{\Phi_a}^{\Phi_b} |\psi(\Phi, t)|^2 d\Phi$ [22]. The probability density function for the charge $|\psi(Q, t)|^2$ gives us the probability distribution for a measurement of charge across the capacitor. We can transform between flux and charge just as we did between position and momentum in the past.

$$\psi(\Phi, t) = \frac{1}{\sqrt{2\pi\hbar}}\int_{-\infty}^{\infty} \psi(Q, t)\exp(-jG\Phi/\hbar)\mathrm{d}Q$$

We have a Heisenberg uncertainty relation between our conjugate variables.

$$(\Delta\Phi)(\Delta Q) \geq \frac{\hbar^2}{2}$$

When the circuit is in the state $|\psi\rangle$, we have:

$$\langle\psi|\,(\Phi - \langle\psi|\Phi|\psi\rangle)^2\,|\psi\rangle\,\langle\psi|\,(Q - \langle\psi|Q|\psi\rangle)^2\,|\psi\rangle \geq \frac{|\,\langle\psi|\,[\Phi, Q]\,|\psi\rangle\,|^2}{2} = \frac{\hbar^2}{2}$$

We even reproduce the Schrödinger equation by restating the Hamiltonian slightly.

$$H = \frac{1}{2C}\left(\frac{\hbar}{i}\right)^2 \frac{\mathrm{d}^2}{\mathrm{d}\Phi^2} + U(\Phi)$$

The impedance z_c gives the magnitude of the inductive and capacitive impedance at the natural frequency $\frac{1}{2\pi\sqrt{LC}}$ of the oscillator. This natural frequency is the frequency at which the circuit oscillates in the absence of any damping. It also gives us the spacing of our quantum levels.

For reasons we will get to in a moment, we typically use a nonlinear inductor called a Josephson Junction in our LC circuits when making superconducting qubits. A Josephson Junction consists of a superconducting island and reservoir separated by an insulating barrier. A quantum mechanical Hamiltonian for the system becomes the following, where we have the charging energy to add one Cooper pair to the island $E_C = \frac{e^2}{2C}$, inductive energy $E_L = \frac{(\Phi_0/2\pi)^2}{L}$, natural frequency ω_r and $\Phi_0 = \frac{h}{2e}$ the superconducting magnetic flux quantum (the smallest amount of magnetic flux allowed by quantum mechanics).

We can quantize the circuit with new conjugate variables in a similar way to what we saw earlier. In this case, however, our variables are the reduced flux as $\phi = 2\pi\Phi/\Phi_0$ and reduced charge as $n = Q\,2e$.

$$[\phi, n] = i$$

Then our Hamiltonian looks the same as the equation for the quantum harmonic oscillator.

$$H = 4E_C n^2 + \frac{1}{2}E_L\phi^2 = \hbar\omega_r\left(a^\dagger a + \frac{1}{2}\right)$$

The quadratic nature of the second term gives us the shape of the harmonic oscillator and the solutions to the eigenvalue problem stated by $H\,|\psi\rangle = E\,|\psi\rangle$ gives us our infinite series of eigenstates.

In practice, we are less interested in a quantum harmonic oscillator with infinite eigenstates and more interested in an anharmonic oscillator. This is because a qubit is a two-level system and an anharmonic oscillator is an n-level system. If we have n energy levels that have similar spacings from their neighbors, we will resonantly drive transitions outside of our qubit subspace when we are trying to drive our qubit.

So, we introduce anharmonicity to the system by replacing the inductor in the circuit with the nonlinear element called a Josephson Junction. This makes the first energy separation between energy levels different from the second, etc.

The JJ can be considered a nonlinear inductor with an energy. The potential energy of the junction is $E_J \cos(\phi)$ where $E_J = \frac{\Phi_0 I_c}{2\pi}$ and $\phi = 2\pi \frac{\Phi}{\Phi_0}$. We can introduce this energy to the Hamiltonian like this.

$$H = \frac{C\dot{\Phi}^2}{2} + \frac{\Phi^2}{2L} - E_J \cos(\phi)$$

The $E_J \cos(\phi)$ detunes the harmonic oscillator so that it is anharmonic. The Hamiltonian of the system then takes the following form up to the fourth order, where ω_q is the qubit frequency (the energy spacing between energy levels 0 and 1) and α gives the difference between the qubit frequency and the second energy spacing.

$$H = \hbar\omega_q a^\dagger a + \frac{\alpha}{2} a^\dagger a^\dagger a a$$

We can see this by performing a truncated power series expansion of the $E_J \cos(\phi)$ term and substituting $\phi = (\frac{2E_C}{E_J})^{\frac{1}{4}}(a^\dagger + a)$. We can do this in the $E_J >> E_C$ regime.

$$E_J \cos(\phi) = \frac{E_J}{2!}\phi^2 - \frac{E_J}{4!}\phi^4 + \ldots$$

$$-E_C \frac{\partial^2}{\partial \phi^2} + E_J \cos(\phi) = \hbar\omega_q a^\dagger a + \frac{E_C}{2} a^\dagger a^\dagger a a + \ldots$$

The energy levels in such an anharmonic oscillator are spaced by increasing factors of $\nu = \frac{e^2}{4\pi\hbar C}$ such that transition frequency $\omega_{(n-1)(n)} = \omega_0 + n\nu$. Here C is the capacitance of the LC capacitor. A microwave quanta (also called a photon) added to the system causes a resonant frequency change of $|\nu|$.

As mentioned briefly in the previous chapter, the anharmonicity in transmons reduces algebraically with a power law in E_J/E_C, but the charge dispersion reduces exponentially. So, by operating the qubit at a large E_J/E_C, we combat charge noise with minimal compromise in the anharmonicity [19]. In order to make $E_C << E_J$ we shunt the Josephson Junction with a large capacitor.

There are more superconducting qubit types including phase qubits, flux qubits and others. However, the LC circuit and transmon are the usual entry point to this world.

4.2.2 Photonic Qubits

Optical modes are used to encode qubits in photonic quantum computing variants. These are vector potential solutions to Maxwell's equations. Discrete solutions as appearing in optimal cavities take the following form, where $\mathbf{u_l}(\mathbf{x})$ is an orthonormal set mode function. By the way, when you see $+h.c.$ in textbooks, it means + Hermitian Conjugate, which means that we also add the conjugate transpose but don't write it for the sake of brevity. l are polarization modes, which are the components of the field for plane waves.

$$\mathbf{A}(\mathbf{x},t)=\sqrt{\frac{\hbar}{2\omega_l\varepsilon_0}}\sum_l(c_l\mathbf{u}_l(\mathbf{x})\exp(-i\omega_l t)+h.c.)$$

From Maxwell's[3] equations, we have $\mathbf{B}=\nabla\times\mathbf{A}$ and $\mathbf{E}=-\frac{1}{c}\delta_t\mathbf{A}$. Then we can quantize by replacing c_l with operator a_l.

$$[a_l,a_k^\dagger]=\delta_{lk}I$$

$$[a_l,a_k]=0$$

We can treat $\mathbf{A}$ as a generalized coordinate and then the canonical momentum density $\mathbf{\Pi}(\mathbf{x},t)=\frac{\partial L}{\partial\dot{\mathbf{A}}}=\varepsilon_0\dot{\mathbf{A}}=-\varepsilon_0\mathbf{E}(\mathbf{x},t)$.

$$[A_l(\mathbf{x},t),\Pi_k(\mathbf{x}',t)]=i\hbar\delta(\mathbf{x}-\mathbf{x}')\delta_{lk}$$

$$\mathbf{\Pi}(\mathbf{x},t)=i\sqrt{\frac{\hbar\omega_l\varepsilon_0}{2}}\sum_l(-a_l\mathbf{u}_l(\mathbf{x})\exp(-i\omega_l t)+h.c.)$$

We use propagating modes, the plane wave solutions of Maxwell's equations [24]. There are two, the electric and magnetic solutions. Here $\mathbf{k}$ is the wave vector in the direction of propagation, $|\mathbf{k}|=\frac{2\pi}{\lambda}$, $\mathbf{r}$ is position and σ is the polarization. You should recall these concepts from the first chapter.

$$\mathbf{E}_{\mathbf{k},\sigma}=\mathrm{Re}[\mathbf{E}_\sigma\exp(i(\mathbf{k}\cdot\mathbf{r}-\omega t))]$$

$$\mathbf{B}_{\mathbf{k},\sigma}=\mathrm{Re}[\mathbf{B}_\sigma\exp(i(\mathbf{k}\cdot\mathbf{r}-\omega t))]$$

As you know, these electromagnetic waves are transverse, meaning that the dot product of $\mathbf{k}$ with either of $\mathbf{B}_\sigma$ or $\mathbf{E}_\sigma$ is zero. $\mathbf{B}_\sigma$ and $\mathbf{E}_\sigma$ and orthogonal with $\mathbf{k}$ and are also linearly independent of one another.

[3] Planck's fundamental contribution to the birth of quantum mechanics stemmed from the classical law of heat radiation which was developed as a result of Maxwell's description of electromagnetic radiation [23].

We can check that these are solutions to Maxwell's equations by substituting them in to each one [25]. Recall Maxwell's equations, with E the electric field, B the magnetic field, ϱ_f the electric charge density, J the current density, μ_0 the vacuum permeability and ε_0 the vacuum permittivity. $D = \varepsilon_0 E$. These equations state essentially that the electric field induced by an electric charge converges from positive toward negative charge, the divergence of the magnetic field is always zero, a changing magnetic field produces a circulating electric field and vice versa.

$$\begin{aligned}
\nabla \cdot D &= \varrho_f \\
\nabla \cdot B &= 0 \\
\nabla \times E &= -\frac{\partial B}{\partial t} \\
\nabla \times B &= \mu_0 (J + \varepsilon_0 \frac{\partial E}{\partial t})
\end{aligned}$$

Say our wave is such that the electric field oscillates in the y plane and the magnetic in the z plane. Then $r = x$ and we can substitute into the first equation.

$$\begin{aligned}
\nabla \cdot D &= \left\{ \frac{\partial}{\partial r}\mathbf{r} + \frac{\partial}{\partial y}\mathbf{y}\frac{\partial}{\partial z}\mathbf{z} \right\} \varepsilon_0 E_\sigma \exp(i(\mathbf{k}\cdot\mathbf{r} - \omega t))\mathbf{y} \\
&= \varepsilon_0 [i\mathbf{k}E_\sigma \exp(i(\mathbf{k}\cdot\mathbf{r} - \omega t))\mathbf{r}\cdot\mathbf{y}] = 0
\end{aligned}$$

$\mathbf{r}\cdot\mathbf{y} = 0$ therefore the form for $\mathbf{E}_{\mathbf{k},\sigma}$ satisfies the first Maxwell's equation. The form for $\mathbf{B}_{\mathbf{k},\sigma}$ similarly satisfies the second equation. We can also check that they satisfy the third equation.

$$i\mathbf{k}E_\sigma \exp(i(\mathbf{k}\cdot\mathbf{r} - \omega t))\mathbf{z} = i\omega B_\sigma \exp(i(\mathbf{k}\cdot\mathbf{r} - \omega t))\mathbf{z}$$

This equation is satisfied when $B_\sigma = \frac{\mathbf{k}}{\omega} E_\sigma$ and enforces a relationship between our electric and magnetic fields.

Finally, we can substitute into the fourth equation to learn:

$$-i\frac{\mathbf{k}}{\mu_0} B_\sigma \mathbf{y} = -i\omega\varepsilon_0 E_\sigma \mathbf{y}$$

This combined with $B_\sigma = \frac{\mathbf{k}}{\omega} E_\sigma$ implies for us a relationship between $\mathbf{k}$, ω, μ_0 and ε_0: $\mathbf{k}^2 = \omega\sqrt{\mu_0\varepsilon_0}$. Therefore when this relationship holds, the forms for $E_{\mathbf{k},\sigma}$ and $B_{\mathbf{k},\sigma}$ satisfy Maxwell's equations.

Generally, the stationary spatiotemporal profiles of the field are linearly independent solutions of Maxwell's equations like we have shown. These are called optical modes, and the excitations of each of them are quantized as a quantum harmonic oscillator. These modes are vessels for quantum states.

$$H = \frac{1}{2}\int d^3x[\varepsilon_0|\mathbf{E}|^2 + \frac{1}{\mu_0}|\mathbf{B}|^2]$$
$$= \sum_l \hbar\omega_l\left(a_l^\dagger a_l + \frac{1}{2}\right)$$

A qubit can be encoded in two independent spatiotemporal degrees of freedom in the electromagnetic field. For example, we can encode our states in the Fock basis, or photon counting basis [2, p.288]. In this basis the kets are labeled by the number of photons in the mode.

$$|n\rangle \; ; \; n = 0, \ldots, \infty$$
$$a^\dagger |n\rangle = \sqrt{n+1}\,|n+1\rangle$$
$$a\,|n\rangle = \sqrt{n}\,|n-1\rangle$$
$$a\,|0\rangle = 0$$
$$n = a^\dagger a$$
$$n\,|n\rangle = n\,|n\rangle$$

However, these occupation number qubits can travel too quickly to keep track of. Before traveling, such a qubit will have a state $\alpha\,|0\rangle + \beta\,|1\rangle$. After traveling, $\alpha\,|0\rangle + \beta\exp(-i\omega t)\,|1\rangle$.

$$H = \hbar\omega n$$
$$U = \exp(-iHt/\hbar)$$

So, we can use an alternative encoding called the dual rail qubit [2, p.288], which encodes a qubit in two physical modes. The "0" state is $|1\rangle\,|0\rangle$ or $|1, 0\rangle$, and the "1" state is $|0\rangle\,|1\rangle$ or $|0, 1\rangle$. The time becomes a global rather than a relative phase so the exact time doesn't need to be taken into account when considering your qubit's state.

More alternative qubit types are time-bin qubits, which use temporal rather than spatial modes, i.e., we don't label our kets by the photon occupation of a waveguide, but by the photon occupation of a particular time window. We can measure these photons using different photo-detector setups.

Finally, we can use polarization qubits which are familiar to us by now. We have not however considered them as traveling modes. To this end, we can think about plane waves with polarization.

$$A(\mathbf{x}, t) = \text{Const.}\int d^3\mathbf{k}\sum_{\sigma\in 0,1}(a_\sigma(\mathbf{k})\mathbf{e}_\sigma(\mathbf{k})\exp(i\mathbf{k}\mathbf{x})\exp(-i\omega(\mathbf{k})t) + h.c)$$
$$[a_\sigma(\mathbf{k}), a_{\sigma'}(\mathbf{k}')] = \delta_{\sigma,\sigma'}\delta(k - k')$$

We choose a set of functions $\{\phi_i(k)\}_i$; $i = 1\ldots,\infty$ that form a complete orthonormal basis.

$$\int d^3\mathbf{k}\phi_i^*(\mathbf{k})\phi_j(\mathbf{k}) = \delta_{ij}$$

$$\sum_{i=1}^{\infty}\phi_i^*(\mathbf{k})\phi_j(\mathbf{k}) = \delta(k-k')I$$

Then by a Fourier transform we have the following, where $a_\sigma(\mathbf{k})$ is our plane wave.

$$c_{i\sigma} = \int d^3\mathbf{k}\phi_i^*(\mathbf{k})a_\sigma(\mathbf{k})$$

Another approach to quantum computing with optics is known as Continuous Variable Quantum Computing (CVQC) [26]. CVQC replaces the qubit as the unit of quantum information with a "qumode". A qumode has the following representation, where we have a continuum in the place of the discrete coefficients we would be used to seeing. The $|x\rangle$ are still eigenstates.

$$|\psi\rangle = \int \mathrm{d}x\psi(x)\,|x\rangle$$

Instead of a finite dimensional Hilbert space we make use of an infinite dimensional Hilbert space. A CVQC quantum state is Gaussian and has the following form, where H is a Bosonic Hamiltonian.

$$|\psi\rangle = \exp(-itH)\,|0\rangle$$

In quantum optical quantum computers, when we model the optical modes as we have done as a quantum harmonic oscillator, we define conjugate variables for each mode. When discussing this type of optical quantum computer, it is typical to discuss a phase space, though this is a valid picture in which to discuss quantum mechanics more broadly as well. A phase space formulation of quantum mechanics was developed independently by Groenewold [27] and Moyal [28] in 1946 and 1949, respectively. It allows us to define a quasiprobability distribution which describes our quantum system instead of a wavefunction, state vector or density matrix.

Phase space distributions are real-valued instead of complex like the wavefunction. The phase space description of a particle's position and momentum, for example, is such that we can integrate the Wigner function $W(x, p)$ (a representation of the quasiprobability) to find the probability of finding the particle on a particular interval.

$$P[a \le X \le b] = \int_a^b \int_{-\infty}^{\infty} W(x, p)\mathrm{d}p\mathrm{d}x$$

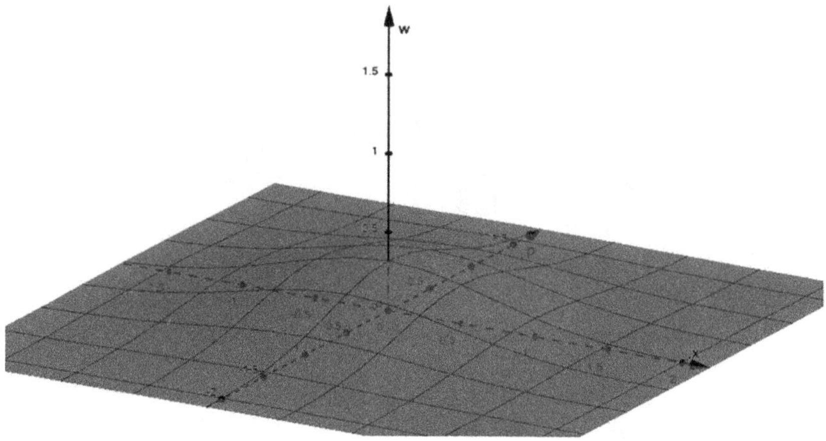

Fig. 4.3 Ground state Wigner function

$$W(x,p) = \frac{1}{2\pi}\int \exp(ivp)\left\langle x - \frac{v}{2}|\varrho|x + \frac{v}{2}\right\rangle \mathrm{d}v$$
$$= \frac{1}{2\pi}\int \exp(ivp)\psi^*(x - \frac{v}{2})\psi(x + \frac{v}{2})\mathrm{d}v$$

Here ϱ is a density matrix. $W(x, p)$ is a quasiprobability so that $\int\int W(x, p)\mathrm{d}x\mathrm{d}p = 1$. This transform between the two is called the inverse Wigner transform [29]. The negativity of the Wigner function is a known measure of how "quantum" a system is. Some states are less classical than others. For example, the ground state $|0\rangle_n$ would be represented by a Wigner function $W(x, p) = \frac{1}{\pi}\exp(-(x^2 + p^2))$ which expresses the uncertainty in the relationship between x and p due to the uncertainty principle (Fig. 4.3).

The expectation of an observable with operator representation A is

$$\langle A\rangle = \int A(x,p)W(x,p)\mathrm{d}p\mathrm{d}x$$

Operators x_l, p_l describe the amplitudes of the quadratures of the electromagnetic field which are $\cos(\mathbf{k}\cdot\mathbf{r} - \omega_k t)$ and $sin(\mathbf{k}\cdot\mathbf{r} - \omega_k t)$.

$$x_l = \frac{1}{\sqrt{2}}(a_l + a_l^\dagger)$$
$$p_l = \frac{1}{\sqrt{2}}\frac{a_l - a_l^\dagger}{i}$$
$$[x_l, p_l] = i$$

Fig. 4.4 Coherent state

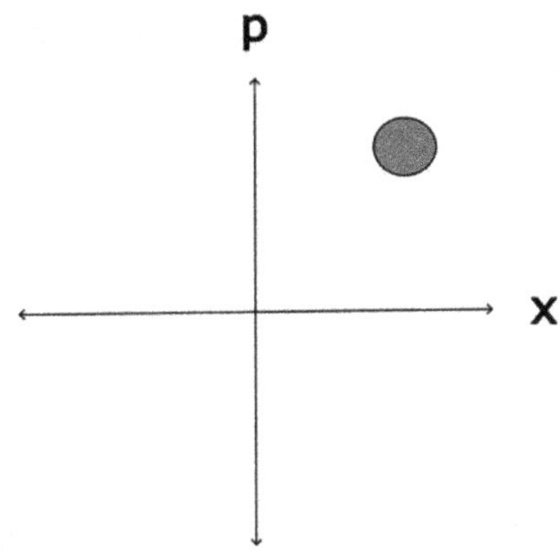

For a plane wave in free space, we have polarization e_l and

$$\mathbf{u}_{kl}(\mathbf{r}) = e_l \frac{1}{\sqrt{v}} \exp(i\mathbf{k}\cdot\mathbf{r})$$

$$\mathbf{E}(\mathbf{r},t) = i\sum_l e_l \sqrt{\frac{\hbar\omega_k}{2\varepsilon_0 v}}[(a_{kl} - a^\dagger_{kl})\cos(\mathbf{k}\cdot\mathbf{r} - \omega_k t) + i(a_{kl} + a^\dagger_{kl}) sin(\mathbf{k}\cdot\mathbf{r} - \omega_k t)]$$

For a coherent state such as that produced by a laser, we have $|\lambda\rangle = \exp(\lambda a^\dagger - \lambda^* a)\,|0\rangle_n$. The phase space representation is graphically a circle with area $\hbar$ in the xp plane (Fig. 4.4).

$$\begin{aligned}\frac{\langle p\rangle}{\sqrt{2}} &= \frac{1}{2}\langle\lambda|\,\frac{a - a^\dagger}{i}\,|\lambda\rangle \\ &= \lambda_I \\ \frac{\langle x\rangle}{\sqrt{2}} &= \frac{1}{2}\langle\lambda|\,(a + a^\dagger)\,|\lambda\rangle \\ &= \lambda_R\end{aligned}$$

A coherent state is a displacement of the vacuum state $|\lambda\rangle = D(\lambda)\,|0\rangle$. In quantum optics, the vacuum state $|0\rangle$ indicates the complete lack of any photons. λ is a complex number, and this displacement operator is $D(\lambda) = \exp(\lambda a^\dagger - \lambda^* a)$.

By the Baker–Campbell–Hausdorff[4] lemma, we have $\exp(A+B)=\exp(A)\exp(B)\exp(-\frac{1}{2}[A,B])$ when $[A,[A,B]]=[B,[A,B]]=0$.

$$\begin{aligned}
[\lambda a^\dagger, -\lambda^* a] &= -\lambda a^\dagger \lambda^* a + \lambda^* a \lambda a^\dagger \\
&= -\lambda\lambda^* a^\dagger a + \lambda^*\lambda a a^\dagger \\
&= -\lambda\lambda^* a^\dagger a + \lambda^*\lambda(1+a^\dagger a)) \because a a^\dagger = 1 + a^\dagger a \\
&= \lambda^*\lambda + a^\dagger a(\lambda^*\lambda - \lambda^*\lambda) \\
&= \lambda^*\lambda \\
&= |\lambda|^2 \\
\therefore D(\lambda) &= \exp(\lambda a^\dagger)\exp(-\lambda^* a)\exp(-|\lambda|^2/2) \\
D(\lambda)^\dagger &= D(\lambda)^{-1} = D(-\lambda) \\
D(\lambda)^\dagger a^\dagger D(\lambda) &= a^\dagger + \lambda^*
\end{aligned}$$

and the displacement operator gets its name from this final property:

$$D(\lambda)^\dagger a D(\lambda) = a + \lambda.$$

We can show that this is the case by the following derivation.

Example: Derivation of the Displacement Operator's Effect

$$\begin{aligned}
D^\dagger(\lambda) a D(\lambda) &= a + \lambda \\
&= D^\dagger(\lambda) D(\lambda) a - D^\dagger(\lambda)[D(\lambda), a]
\end{aligned}$$

$$\begin{aligned}
[a, D(\lambda)] &= [a, \exp(\lambda a^\dagger - \lambda^* a)] \\
&= [a, \exp(\lambda a^\dagger)\exp(-\lambda^* a)\exp(-|\lambda|^2/2)] \\
&= [a, \exp(\lambda a^\dagger)]\exp(-\lambda^* a)\exp(-|\lambda|^2/2) \because [a,a]=0 \\
&= (a\exp(\lambda a^\dagger) - \exp(\lambda a^\dagger)a)\exp(-\lambda^* a)\exp(-|\lambda|^2/2) \\
&= [a\sum_{n=0}^{\infty}\frac{(\lambda a^\dagger)^n}{n!} - \sum_{n=0}^{\infty}(\frac{(\lambda a^\dagger)^n}{n!})a]\exp(-\lambda^* a)\exp(-|\lambda|^2/2) \\
&= (\sum_{n=0}^{\infty}\frac{\lambda^n}{n!}[a, a^{\dagger n}])\exp(-\lambda^* a)\exp(-|\lambda|^2/2)
\end{aligned}$$

[4] In addition to his works in mathematics, Hausdorff began publishing philosophy under a pen name "Paul Mongré" in 1897 [30].

$$\begin{aligned}
[a, a^{\dagger n}] &= a(a^{\dagger})^n - (a^{\dagger})^n a \\
&= aa^{\dagger}(a^{\dagger})^{n-1} - a^{\dagger}a(a^{\dagger})^{n-1} \\
&+ a^{\dagger}aa^{\dagger}(a^{\dagger})^{n-2} - (a^{\dagger})^2 a(a^{\dagger})^{n-2} \\
&+ (a^{\dagger})^2 aa^{\dagger}(a^{\dagger})^{n-3} - \dots \\
&= [a, a^{\dagger}](a^{\dagger})^{n-1} + a^{\dagger}[a, a^{\dagger}](a^{\dagger})^{n-2} + (a^{\dagger})^2[a, a^{\dagger}](a^{\dagger})^{n-3} + \dots \\
&= n(a^{\dagger})^{n-1} \because [a, a^{\dagger}] = 1
\end{aligned}$$

$$\begin{aligned}
&\left(\sum_{n=0}^{\infty} \frac{\lambda^n}{n!}[a, a^{\dagger n}]\right) \exp(-\lambda^* a)\exp(-|\lambda|^2/2) \\
&= \left(\sum_{n=0}^{\infty} \frac{\lambda^n}{n!} n(a^{\dagger})^{n-1}\right) \exp(-\lambda^* a)\exp(-|\lambda|^2/2) \\
&= \left(\lambda \sum_{n=1}^{\infty} \frac{(\lambda a^{\dagger})^{n-1}}{(n-1)!}\right) \exp(-\lambda^* a)\exp(-|\lambda|^2/2) \because a^{\dagger n} = 1 \text{ when } n = 0 \\
&= \lambda \exp(\lambda a^{\dagger})\exp(-\lambda^* a)\exp(-|\lambda|^2/2) \\
&= \lambda D(\lambda)
\end{aligned}$$

$$\begin{aligned}
&D^{\dagger}(\lambda)D(\lambda)a - D^{\dagger}(\lambda)[D(\lambda), a] \\
&= D^{\dagger}(\lambda)D(\lambda)a + \lambda D^{\dagger}(\lambda)D(\lambda) \\
&= a + \lambda \because D^{\dagger}(\lambda)D(\lambda) = 1
\end{aligned}$$

A coherent state is an eigenstate of a.

$$\begin{aligned}
D(\lambda)^{\dagger} a \left|\lambda\right\rangle &= D(\lambda)^{\dagger} a D(\lambda) \left|0\right\rangle \\
&= (a + \lambda) \left|0\right\rangle \\
&= \lambda \left|0\right\rangle \\
D(\lambda)D(\lambda)^{\dagger} a \left|\lambda\right\rangle &= D(\lambda)\lambda \left|0\right\rangle \\
\therefore a \left|\lambda\right\rangle &= \lambda \left|\lambda\right\rangle
\end{aligned}$$

The coherent state's circle in the xp plane has a width of Δx and height of Δp.

$$
\begin{aligned}
(\Delta x)^2 &= \langle\lambda|\, x^2 \,|\lambda\rangle - \langle\lambda|\, x \,|\lambda\rangle^2 \\
&= \frac{1}{2}\, \langle\lambda|\, (a+a^\dagger)^2 \,|\lambda\rangle - [\frac{1}{\sqrt{2}}(\lambda+\lambda^*)]^2 \\
&= \frac{1}{2}\, \langle\lambda|\, (a^2 + aa^\dagger + a^\dagger a + a^{\dagger^2}) \,|\lambda\rangle - \frac{1}{2}(\lambda+\lambda^*)^2 \\
&= \frac{1}{2}[\lambda^2 + 1 + 2|\lambda|^2 + \lambda^{*^2}] - \frac{1}{2}(\lambda^2 + 2|\lambda|^2 + \lambda^{*^2}) \\
&= \frac{1}{2}
\end{aligned}
$$

Similarly, $(\Delta p)^2 = \frac{1}{2}$ and $(\Delta x)(\Delta p) = \frac{1}{2}$. For any physical state, $(\Delta x)(\Delta p) \geq \frac{1}{2}$. Therefore, coherent states are minimum uncertainty relation states.

Squeezed states $|\xi\rangle$ are of the form $\exp(\frac{1}{2}(\xi^* z^2 - \xi a^{\dagger^2}))$ and are graphically ellipses in the xp plane with area proportional to $\hbar$ and $\Delta x \neq \Delta p$ (Fig. 4.5).

Similar to the displacement operator, we can define a squeezing operator.

$$
\begin{aligned}
S^\dagger(\varepsilon) &= S^{-1}(\varepsilon) = S(-\varepsilon) \\
S^\dagger(\varepsilon) a S(\varepsilon) &= a\cosh(r) - a^\dagger \exp(2i\phi)\sinh(r) \\
S^\dagger(\varepsilon) a^\dagger S(\varepsilon) &= a^\dagger \cosh(r) - a\,\exp(-2i\phi)\sinh(r)
\end{aligned}
$$

Considering the rotated amplitudes $y_1 = \frac{1}{\sqrt{2}}(\exp(-i\phi)a + \exp(i\phi)a^\dagger)$ and $iy_2 = \frac{1}{\sqrt{2}}(\exp(-i\phi)a - \exp(i\phi)a^\dagger)$,

$$
S^\dagger(\varepsilon)(y_1 + iy_2)S(\varepsilon) = y_1\exp(-r) + i\frac{1}{2}\exp(r)
$$

Fig. 4.5 Squeezed state

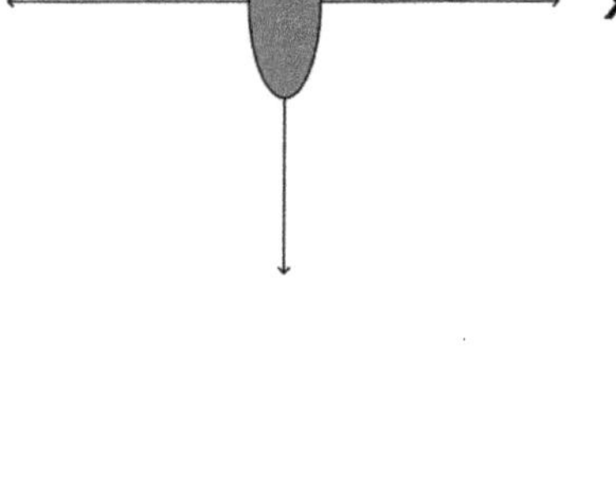

Here $r = |\varepsilon|$ is the squeezing factor. $(\Delta y_1)^2 = \frac{1}{2}\exp(-2r)$, $(\Delta y_2)^2 = \frac{1}{2}\exp(2r)$ and $(\Delta y_1)(\Delta y_2) = \frac{1}{2}$.

A thermal state near the center of the Bloch sphere with ϱ proportional to $\exp(-\frac{H}{KT})$ is a circle with radius T in phase space (Fig. 4.6).

The first excited state $|1\rangle_n$ has a nonzero ω component. Below is the graph for p = 0 (Fig. 4.7).

Instead of the gates we are used to dealing with in gate model quantum computing, CVQC makes use of displacements and squeezing, etc., in phase space to do computing.

4.2.3 Quantum Annealers

D-Wave's annealers are useful for minimizing functions by finding minima in an energy landscape [5]. This is done by turning on a transverse field and then gradually reducing its strength to allow the quantum state to settle in a low-energy state. These annealers have different architectures with different connectivities but are based on the transverse Ising model.

$$H(\sigma) = \sum_{i=0}^{N-1} h_i \sigma_i + \sum_{i=0}^{N-2} \sum_{j=i+1}^{N-1} J_{ij} \sigma_i \sigma_j$$

Here the σ corresponds to physical qubits, h_i is local magnetic fields and J_{ij} is interqubit coupling strengths. These coefficients are configured via a process called minor embedding to encode the dynamics of the specific problem to be solved. Problems should be expressed as quadratic unconstrained Boolean optimizations before they are minor-embedded. Many NP-hard problems can be encoded in these hardware topologies, such as the two-dimensional 8 qubit bipartite graph that some D-Wave annealers use.

Fig. 4.6 Thermal state

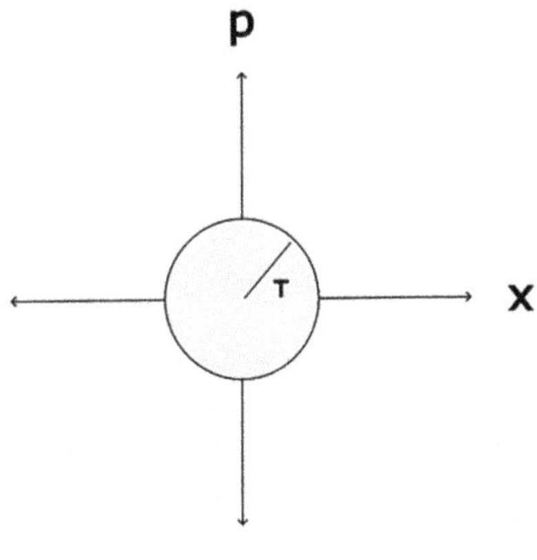

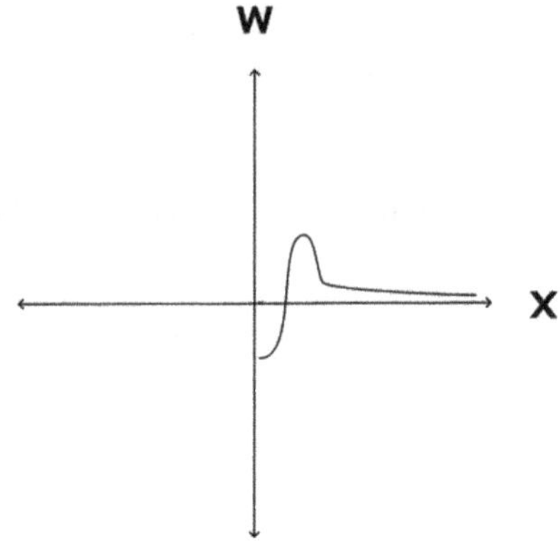

Fig. 4.7 Excited state

This transverse Ising[5] model is actually a special case of the full Hamiltonian for D-Wave's qubits, which looks like this.

$$H = -\frac{1}{2}\sum_{i=1}^{N}(\Delta\sigma_x + \varepsilon_i\sigma_z) + \sum_{i<j} J_{ij}\sigma_{iz}\sigma_{jz}.$$

Here Δ is controllable by the transverse magnetic field, and ε and J_{ij} are also tunable. See that when $\Delta \to 0$, we have the programmable transverse Ising model.

The J-coupling is realized by a mutual inductance between qubits. This is provided by a DC Squid coupler. A DC Squid is like a single Josephson Junction with a flux-tunable critical current (Fig. 4.8).

The DC Squid makes use of interference between phases ϕ_1 and ϕ_2. The current is maximized when the two are in-phase s.t. $\phi_2 = \phi_1 - 2\pi\frac{\Phi}{\Phi_0} = (2n+1)\pi - \phi_1$. Here $\phi_2 = \phi_1 - 2\pi\frac{\Phi}{\Phi_0}$ is a relationship that can be found by solving for the phase difference in the arms of the DC Squid using path integrals such that $J_S = 0$, and $\phi_2 = (2n+1)\pi - \phi_1$ is the special case where ϕ_1 and ϕ_2 are in-phase. Then assuming $I_{c1}, I_{c2} = I_c$,

[5] After developing his much-studied Ising model, Ernst Ising disconnected completely from scientific life and did not realize until years later that his work, since cited by Heisenberg, had become widely known [31].

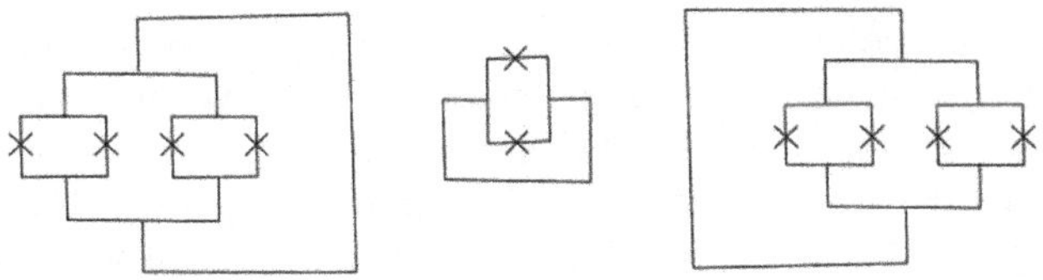

Fig. 4.8 DC Squid coupling (illustration by Joan Weng)

$$I = I_{c1}\sin(\phi_1) + I_{c2}\sin\left(\phi_1 - 2\pi\frac{\Phi}{\Phi_0}\right)$$
$$2\phi_1 = (2n+1)\pi + 2\pi\frac{\Phi}{\Phi_0}$$
$$I_{\max} = 2I_c|\sin\left(\left(\frac{2n+1}{2} + \frac{\Phi}{\Phi_0}\right)\pi\right)|$$
$$I_{\max} = 2I_c|\cos\left(\pi\frac{\Phi}{\Phi_0}\right)|$$

This realizes our coupler. The qubits themselves are "flux qubits" with the following Hamiltonian [5]:

$$H = 4E_C n^2 - E_J\cos(\phi) + \frac{1}{2}E_L(\phi - \phi_x)^2.$$

Here $E_C = \frac{e^2}{2C}$, $E_J = \frac{\Phi_0}{2\pi}I_c$, $E_L = \frac{1}{L}(\frac{\Phi_0}{2\pi})^2$. The E_J, E_L terms are together considered the potential energy. The result is a double-well shape when $\phi_x = \pi$ and $\beta_L = \frac{E_J}{E_L} = \frac{2\pi}{\Phi_0}LI_c > 1$. A qubit is typically formed by treating a low-energy state in the left well as $|0\rangle$ and a low-energy state in the right well as $|1\rangle$ or vice versa. The state on the right typically has approximately half flux quantum more flux than the one on the left. The height of the barrier between the wells is controllable by β. The asymmetry of the wells is controllable by the external flux ϕ_x.

Using a DC Squid in place of the JJ can allow us to tune I_C, ε, β and therefore the tunneling between the wells. However, we also want to account for fabrication variation so we replace the two JJs in the DC Squids by *another* pair of JJs. This gives us current control of both.

4.2.4 Nuclear Magnetic Resonance

An experimental NMR setup is based on spectrometer technology which is very mature and has been developed since 1946. A liquid Helium jacket is further enclosed in a liquid nitrogen dewar. The sample molecular bath is lowered into the very center between small superconducting coils which are used to control and measure the quantum state. The nuclear spin can be changed in around one nanosecond with errors of around 1 part per 10^{12} in qubit rotation.

C_{13}, H_1, F_{19} and N_{15} nuclei are spin-$\frac{1}{2}$, for example, [2, p.328]. Trichloroethylene has also historically been used for the molecular bath, which was also used as a cleaning product and to make decaf coffee before it was found to be carcinogenic.

Instead of an electron spin, NMR uses the nuclear spin of atoms in a molecular bath. The bath itself is in a strong background magnetic field. The coil generating the magnetic field might have as many as 10^4 windings and have 100 amps running through it.

Qubit addressability in NMR is provided by the widely separated ω_0 of the nuclei. The local environments of the nuclei provide chemical shifts of the ω_0 of 600 900 Hz, i.e. $\omega_0(H) = 500\ MHz$ and $\omega_0(C_{13}) = 125\ MHz$ at 11.7T [2, p.325].

The interaction of a spin **u** with the electromagnetic field is given by $H = \boldsymbol{\mu}\mathbf{B}$. The field in the case of NMR is principally in the z-axis and generated by the large superconducting magnet, but smaller transversely oriented Helmholtz coils create smaller oscillating field components in the xy plane [2, p.325].

$$\mathbf{B} = B_0\mathbf{z} + B_1(\mathbf{x}\cos(\omega t) + \mathbf{y}\sin(\omega t))$$

This Hamiltonian can also be written in terms of the Pauli matrices.

$$H = \frac{\omega_0}{2}Z + g(X\cos(\omega t) + Y\cos(\omega t))$$

We have a rotating frame Hamiltonian [2, p. 327] that gives our dynamics in the rotating frame.

$$H = \frac{\omega_0 - \omega}{2}Z + gX$$

Interaction between qubits is facilitated by interaction between the nuclear spins mediated by an electron cloud which is introduced to provide this medium. We need this electron cloud because dipole-dipole interactions average to zero.

Two-qubit gates are mediated by electrons and we get an effective J-coupling.

$$H_J = \frac{\pi}{2}J\sigma_z^{(1)}\sigma_z^{(2)}$$

This is a specific case of the more general Hamiltonian for electron-mediated couplings. In this case, we consider scalar instead of tensor J, which is appropriate in liquids with weak couplings and/or nuclear spins with differing precession frequencies. The more general Hamiltonian is the following [2, p.328]:

$$H_{1,2}^J = \frac{\hbar J}{4}\boldsymbol{\sigma}_1 \cdot \boldsymbol{\sigma}_2 = \frac{\hbar J}{4}Z_1Z_2 + \frac{\hbar J}{8}(\sigma_+\sigma_- + \sigma_-\sigma_+).$$

J might be in the range of hundreds of Hertz for adjacent spins in a molecule and tens of Hertz for non-adjacent spins, for example.

4.2.5 Ion Traps

In ion trap-based quantum computing, the qubits are the electronic states of ions, which are held in a chain in an electromagnetic trap such as a Paul trap. Ion trap QC was originally considered feasible due to its similarity to atomic clock technology. The ions are never completely stationary, but move back and forth and repel each other as they do so. The motional states of the traps are used as a bus for coupling the qubits. Ions of Sr, Hg and Yb can be used, for example.

As discussed in the opening of the book, we need something like a stable potential to hold our ions in our trap. However, according to Earnshaw's Theorem, the closest thing that we can achieve is a saddle point [2, p. 309]. What we do therefore is rotate the potential by varying the signals we apply to our electrodes. This creates a radiofrequency potential of the form $\Phi_{rf} = (V_0 cos\Omega_T t + U_r)(1 + (x^2 \; y^2)/R^2)/2$ [2, p. 311].

Our potential is restrictive in the xy plane but allows the ions to space themselves out in the z dimension according to their Coulomb repulsion. This is called a "Coulomb chain" or "Coulomb crystal". So, each ion sits in a harmonic z potential. The vibrational motion of these ions is harmonic and is quantized as a quantum harmonic oscillator. The more ions there are, the smaller these oscillations are.

$$
\begin{aligned}
H_0 &= \frac{\varrho^2}{2m} + \frac{m}{2}\omega^2 z^2 + \frac{\hbar\omega_0}{2}\sigma_z \\
&= \hbar\omega\left(a^\dagger a + \frac{1}{2}\right) + \frac{\hbar\omega_0}{2}\sigma_z
\end{aligned}
$$

Here we have a harmonic oscillator $\hbar\omega(a^\dagger a + \frac{1}{2})$ and a qubit $\frac{\hbar\omega_0}{2}\sigma_z$. A z-axis laser can cause an interaction between the motional oscillation and the qubit state. We can transfer the state of an ion to the motional phonon state and then to another ion, achieving a coupling of the two ions.

This is a particular case of the more general Hamiltonian which includes an x and a y component [2, p. 311] with N ions of mass M. This more general Hamiltonian is of interest because some traps will alternatively restrict motion in z so that a Coulomb crystal forms in the xy plane.

$$
H = \sum_{i=1}^{N} \frac{M}{2}\left(\omega_x^2 x_i^2 + \omega_y^2 y_i^2 + \omega_z^2 z_i^2 + \frac{|\mathbf{p}_i|^2}{M^2}\right) + \sum_{i=1}^{N}\sum_{j>i} \frac{e^2}{4\pi\varepsilon_0 |\mathbf{r}_i - \mathbf{r}_j|}
$$

In the interaction picture, our xy-restricted Hamiltonian has three components, $H_I = H_{\text{carrier}} + H_{\text{red}} + H_{\text{blue}}$. The red and blue components describe the coupling of the spin state of the ion to its motional state [2, p.318]. Each of H_{blue}'s and H_{red}'s two terms corresponds to motional sideband transitions that involve the creation or annihilation of one phonon.

We can represent that quantum state of a trapped ion and the trap as $|n, m\rangle$ where n is our spin state and m are our phonons. Then the sidebands involve the following transitions (Fig. 4.9).

The H_{carrier} component changes our qubit for $\omega_0 \doteq \omega_l$. Here ϕ is the electromagnetic potential, ω is our phonon or trap frequency and ω_0 is our ion resonant frequency.

$$H_{\text{carrier}} = \hbar\Omega[\sigma_+ \exp(i(\omega_0 - \omega_l)t)\exp(-i\phi) + \sigma_+ \exp(-i(\omega_0 - \omega_l)t)\exp(i\phi)]$$

H_{red} excites our qubit, which relaxes our trap.

$$\begin{aligned} H_{\text{red}} = & -i\hbar\Omega N[\sigma_+ a \exp(-i(\omega_l - \omega_0 + \omega)t)\exp(i\phi) \\ & + \sigma_+ a^\dagger \exp(i(\omega_l - \omega_0 + \omega)t)\exp(-i\phi)] \end{aligned}$$

H_{blue} excites both our qubit and trap at $\omega_0 \doteq \omega_l + \omega$.

$$\begin{aligned} H_{\text{blue}} = & -i\hbar\Omega N[\sigma_+ a^\dagger \exp(-i(\omega_l - \omega_0 - \omega)t)\exp(i\phi) \\ & + \sigma_+ a \exp(i(\omega_l - \omega_0 - \omega)t)\exp(-i\phi)] \end{aligned}$$

A control laser composed of photons interacts with an ion in the trap by adding or subtracting units of angular momenta from the ion. In the previous chapters and the introduction to this one we have mostly discussed two-level systems and spin-1/2 particles. However, an ion has both a spin-1/2 electron providing its overall charge and a nuclear spin which is spin-3/2. A photon interacting with the ion does not distinguish between these spins and in an effort to describe the interaction that takes place it is best for us to describe the total angular momentum of the system [2, p.313].

As we saw a preview of in Chap. 2, the angular momentum operators for a spin-3/2 particle are:

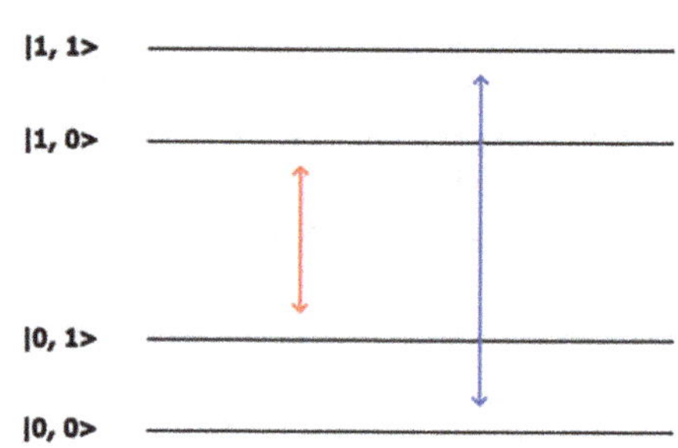

Fig. 4.9 Ion trap levels

$$L_z = \frac{\hbar}{2}\begin{bmatrix} 3 & 0 & 0 & 0 \\ 0 & 1 & 0 & 0 \\ 0 & 0 & -1 & 0 \\ 0 & 0 & 0 & -3 \end{bmatrix}$$

$$L_y = \frac{\hbar}{2}\begin{bmatrix} 0 & -i\sqrt{3} & 0 & 0 \\ i\sqrt{3} & 0 & -2i & 0 \\ 0 & -2i & 0 & -i\sqrt{3} \\ 0 & 0 & i\sqrt{3} & 0 \end{bmatrix}$$

$$L_x = \frac{\hbar}{2}\begin{bmatrix} 0 & -\sqrt{3} & 0 & 0 \\ -\sqrt{3} & 0 & -2 & 0 \\ 0 & -2 & 0 & -\sqrt{3} \\ 0 & 0 & -\sqrt{3} & 0 \end{bmatrix}.$$

Our total angular momentum F is a sum $F = S \otimes I + I \otimes L$ of the spin S and nuclear spin L. We have $F^2 = S^2L^2 + 2S_zL_z + S_+L_- + S_-L_+$ defined in similar fashion to L^2 of chapter 2, and following the same process as we did there we can find the basis states $|fm\rangle$ such that $F^2\,|fm\rangle = \lambda\,|fm\rangle$ and $F_z\,|fm\rangle = m_F\,|fm\rangle$ where $\lambda = f(f+1)\hbar^2$ in order to construct our basis [32, p.223–229].

We have two choices for the basis states. The first choice is that of the simultaneous eigenstates of S^2, L^2, S_z and L_z which we will denote $|sl; m_s m_l\rangle$. These operators are mutually compatible observables since they all commute.

$$\begin{aligned} S^2\,|sl; m_s m_l\rangle &= s(s+1)\hbar^2\,|sl; m_s m_l\rangle \\ S_z\,|sl; m_s m_l\rangle &= m_s\hbar\,|sl; m_s m_l\rangle \\ L^2\,|sl; m_s m_l\rangle &= l(l+1)\hbar^2\,|sl; m_s m_l\rangle \\ L_z\,|sl; m_s m_l\rangle & m_l\hbar\,|sl; m_s m_l\rangle \end{aligned}$$

The second choice is to use the simultaneous eigenstates of S^2, L^2, F^2 and F_z denoted $|s,l;\,fm\rangle$. These are often also denoted simply $|fm\rangle$ as done in [2].

$$\begin{aligned} S^2\,|s,l;\,fm\rangle &= s(s+1)\hbar^2\,|s,l;\,fm\rangle \\ L^2\,|s,l;\,fm\rangle &= l(l+1)\hbar^2\,|s,l;\,fm\rangle \\ F^2\,|s,l;\,fm\rangle &= f(f+1)\hbar^2\,|s,l;\,fm\rangle \\ F_z\,|s,l;\,fm\rangle &= m\hbar\,|s,l;\,fm\rangle \end{aligned}$$

Since F^2 and S_z do not commute and F^2 and L_z do not commute, these two choices represent the largest sets of compatible observables that we may choose.

When we solved for the computational basis states in Chap. 2, it involved finding a relationship between a λ_{min} and λ_{max} and reasoning about the λ in between. In this case, we can relate our two choices of bases to find recursion relations involving the quantum numbers.

Relating the two bases causes us to consider Clebsch–Gordon coefficients $\langle sl;\, m_s m_l | s, l;\, fm\rangle$ which provide a change of basis in that they help us to form a real unitary transformation matrix:

$$|s, l;\, fm\rangle = \sum_{m_s} \sum_{m_l} |sl;\, m_s m_l\rangle \langle sl;\, m_s m_l | s, l;\, fm\rangle$$

Since a real unitary matrix is also orthonormal, we know:

$$\sum_{f} \sum_{m} \langle sl;\, m_s m_l | s, l;\, fm\rangle \left\langle sl;\, m'_s m'_l | s, l;\, fm\right\rangle = \delta_{m_s m'_s} \delta_{m_l m'_l}$$

$$\sum_{m_s} \sum_{m_l} \langle sl;\, m_s m_l | s, l;\, fm\rangle \left\langle sl;\, m_s m_l | s, l;\, f'm'\right\rangle = \delta_{ff'} \delta_{mm'}$$

We also have raising and lowering operators $F_+ \, |f, m\rangle = \sqrt{(f-m)(f+m+1)}\hbar \, |f, m+1\rangle$ and $F_- \, |f, m\rangle = \sqrt{(f+m)(f-m+1)}\hbar \, |f, m-1\rangle$ [32, p. 197] so that:

$$F_\pm \, |sl;\, fm\rangle = (S_\pm + L_\pm) \sum_{m_s} \sum_{m_l} |sl;\, m_s m_l\rangle \langle sl;\, m_s m_l | s, l;\, fm\rangle$$

$$\sqrt{(f \mp m)(f \pm m + 1)} \, |sl;\, f, m \pm 1\rangle = \sum_{m_s} \sum_{m_l} (\sqrt{(s \mp m_s)(s \pm m_s + 1)} \, |sl;\, m_s \pm 1, m_l\rangle$$
$$+ \sqrt{(l \mp m_l)(l \pm m_l + 1)} \, |sl;\, m_s, m_l \pm 1\rangle) \times \langle sl;\, m_s m_l | s, l;\, fm\rangle$$

In order to relate this equation to our Clebsch–Gordon coefficients we may multiply on both sides by $\left\langle sl;\, m'_s m'_l \right|$. Then since we know $\langle a|b\rangle = \delta_{ab}$ for a, b belonging to any orthonormal set, we can see that the terms on the right-hand side will vanish whenever $m'_s \neq m_s \pm 1$, $m'_l \neq m_l$ for the first term and $m'_s \neq m_s$, $m'_l \neq m_l \pm 1$ for the second term. This simplifies our equation, and we get the following recursion relation:

$$\begin{aligned} &\sqrt{(f \mp m)(f \pm m + 1)} \, \langle sl;\, m_s m_l | sl;\, f, m \pm 1\rangle \\ &= \sqrt{(s \mp m_s + 1)(s \pm m_s)} \, \langle sl;\, m_s \mp 1, m_l | sl;\, fm\rangle \\ &+ \sqrt{(l \mp m_l + 1)(l \pm m_l)} \, \langle sl;\, m_s, m_l \mp 1 | sl;\, fm\rangle . \end{aligned}$$

This recursion relation can then be used to solve for the eigenbasis of the total angular momentum, which is usually given in terms of the joint eigenstates of the square of the orbital angular momentum L^2 and generator of rotations about the azimuthal axis L_z. The joint eigenstates of L^2 and L_z are called spherical harmonics $Y_l^m(\theta, \phi)$, and the eigenstates of F^2 are called spinor spherical harmonics $\mathsf{Y}_l^{f=l\pm1/2,m}(\theta, \phi)$.

$$
\begin{aligned}
Y_l^m(\theta,\phi) &= (-1)^m \sqrt{\frac{2l+1}{4\pi}\frac{(l-m)!}{(l+m)!}} P_l^m(cos\phi)\exp(im\phi) \\
P_l^{-m} &= (-1)^m \frac{(l-m)!}{(l+m)!} P_l^m \\
\mathsf{Y}_l^{f=l\pm1/2,m}(\theta,\phi) &= \pm\sqrt{\frac{l\pm m+1/2}{2l+1}} Y_l^{m-1/2}(\theta,\phi)\,|0\rangle \\
&+ \sqrt{\frac{l\pm m+1/2}{2l+1}} Y_l^{m+1/2}(\theta,\phi)\,|1\rangle \\
&= \frac{1}{\sqrt{2l+1}} \begin{bmatrix} \pm\sqrt{l\pm m+1/2}Y_l^{m-1/2}(\theta,\phi) \\ \sqrt{l\mp m+1/2}Y_l^{m+1/2}(\theta,\phi) \end{bmatrix}
\end{aligned}
$$

Our mutually compatible observables are defined in space and may be expressed in spherical coordinates. These are the coordinates used in the above equations. For example,

$$
\begin{aligned}
L^2 &= \hbar^2(-r^2\nabla^2 + \left(r\frac{\partial}{\partial r}+1\right) r\frac{\partial}{\partial r}) \\
&= \hbar^2\left(-\frac{1}{sin\theta}\frac{\partial}{\partial\theta}sin\theta\frac{\partial}{\partial\theta} - \frac{1}{sin^2\theta}\frac{\partial^2}{\partial\phi^2}\right) \\
L_z &= -i\hbar\left(x\frac{\partial}{\partial y} - y\frac{\partial}{\partial x}\right) \\
&= -i\hbar\frac{\partial}{\partial\phi}.
\end{aligned}
$$

A thorough treatment and derivation of these spin angular functions can be found in [32, p. 229–232].

4.2.6 Spin Qubits

Spin qubits are qubits defined using the spin of a particle. These include semiconductor spin qubits, which are defined in a semiconductor by trapping a single hole or electron, and radiation damage centers, where "defects" are implanted artificially into a lattice by exposure to radiation. Similarly to NMR setups, we are concerned with qubits whose levels are derived from Zeeman splitting induced by an external magnetic field. A spin-$\frac{1}{2}$ particle in a magnetic field has the following Zeeman Hamiltonian, this time expressed in terms of the Landé g-factor:

$$H = \frac{1}{2} g\mu_B \mathbf{B} \cdot \boldsymbol{\sigma}.$$

The g-factor relates the Larmor precession frequency of our qubit to the oriented magnetic field. An effective g-factor is given by $g* = hf_Q/(\mu_B B)$. Here f_Q is the Larmor frequency, h is Planck's constant and μ_B is the Bohr magneton. In spin qubits we are especially interested in this g-factor. The g-factor depends greatly on material homogeneity and can differ between sites in a semiconductor spin qubit processor, yielding qubits that behave differently from each other. We call a g-tensor with local variation anisotropic. It is an important area of study to develop materials which yield predictable g-factors and to learn how to take advantage of "sweet spots" where the magnetic field is oriented such that the operation of a qubit is optimally fast. By varying our magnetic field angle we can investigate the anisotropy of our qubit's coupling to nuclear spins and the qubit drive mechanism. For example, Salfi predicted sweet spots in Si in 2016 [33], which were later verified. These types of sweet spots are still being investigated as late as 2024 [34, 35].

The g-factor is most generally a tensor, and the Zeeman Hamiltonian has an expanded form [36].

$$H = \frac{\mu_B}{\hbar} B_0 \begin{bmatrix} \cos\phi \sin\theta \ \sin\phi \cos\theta \ \cos\theta \end{bmatrix} \begin{bmatrix} g_x & 0 & 0 \\ 0 & g_y & 0 \\ 0 & 0 & g_z \end{bmatrix} \begin{bmatrix} \mathbf{S_x} \\ \mathbf{S_y} \\ \mathbf{S_z} \end{bmatrix}$$

Here ϕ, θ give the orientation of the magnetic field. The effective g-factor at a given magnetic field orientation is $g^* = \sqrt{g_x^2 \sin^2\theta \cos^2\phi + g_y^2 \sin^2\theta \sin^2\phi + g_z^2 \cos^2 \cos\theta}$.

Generally, H is minimized when $\mathbf{B}$ and $\boldsymbol{\sigma}$ are anti-parallel and it is maximized when they are parallel. Any relative orientation that is neither parallel nor anti-parallel yields spin precession about $\mathbf{B}$. A typical Larmor frequency for a spin qubit might be around $f_Q = 2\pi \cdot 5$ GHz.

Zeeman splitting [32, p. 336] yields the energy separation of the energy eigenstates of the spin qubit which are given by:

$$E = \pm \frac{g\mu_B B}{2}$$

There are a cubic and a diamagnetic term in the Zeeman Hamiltonian which are usually neglected, but are explored in [37]. These terms however come into play when we consider the anisotropy (without them the anisotropy can go to infinity).

The energy level spacing is δ_s and ideally we want $\delta_s > KT$ so that the qubit can have a well-defined state.

There are multiple varieties of spin qubits, including semiconductor quantum dots, and qubits bound to impurities. These are different approaches to isolating single charge carriers.

In quantum dots, there is a DC bias on the top gate which we use to accumulate or isolate a hole or electron. Lateral quantum dots use this top gate voltage to confine a particular charge carrier in a two-dimensional electron gas (2DEG) or two-dimensional hole gas (2DHG) which exists in a quantum well usually tens of nanometers beneath the surface of the substrate. This quantum well itself must have a thickness comparable to the Fermi wavelength so that the z-confinement is such that we have a 2DEG or 2DHG and not a 3D gas.

The effective mass of the carrier determines in part the size we need for our quantum dots. The effective mass in Si is large, for example. This means that quantum dots need to be tens of nanometers in size. This footprint is small and difficult to fabricate. Using holes in Germanium conversely allows us to fabricate dots 50 to 100 nm in size. There is a trade-off between ease of fabrication and scalability. It is harder to fit a large number of dots on a chip if they have a large footprint.

Loading an nth charge onto a dot requires crossing an energy barrier of $\frac{n^2e^2}{2C}$. This is the Coulomb blockade. The quantum dot and current source can be modeled by a tunnel resistance and a capacitor. The energy stored in the capacitor E_C can be made larger by making the capacitance C smaller. C depends on the volume of the island.

$$E_C = \frac{e^2}{2C} = \frac{e^2}{4\pi\varepsilon V^{\frac{1}{3}}}$$

These charges can either be electrons or holes. This depends on the quantum well material. For example, a GeSi/Ge/GeSi quantum well will confine holes rather than electrons. Electron spin qubits are simpler, the qubit possibly being encoded directly into the spin of the particle which is spin-1/2. The electron Hamiltonian $H = \frac{\hbar^2 k^2}{2m_{eff}}$ is compact in comparison to the hole Hamiltonian $H = \frac{\hbar^2}{2m_{eff}}[(\gamma_1 + \frac{5}{2}\gamma_s)k^2 - 2\gamma_s(\mathbf{k} \cdot \mathbf{J})^2]$ [10, 38]. Here the γs are Luttinger parameters, and k is crystal momentum. This is the (simplified) Hamiltonian for Germanium. Some of the complication for hole spins comes from the fact that we have light and heavy holes, which have different effective masses. The effective mass of heavy holes is heavy out of the plane and along the quantization axis. We can get this from the Luttinger parameters. Heavy holes are lighter in the plane, and they are easier to confine in the plane. The mixing of heavy and light holes is zero at zero magnetic field. Some band structure details follow [39]. The light and heavy holes are each doubly degenerate, meaning that we can model them together at minimum with a 4 band model. The Luttinger–Kohn model is a 6 band model. This comes from the P-type valence band's symmetry being 3-times degenerate, which is doubled because of the spin quantum number. The Kane model is another, which is 8 band because it considers the S-type conduction band. All of these models are derived from the 30 band model by rescaling parameters using Schrieffer–Wolff transformations. The Hamiltonians of the quantum wells in any of these models are written using kp perturbation theory with a confining potential.

Heavy and light holes have different energies, and when they meet in an energy level diagram, we get an anti-crossing. The lowest energy states are the heavy hole states. The ground state is not purely heavy holes however, because spin orbit coupling and the Zeeman effect introduce a small admixture of light holes.

We do not encode the qubit into the spin of a particle directly when it comes to hole spin qubits, but rather into a pseudo-spin. The ground state for a single hole confined in a quantum dot is a degenerate Kramers doublet at zero magnetic field [10].

Kramers theorem [40] states that the energy levels of a time-reversal symmetric system with half integer spin are at least doubly degenerate, meaning that for every energy eigenstate there is at least one more eigenstate with the same energy. In solid-state physics, the time-reversal operator operates on a Bloch state $|n, k, \sigma\rangle$ of band index n, crystal momentum k and spin σ by inverting the signs on k and σ so that by Kramers theorem for every $|n, k, \sigma\rangle$ there is a $|n, -k, -\sigma\rangle$ with the same energy. Such a doublet is split in energy when there is an external magnetic field. This becomes a natural basis into which we can encode a qubit. For heavy holes we have $J = \frac{3}{2}$ and $J_z \pm \frac{3}{2}$ and the hole wave function is a superposition of the $|0\rangle$ and $|1\rangle$ spin states with the orbital degrees of freedom. We have seen similar physics before! $J = 3/2$ systems have electric and elastic quadrupoles [37] and the physics and microscopic mechanisms of corresponding physical systems are not fully understood yet.

For light holes we have $J = \frac{3}{2}$ and $J_z \pm \frac{1}{2}$. The confinement potential we use to isolate a single hole itself causes the pseudo-spin to appear in the z direction since the z component of **k** becomes smaller than the components in the xy plane.

We do not stop here with qubit encodings. There are many encodings including Loss–DiVincenzo, Flopping Mode, Quadrupole, Flip-Flop, Spin-Charge, Hybrid, RX / AEON, Exchange Only, QUEX, Singlet-Singlet and Singlet-Triplet encodings [41]. The simplest is the Loss–DiVincenzo encoding [7].

In a Loss–DiVincenzo qubit, the Pauli operators are encoded directly into the spin operators of the spin-1/2 electron which is electrically isolated and confined. Two spin qubits interact with each other via the Heisenberg exchange mechanism. The unitary evolution operator of this Hamiltonian is given by $U(t_2 - t_1)$.

$$H = J(\sigma_x^{(2)} \otimes \sigma_x^{(1)} + \sigma_y^{(2)} \otimes \sigma_y^{(1)} + \sigma_z^{(2)} \otimes \sigma_z^{(1)})$$
$$U(t_2 - t_1) = \exp(-iH(t_2 - t_1)/\hbar)$$

The Hamiltonian includes a single-electron Zeeman term and a Heisenberg exchange term [41].

$$H(t) = \frac{1}{4}\sum_{\langle i,j\rangle} J_{ij}(t)\boldsymbol{\sigma}_i \cdot \boldsymbol{\sigma}_j + \frac{1}{2}\sum_i g_i \mu_B \mathbf{B}_i \cdot \boldsymbol{\sigma}_i$$

In reality we will also consider a hyperfine term of the form $\sum_{i,j} A_{i,j,k} I_{i,k} S_{j,k}$ where $A_{i,j,k}$ is the hyperfine coupling between nuclear spin i and spin j in direction k. $I_{i,k}$ are nuclear spins and S are Pauli spin operators.

The next least complicated encoding of a qubit is the singlet-triplet qubit which encodes a qubit into two spins [42, 43]. The singlet-triplet is so-named because of its connection to singlet and triplet states. The singlet state involves two particles who share an energy level in a potential well and due to the Pauli exclusion principle therefore have spin quantum numbers with opposite signs. Their total angular momentum is zero. A triplet state on the other hand involves two particles $|s_1 m_1\rangle$ and $|s_2 m_2\rangle$ in different energy levels whose spins add to a total s of 1 and whose projection onto z m can therefore take values in $\{-1, 0, 1\}$. If we label the $\left|\frac{1}{2}, +\frac{1}{2}\right\rangle$ state as $|\uparrow\rangle$ and $\left|\frac{1}{2}, -\frac{1}{2}\right\rangle$ as $|\downarrow\rangle$, then the three states with total spin angular momentum equaling 1 are $|\uparrow_1\uparrow_2\rangle = |T_+\rangle$, $|\downarrow_1\downarrow_2\rangle = |T_-\rangle$ and $\frac{1}{\sqrt{2}}(|\uparrow_1\downarrow_2\rangle + |\downarrow_1\uparrow_2\rangle) = |T_0\rangle$ which are related to the total angular momentum by Clebsch–Gordon coefficients. The anti-symmetric two-particle spin configuration $\frac{1}{\sqrt{2}}(|\uparrow_1\downarrow_2\rangle - |\downarrow_1\uparrow_2\rangle)$ describes a singlet state also denoted $|S\rangle$.

We can encode the computational $|0\rangle$ and $|1\rangle$ states of the singlet-triplet qubit into the $\frac{1}{\sqrt{2}}(|\uparrow_1\downarrow_2\rangle - |\downarrow_1\uparrow_2\rangle)$ state and the $\frac{1}{\sqrt{2}}(|\uparrow_1\downarrow_2\rangle + |\downarrow_1\uparrow_2\rangle)$ state of the particles respectively. When we operate a singlet-triplet qubit, we understand that no more than one energy level is accessible in each quantum dot. A triplet state therefore cannot form in one quantum dot, but must involve both. The Pauli X, $Y$and Z operators become $S_z^{(1)} - S_z^{(2)}$, $2z \cdot \mathbf{S}^{(2)} \times \mathbf{S}^{(1)}$ and $2(S_z^{(1)} S_z^{(2)} - \mathbf{S}^{(1)} \cdot \mathbf{S}^{(2)})$, respectively. One can check that these encodings respect the Pauli algebra.

The Hamiltonian in the encoded subspace is the following [41]. $\Delta(g^* B_z)$ is the effective magnetic field difference between the dots.

$$H = J_{12}\frac{\sigma_z}{2} + \mu_B \Delta(g^* B_z)\frac{\sigma_x}{2}$$

Another choice is to encode the computational basis states into the singlet state and *another* triplet state, such as $|\uparrow\uparrow\rangle$. If this is done, our Pauli operators map differently onto the spin operators so that Pauli X, Y and Z are realized by $(S_x^{(2)} - S_x^{(1)})/\sqrt{2} + \sqrt{2}(S_z^{(1)} S_x^{(2)} - S_x^{(1)} S_z^{(2)})$, $(S_y^{(1)} - S_y^{(2)})/\sqrt{2} + \sqrt{2}(S_y^{(1)} S_z^{(2)} - S_z^{(1)} S_y^{(2)})$ and $-(S_z^{(1)} - S_z^{(2)})/2 - \mathbf{S}^{(1)} \cdot \mathbf{S}^{(2)} - S_z^{(1)} S_z^{(2)}$, respectively [41].

The flip-flip qubit also encodes the qubit into two spins, using $|\uparrow\downarrow\rangle$ and $|\downarrow\uparrow\rangle$ as eigenstates. The Pauli operator encodings have overlap with those of the singlet-triplet qubit, and indeed, the flip-flop qubit is a rotation of the singlet-triplet qubit about the y axis of the Bloch sphere [41]. Pauli X is realized by $2(\mathbf{S}^{(1)} \cdot \mathbf{S}^{(2)}) - S_z^{(1)} S_z^{(2)}$, Y by $2\mathbf{z} \cdot \mathbf{S}^{(2)} \times \mathbf{S}^{(1)}$ and Z by $S_z^{(1)} - S_z^{(2)}$.

Other encodings make use of more than two particles to encode a qubit. When respecting the rules of angular momentum addition (we have seen the orbital + spin angular addition in this chapter and refer the reader to [32, p. 223–229] for the general theory), the total spin Hilbert space of three spin-1/2 particles decomposes into a direct sum of one spin-3/2 subspace and two spin-1/2 subspaces [41]. The two spin-1/2 subspaces are used to encode the qubit. There are two copies of the qubit with different $m = \pm 1/2$. The computational basis states $|0\rangle$ and $|1\rangle$ map to $|S\uparrow\rangle$ and $\sqrt{2/3}\,|T_+\rangle\,|\downarrow\rangle - \sqrt{1/3}\,|T_0\rangle\,|\uparrow\rangle$ [44] where the singlet and triplet states are the states of the first two qubits in the three-qubit system. m can be polarized by high magnetic fields or can be ignored. Our Pauli operators X, Y, Z map to $2(\mathbf{S}^{(2)} - \mathbf{S}^{(1)}) \cdot \mathbf{S}^{(3)}/\sqrt{3}$, $4(\mathbf{S}^{(1)} \times \mathbf{S}^{(2)} \cdot \mathbf{S}^{(3)})/\sqrt{3}$ and $2((\mathbf{S}^{(1)} + \mathbf{S}^{(2)}) \cdot \mathbf{S}^{(3)} - 2\mathbf{S}^{(1)} \cdot \mathbf{S}^{(2)})/3$ [41].

There is also a four-qubit case of the exchange-only qubit in which case $m = 0$ in the $S = 0$ subspace [41]. The computational basis states are encoded into $|S\rangle\,|S\rangle$ and $(|T_+\rangle\,|T_-\rangle + |T_-\rangle\,|T_+\rangle - |T_0\rangle\,|T_0\rangle)/\sqrt{3}$. The paulis become realized by $2(\mathbf{S}^{(1)} \times \mathbf{S}^{(2)} \cdot \mathbf{S}^{(3)} \times \mathbf{S}^{(4)} + (\mathbf{S}^{(2)} - \mathbf{S}^{(1)}) \cdot (\mathbf{S}^{(3)} - \mathbf{S}^{(4)})/4)/\sqrt{3}$, $(\mathbf{S}^{(1)} \times \mathbf{S}^{(2)} \cdot (\mathbf{S}^{(3)} - \mathbf{S}^{(4)}) + \mathbf{S}^{(3)} \times \mathbf{S}^{(4)} \cdot (\mathbf{S}^{(1)} - \mathbf{S}^{(2)}))/\sqrt{3}$ and $2(\mathbf{S}^{(1)} \times \mathbf{S}^{(4)} \cdot \mathbf{S}^{(2)} \times \mathbf{S}^{(3)} + \mathbf{S}^{(1)} \times \mathbf{S}^{(3)} \cdot \mathbf{S}^{(2)} \times \mathbf{S}^{(4)})/3 + ((\mathbf{S}^{(1)} - \mathbf{S}^{(4)}) \cdot (\mathbf{S}^{(3)} - \mathbf{S}^{(2)}) + (\mathbf{S}^{(1)} - \mathbf{S}^{(3)}) \cdot (\mathbf{S}^{(4)} - \mathbf{S}^{(2)}))/6$.

The other qubit types listed, such as resonant-exchange qubits, involve different ways to operate the qubits we have already discussed. However, we do not cover qubit operation here but leave it to another chapter.

Different approaches to building spin qubits are at different levels of maturity. For example, of all of these qubit encodings only the Loss–DiVincenzo qubit and Singlet-Triplet have been realized using hole spin qubits in Germanium.

4.3 Case Study: Building a Spin Qubit

Let's drill down into a specific example of how one might practically go about building one of our types of qubits: a semiconductor spin qubit. Spin qubits are isolated using nanoscale devices akin to transistors. These devices, like all micro- and nanoscale electrical devices, are fabricated in a cleanroom. You may have heard of the cleanrooms of TSMC and GlobalFoundries, two industrial scale chip foundries where chips including the microprocessors that run your phones, tablets and other computers are made. These foundries are geared toward mass production. On the other hand, quantum devices are not mass produced. Instead, prototypes are fabricated in smaller quantities by research labs to experiment with new designs and techniques. However, there are certain efforts being made to ensure that spin qubits' fabrication steps are industry-compatible. This should be possible in theory by virtue of their being based on semiconductor heterostructures [45–48].

What are heterostructures? A heterostructure is just that: a non-homogeneous structure made of different materials. In our case, we are interested in structures of metals and insulators. In most semiconductor devices, we are interested in controlling the flow of current through the device. When it comes to single electron/hole transistors and qubits, we are interested in even more control. We aim to gain control over individual charge carriers, their positions and intrinsic angular momenta.

DiVincenzo and Loss first proposed encoding qubits into the spins of single charge carriers isolated in quantum dots in their 1998 paper [7]. There have since been many different proposals for quantum dot based qubits [49]. Semiconductor spin qubits may be based on different materials, as long as charges can be selectively accumulated or depleted in a quantum well grown in that material. Quantum dots have been based on Silicon, Gallium Arsenide and Germanium, for example. There are several factors to consider when choosing a material from these.

A strong magnetic field is needed to create a Zeeman splitting sufficient to define the two levels of our qubit in GaAs since the energy splitting between spin states in GaAs is very weak. The energy splitting between spin states is also relatively low in Silicon, so we run into the same problem with Si. Germanium is the only one of the three with a strong spin splitting between spin states that allows us to use low magnetic fields.

Arsenic and Gallium also both have nonzero nuclear spin, so there are relatively strong hyperfine interactions between electrons/holes and nuclei in GaAs. These hyperfine interactions cause spin decoherence and lower the lifetime of our qubit. Silicon is a better choice in this regard, as there is no nuclear spin in Si. Germanium also has no nuclear spin in its most common isotope.

Spin Orbit Interaction (SOI), as we will learn, is a necessary ingredient to allow us to electrically control our qubit. There is a trade-off here, however, as a strong SOI is also a source of noise. In any case, SOI is stronger in GaAs and Ge than it is in Si meaning that operations in GaAs and Ge can be faster.

The effective mass is relatively large in Silicon, meaning that our quantum dots will need to be tens of nanometers in size. These will be challenging to fabricate as Electron Beam Lithography (EBL) systems will struggle to calibrate a beam size much less than 10 nm. The effective mass is smaller in GaAs and Ge, so quantum dots can be a bit larger. There is also a trade-off here, however, as smaller dots will allow in the end for a greater density of qubits and have positive implications for scaling.

Silicon is the dominant semiconductor material in industry, but GaAs is used for certain high-speed and optoelectronic applications. There may be a path to integration with existing industrial processes in that case with either material.

The material quality has several more effects on the qubits themselves. For example, inhomogeneity in the material can lead to highly anisotropic g-factors for the qubits. This results in different operation speeds per site and different optimal field angles per site as well. This is difficult to overcome since a global field is usually applied, and it cannot be in the operational sweet spot for every qubit. The roughness of the material also effects carrier mobility. Carrier mobility is related to the material's resistivity, conductivity and density. We want to be able to create a conductive channel in the material and to be able to turn this channel on an off with an applied voltage. This is a pre-requisite to forming quantum dots.

The way we form quantum dots is to begin by accumulating carriers in the quantum well if the material is not already doped. If it is highly doped, this step is not necessary. We can do this by applying a voltage to a global top gate. The quantum well itself should be a 10nm layer of conductive material grown on a buffer layer and topped with a barrier. We choose 10nm since this is roughly the Fermi wavelength and we want to confine our carriers in the z direction. The buffer layer allows us to control the spread of dislocations due to lattice spacing differences between the substrate and the growth materials. We want to end the buffer with a mixture that the pure quantum well can be grown on without any dislocations at the material interface. These buffers can be linearly graded, introducing the desired quantum well material gradually (linearly) in the growth thickness. They can also be reverse-graded, starting from a pure material and then backing off to the desired concentrations. High-temperature annealing can be applied after the deposition of these layers can improve atomic layer uniformity fine features including roughness, and therefore carrier mobility.

When we characterize a material for use as a quantum well, we are interested in the carrier mobility, sheet density and the percolation density. The sheet density gives the number of charge carriers you have in a cross-section of the material. The percolation density is the minimum density you need to get conduction through the material. Therefore, the percolation density characterizes disorder in the material, as it is the minimum density required to overcome the disorder landscape. If our material is very homogeneous, then our disorder will be lower and so will our percolation density, which is ideal.

However, we do not want so much conduction that we cannot turn it off. Background conduction and parallel conduction paths through the substrate can be a problem. We want to be able to tune our SET so precisely that exactly one carrier is isolated within the quantum well, in the end. This is next to impossible if there's no way to control the current.

This brings us to the next steps after a two-dimensional hole electron gas 2DEG or hole gas 2DHG is accumulated. There are different approaches in the literature, some of them more systematic than others. McJunkin, for example, breaks dot tuning down into the following steps [50].

- There's an offset on sources—need to calibrate.
- Create 2DEG (or 2DHG) under everything.
- Next pinch off channels with screening gate.
- See that biases only effect the right currents: channels are decoupled.
- Tie all the gates over each channel together, test turn on of each channel with voltage.
- Next deplete (another screening gate) to get a narrow channel.
- Then increase voltage on the plungers to pull in individual carriers.
- Play with barriers and plungers to get dots.
- Look for loading lines in the currents.

There is a lot to unpack here, and it is dependent upon the design of the specific device that was fabricated.

4.4 Tutorial: K-Layout

In this chapter, instead of providing sample Python code, we provide this tutorial on using K-Layout. Let's take a closer look at the quantum dot device that was fabricated and measured by McJunkin et al. The considerations that would go into the choice of materials to use for such a device have been discussed. But what would it take to design the quantum device itself?

In nanofabrication, Computer-Aided Design (CAD) programs are used to create precise designs. One such tool which is free and open source (under GNU public license) is K-Layout. K-Layout differs from tools like Auto CAD in that it is specialized for micro- and nanoscale designs. It is an editor for .GDS files, a format that is used widely and is taken as an input format for EBL systems, for example, which will turn our designs into patterns on a chip.

When you first download K-Layout and open the editor, the default window size is 2 microns (by default). This gives you an idea of the target audience for the tool. A helpful scalebar is present at all times when you are working on a design in the editor.

The tools which you will use during the creation of a design include the select tool, move tool, ruler, add tool, polygon tool, box tool, text entry mode, path tool, instance tool and partial tool. Before we dive into the use of each of these, however, let's get familiar with the structure of a layout and the main window in K-Layout.

The main window is segmented into three vertical segments. One is labeled "Cells", one shows our layout itself (at the beginning, just the grid over which we will define our layout) and the final segment is labeled "Layers". These introduce for us the central concepts behind any good layout.

Any useful design system has some concept of the separation of concerns, whether it be the high-level design of a piece of software or, in this case, a piece of hardware. Separation of concerns means you can divide your design into pieces which each answer a specific question about your system, addressing a specific concern. Each piece of the design, called a module from here on, should not address the concerns addressed in any other module, and wherever we we need to reference a particular property of the system, we should always know there is exactly one place to reference that property: one answer to our question about the system, one module that addresses this concern. This is all to avoid the situation where we have multiple sources of truth which may conflict if one or the other is more consistently maintained and updated than the others. We are also interested in saving time and effort. It is easier to update one module than it is to update twelve similar modules when a design change is required.

In software design, you may be familiar with thinking of your modules as classes or packages. In nanoelectronics design, they are cells. When you create a cell, you are shown a new grid over which to define a new layout. Once this cell is defined, it will populate your cell library. The instance tool can then be used to populate your top-level cell (TOP by default) with several instances of cells selected from the cell library. This might look like designing a cell per device on a multi-device chip, and then instantiating each cell multiple times to make optimal use of the surface area of your chip, and to account for yield. The fact is, even world-class foundries spend a lot of time thinking and worrying about yield, the percentage of devices fabricated which actually work, which is never 100%.

Conceptually, that covers cells. Layers are the other important thing to understand before we begin to design a layout. Nanoelectronic devices are fabricated in two-dimensional layers. This is either by etching away materials in 2D patterns or depositing materials in 2D patterns, mainly. If you zoom in, the layers will build up on one another in a way that is not strictly two-dimensional, as the cumulative height of the layers at any (x, y) coordinate will offset the starting height of the material in the next layer at that coordinate. Some processes are conformal, meaning that material is etched or deposited, for example, in a way that has an equal thickness at all coordinates no matter the topography of the preceding layers. Not all processes are conformal. But all this is outside the realm of our layout, which does not capture any information about thicknesses or the depths of etches. The layout is only concerned with the 2D view, and in most cases, we are designing a 2D mask for each layer of material. These are the layers referenced in the K-Layout UI.

A layer can be created by right clicking under "Layers". You can give the layer a meaningful name here, or a number. When there are multiple layers, the active layer can be selected from the "Layers" panel where they will appear in a list. For our design, we will be creating a similar device to that presented on page 128 of McJunkin's thesis and used as an example for device tuning therein. Whereas they present a triple dot device, we will design a simpler double quantum dot device. We will have six layers: barriers, plungers and reservoirs, ohmics, gate bond pads, ohmic bond pads and screening gates. We can go ahead and create these layers straight away.

Next we can think about the design of our device. It really contains three devices: a charge sensor and two quantum dots. However, there is an odd number of barriers around the dots so breaking the two dots into two instances of a cell doesn't really work. On the other hand, if we don't include the barriers in our dot cell then the cell might only contain one polygon, which might defeat the purpose of having a cell. So, we can define a Double Quantum Dot (DQD) cell that contains both dots. The charge sensor is also a complete device that we can separate out into a cell. A final cell can combine the two we've already discussed into the complete charge sensor + DQD design. Then we can instantiate as many instances of the design as we need in the final top-level layout and if we ever want to edit our design, we need only edit it once and the update will take effect in all the instances across the entire layout. We can create these cells now, in a similar fashion to how we created our layers. Right click under "Cells" and select "New Cell". Give each cell a descriptive name (Fig. 4.10).

To edit the contents of a cell, right click its name in the left panel and select "Show As New Top". Select the "plungers and reservoirs" layer by clicking on it. Do this for the DQD cell you created. We are now ready to draw our first polygon.

The first thing to think about when drawing a polygon is its shape and size. In the thesis we are basing our recreation on, the dot gates are said to have narrow <100 nm widths [50]. We can be even more precise, as some scanning electron microscope (SEM) images with scalebars are provided of devices similar to what we want to design. In order to be precise about our work, we can import the SEM image into a graphics editor like the free and open-source InkScape or similar and use a line to measure the exact number of pixels that make up this scale bar. Then, by drawing more lines of the right widths next to other features of the image, we can compare these lines' pixel lengths to that of the scalebar to get their lengths as ratios of the scalebar length that we know is 250nm.

Doing this tells us that the dots are about 50nm in width and 100nm in height. We are mainly interested in heights and widths here, but it is also good to keep in mind the thicknesses of the layers as we create our layers, so that we are aware of what topographies our conductive traces may need to climb, for example. The thesis says that the bondpads should be around 200nm thick while the gates should be "much

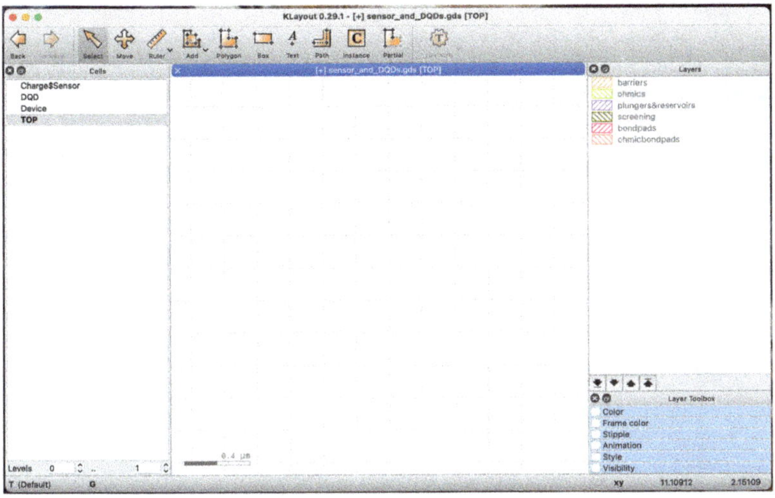

Fig. 4.10 Tutorial setup

thinner". The bondpads need to be thick in order to mechanically withstand the force of wirebonding. Good to keep in mind.

Activate the "Box" tool in the toolbar and draw a 50 × 100 nm box. If getting the size perfect is tricky, you can always make a box of a different size, and then edit it by hitting "escape" to go back to select mode, double clicking on your malformed box, and entering the exact dimensions under "center/size".

Our plunger will be used to trap the charge carrier whose intrinsic angular momentum will be used to encode our qubit. On either side of it, we need barrier gates to keep the charge contained in a restrictive potential well. These have the same dimensions as the plunger, but should be on the "barriers" layer. To get the right polygon, you can click the existing plunger, hit "Ctrl-C" to copy and "Ctrl-V" to paste. Then activate the "Move" tool and move the new polygon to the side. To change the layer of the polygon, you first need to select the "barriers" layer. Then, you can click "Edit → Selection → Change layer" et voila.

Similarly, make another barrier on the other side of the plunger and use the ruler tool to make sure the plunger is centered between them, with maybe 10 to 20 nm of space separating the polygons. Copy and paste the plunger with one barrier to create the second dot, moving it into position on the appropriate side of the existing dot (Fig. 4.11).

Some things not included in the SEM images we have made use of are the bondpads. These are aptly named, as they are metal pads that we can wirebond to. Wirebonding allows us to connect the external world to our chip, and it will be covered in more detail elsewhere. However, they should be around 150–200 microns in length and width in order to accommodate our wirebonds. This is several orders of magnitude larger than our plungers and barriers. No wonder they didn't fit in our SEM image!

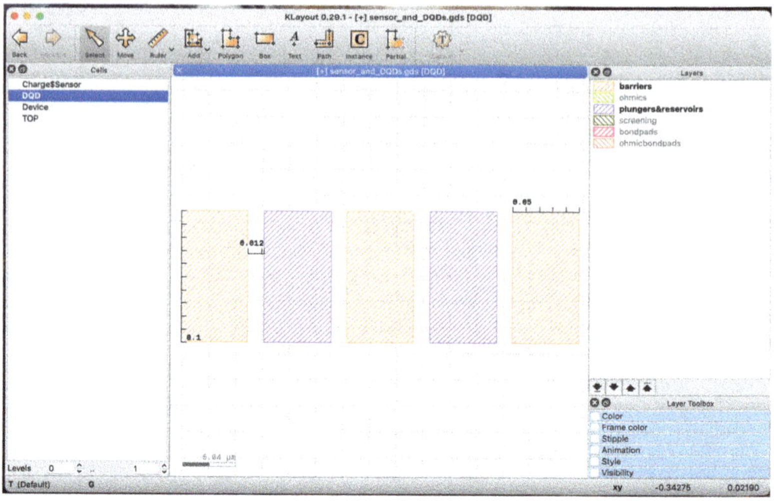

Fig. 4.11 Double quantum dot boxes

We will have to draw the bondpads next and traces to connect them to all of the barriers and plungers, which we all want to control using an external electronics setup. Try to taper your traces to go from the small to the large sizes gradually. We can use the "Polygon" tool for this, clicking once to define each vertex of a tapered trace and twice on the last one to let K-Layout know that we are done defining the shape.

These small- and large-scale features will be fabricated using different processes. We will not divulge the recipes for fabrication processes that would be highly specific to particular devices, but we can point out generally that the smaller features we have drawn would be patterned by an Electron Beam Lithography (EBL) system that has a higher resolution and lower beam size due to the smaller electron wavelength than an optical lithography system. An optical lithography system such a mask aligner or CNC-like maskless aligner would be more than sufficient for the bondpads and in general, any pattern with a critical linewidth of several microns or more.

That said, however, we will want to have another layer for the bondpads since they will be fabbed by a separate process. Hence our "bond pads" layer. Select this layer and draw half your trace. Draw the other half using the appropriate barrier or plungers and reservoirs layer, with enough overlap that the metals from the two depositions will galvanically contact.

If you want to visually clean things up you can merge the "bondpad" halves of your traces with the bondpads themselves by selecting a trace and its corresponding pad together and then clicking "Edit → Selection → Merge shapes". If you place something wrong, you can always edit a polygon a vertex or edge at a time with the "Partial" tool. Give it a try. Activate "Partial" mode and move a vertex around (Fig. 4.12).

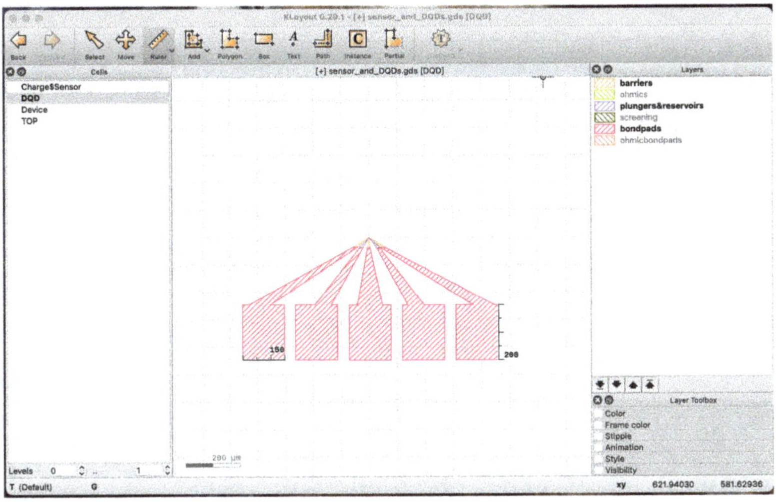

Fig. 4.12 Bond pads

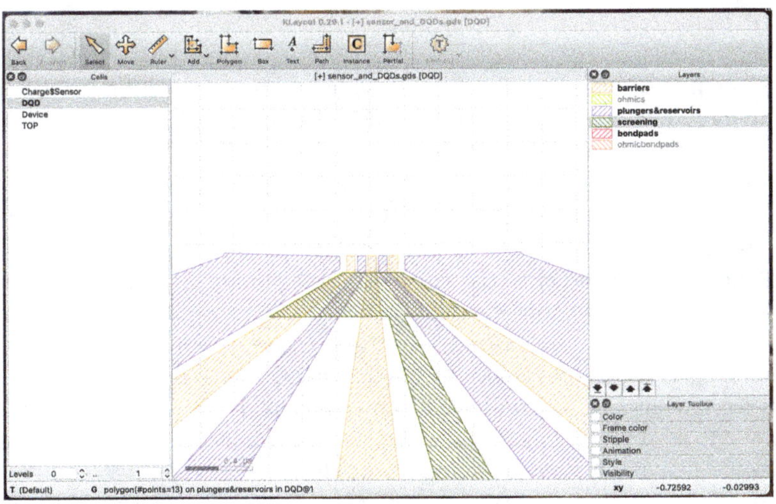

Fig. 4.13 Screening layer

Now we just need to define our reservoirs on the plungers and reservoirs layer and our screening gate on the screening layer. But you know the drill by now, use the polygon tool while working on the appropriate layers! You will have to make room for their bondpads (Fig. 4.13).

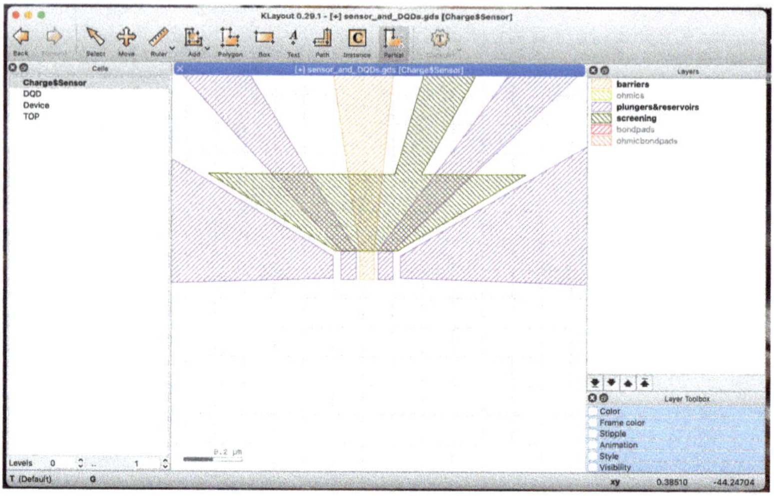

Fig. 4.14 Charge sensor

Making the charge sensor is comparatively easy to what we have just done, since it is essentially a single quantum dot (it has less polygons than the DQD). Using the same tools and know-how, define the SET charge sensor in the appropriate cell by right clicking on the "Charge Sensor" cell and clicking "Show As New Top" (Fig. 4.14).

We still have to combine our DQD and charge sensor into a working device. To do this, we will work in the "Device" cell. Select the "Instance" tool, and you will see in the bottom left corner of the screen a search button. Click that search button and select the DQD cell from the list that appears. You will then be able to place the DQD within the Device layout. Select the charge sensor next, and place it as well. Now separate the charge sensor from the DQDs with a final screening gate that belongs to the Device cell. It will need its own bondpad.

With that, our device design is complete and we can instantiate as many of them as we would like to fabricate in the top-level cell (Fig. 4.15).

When you design such a pattern you will also want to include what are called test structures such as hallbars, Vernier scales and duplicates of critical sections of the design for inspection with a scanning electron microscope (SEM). These allow you to learn about the material properties, the alignment and the outcome/yield, respectively. You might also include simpler field effect transistors (FETs) to validate that you can indeed get a channel type transistor to "turn on" before you try to get your single electron transistors (SETs) or single hole transistors to work.

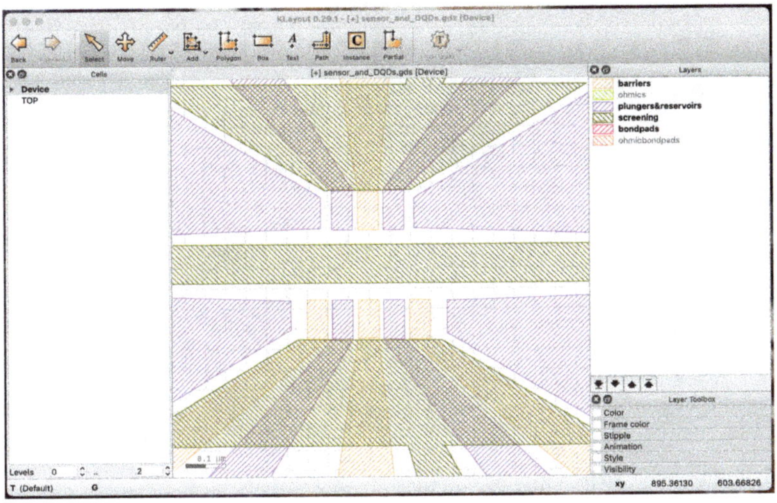

Fig. 4.15 Device

Some of these test structures are tedious to draw by hand exactly and so we recommend a package called GDS Factory to help you create the right polygons. For example, the following snippets generate the corresponding test structures in just a single line of Python each that allow you to provide every relevant parameter (Fig. 4.16 and 4.17).

Fig. 4.16 Lithographic calipers

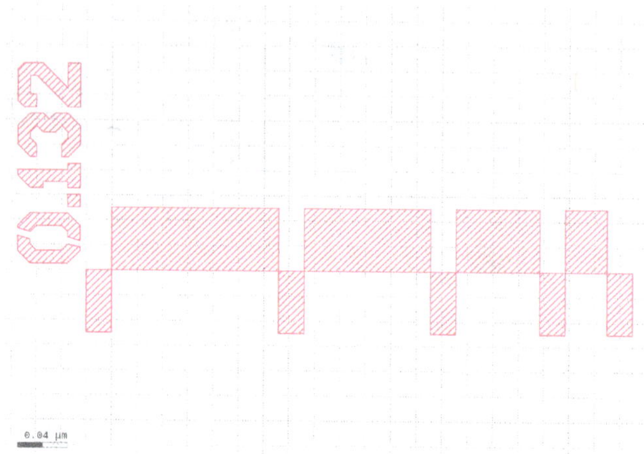

Fig. 4.17 Lithographic steps

Example: Test Structures with GDSfactory

```
import gdsfactory as gf

c = gf.components.litho_calipers(notch_size=(1, 4), notch_spacing=2, num_notches=11,
        offset_per_notch=0.01, row_spacing=0, layer1='1/0', layer2='2/0')
c.write_gds("calipers.gds")
```

```
import gdsfactory as gf

c = gf.components.litho_steps(line_widths=(0.033, 0.066, 0.099, 0.132), line_spacing=0.02, height=0.1,
        layer='1/0')
c.write_gds("steps.gds")
```

References

1. Easttom, C.: Hardware for Quantum Computing. Springer Nature Switzerland, Cham (2024). https://doi.org/10.1007/978-3-031-66477-9
2. Nielsen, M.A., Chuang, I.L.: Quantum Computation and Quantum Information 10th, Anniversary Cambridge University Press, Cambridge; New York (2010)
3. Laflamme, R., Wilson, C.: Quantum information processing devices at university of waterloo (2019). Graduate course
4. Salfi, J.: Quantum information and computation at university of british columbia (2023). Graduate course
5. de Sousa, R.: Quantum devices 1 at university of victoria (2024). Graduate course
6. Scppucci, G., Veldhorst, M., Greplova, E., Andresen, S., Bermejo, P., Chatterjee, A., Crastina, D., Katsaros, G., Li, R., Palyi, A.: Development and applications of germanium quantum technologies at tu delft (2024). Online course
7. Loss, D., DiVincenzo, D.P.: Quantum computation with quantum dots. Phys. Rev. A **57**, 120–126 (1998). https://doi.org/10.1103/PhysRevA.57.120

8. Braibant, S., Giacomelli, G., Spurio, M.: Particles and Fundamental Interactions. Undergraduate Lecture Notes in Physics. Springer Netherlands, Dordrecht (2012). https://doi.org/10.1007/978-94-007-2464-8. http://link.springer.com/10.1007/978-94-007-2464-8
9. Krantz, P., Kjaergaard, M., Yan, F., Orlando, T.P., Gustavsson, S., Oliver, W.D.: A quantum engineer's guide to superconducting qubits. Appl. Phys. Rev. **6**(2), 021318 (2019). https://doi.org/10.1063/1.5089550. http://arxiv.org/abs/1904.06560. ArXiv:1904.06560 [cond-mat, physics:physics, physics:quant-ph]
10. Fang, Y., Philippopoulos, P., Culcer, D., Coish, W.A., Chesi, S.: Recent advances in hole-spin qubits. Mater. Quantum Technol. **3**(1), 012003 (2023). https://doi.org/10.1088/2633-4356/acb87e. https://iopscience.iop.org/article/10.1088/2633-4356/acb87e
11. 5.1: Two-level systems (2022). https://phys.libretexts.org/Bookshelves/Quantum_Mechanics/Essential_Graduate_Physics_-_Quantum_Mechanics_(Likharev)/05%3A_Some_Exactly_Solvable_Problems/5.01%3A_Two-level_Systems
12. Altherr, R., White, E., White, C.: An ion of the highest fidelity. https://embedd]ed.fm/transcripts/483
13. Germanium | Properties, Uses, & Facts | Britannica (2025). https://www.britannica.com/science/germanium
14. Arute, F., Arya, K., Babbush, R., Bacon, D., Bardin, J.C., Barends, R., Biswas, R., Boixo, S., Brandao, F.G.S.L., Buell, D.A., Burkett, B., Chen, Y., Chen, Z., Chiaro, B., Collins, R., Courtney, W., Dunsworth, A., Farhi, E., Foxen, B., Fowler, A., Gidney, C., Giustina, M., Graff, R., Guerin, K., Habegger, S., Harrigan, M.P., Hartmann, M.J., Ho, A., Hoffmann, M., Huang, T., Humble, T.S., Isakov, S.V., Jeffrey, E., Jiang, Z., Kafri, D., Kechedzhi, K., Kelly, J., Klimov, P.V., Knysh, S., Korotkov, A., Kostritsa, F., Landhuis, D., Lindmark, M., Lucero, E., Lyakh, D., Mandrà, S., McClean, J.R., McEwen, M., Megrant, A., Mi, X., Michielsen, K., Mohseni, M., Mutus, J., Naaman, O., Neeley, M., Neill, C., Niu, M.Y., Ostby, E., Petukhov, A., Platt, J.C., Quintana, C., Rieffel, E.G., Roushan, P., Rubin, N.C., Sank, D., Satzinger, K.J., Smelyanskiy, V., Sung, K.J., Trevithick, M.D., Vainsencher, A., Villalonga, B., White, T., Yao, Z.J., Yeh, P., Zalcman, A., Neven, H., Martinis, J.M.: Quantum supremacy using a programmable superconducting processor. **574**(7779), 505–510 (2019). https://doi.org/10.1038/s41586-019-1666-5
15. Madsen, L.S., Laudenbach, F., Askarani, M.F., Rortais, F., Vincent, T., Bulmer, J.F.F., Miatto, F.M., Neuhaus, L., Helt, L.G., Collins, M.J., Lita, A.E., Gerrits, T., Nam, S.W., Vaidya, V.D., Menotti, M., Dhand, I., Vernon, Z., Quesada, N., Lavoie, J.: Quantum computational advantage with a programmable photonic processor. **606**(7912), 75–81 (2022). https://doi.org/10.1038/s41586-022-04725-x
16. Zhong, H.S., Deng, Y.H., Qin, J., Wang, H., Chen, M.C., Peng, L.C., Luo, Y.H., Wu, D., Gong, S.Q., Su, H., Hu, Y., Hu, P., Yang, X.Y., Zhang, W.J., Li, H., Li, Y., Jiang, X., Gan, L., Yang, G., You, L., Wang, Z., Li, L., Liu, N.L., Renema, J.J., Lu, C.Y., Pan, J.W.: Phase-programmable Gaussian Boson sampling using stimulated squeezed light. **127**(18), 180502 (2021). https://doi.org/10.1103/PhysRevLett.127.180502
17. Zhong, H.S., Wang, H., Deng, Y.H., Chen, M.C., Peng, L.C., Luo, Y.H., Qin, J., Wu, D., Ding, X., Hu, Y., Hu, P., Yang, X.Y., Zhang, W.J., Li, H., Li, Y., Jiang, X., Gan, L., Yang, G., You, L., Wang, Z., Li, L., Liu, N.L., Lu, C.Y., Pan, J.W.: Quantum computational advantage using photons. Science **370**(6523), 1460–1463 (2020). https://doi.org/10.1126/science.abe8770
18. King, A.D., Nocera, A., Rams, M.M., Dziarmaga, J., Wiersema, R., Bernoudy, W., Raymond, J., Kaushal, N., Heinsdorf, N., Harris, R., Boothby, K., Altomare, F., Berkley, A.J., Boschnak, M., Chern, K., Christiani, H., Cibere, S., Connor, J., Dehn, M.H., Deshpande, R., Ejtemaee, S., Farré, P., Hamer, K., Hoskinson, E., Huang, S., Johnson, M.W., Kortas, S., Ladizinsky, E., Lai, T., Lanting, T., Li, R., MacDonald, A.J.R., Marsden, G., McGeoch, C.C., Molavi, R., Neufeld, R., Norouzpour, M., Oh, T., Pasvolsky, J., Poitras, P., Poulin-Lamarre, G., Prescott, T., Reis, M., Rich, C., Samani, M., Sheldan, B., Smirnov, A., Sterpka, E., Clavera, B.T., Tsai, N., Volkmann, M., Whiticar, A., Whittaker, J.D., Wilkinson, W., Yao, J., Yi, T.J., Sandvik, A.W., Alvarez, G., Melko, R.G., Carrasquilla, J., Franz, M., Amin, M.H.: Computational supremacy in quantum simulation (2024). https://arxiv.org/abs/2403.00910

19. Koch, J., Yu, T.M., Gambetta, J., Houck, A.A., Schuster, D.I., Majer, J., Blais, A., Devoret, M.H., Girvin, S.M., Schoelkopf, R.J.: Charge insensitive qubit design derived from the cooper pair box. Phys. Rev. A **76**(4), 042319 (2007). https://doi.org/10.1103/PhysRevA.76.042319. http://arxiv.org/abs/cond-mat/0703002. ArXiv:cond-mat/0703002
20. Struik, D.J.: Joseph-Louis Lagrange, comte de l'Empire | French Mathematician, Comte de l'Empire | Britannica. https://www.britannica.com/biography/Joseph-Louis-Lagrange-comte-de-lEmpire
21. 8.2: Legendre Transformation between Lagrangian and Hamiltonian mechanics (2017). https://phys.libretexts.org/Bookshelves/Classical_Mechanics/Variational_Principles_in_Classical_Mechanics_(Cline)/08%3A_Hamiltonian_Mechanics/8.02%3A_Legendre_Transformation_between_Lagrangian_and_Hamiltonian_mechanics
22. Bardin, J.C., Sank, D., Naaman, O., Jeffrey, E.: Quantum computing: an introduction for microwave engineers. IEEE Microwave Mag. **21**(8), 24–44 (2020). https://doi.org/10.1109/MMM.2020.2993475. https://ieeexplore.ieee.org/document/9134701/?arnumber=9134701. Conference Name: IEEE Microwave Magazine
23. Domb, C.: James Clerk Maxwell | Biography & Facts | Britannica (2025). https://www.britannica.com/biography/James-Clerk-Maxwell
24. Griffiths, D.J.: Introduction to Electrodynamics. Prentice-Hall (1981)
25. Mobley, C.: Plane wave solutions, ocean optics web book. https://www.oceanopticsbook.info/view/theory-electromagnetism/level-2/plane-wave-solutions
26. Buck, S., Coleman, R., Sargsyan, H.: Continuous variable quantum algorithms: an introduction (2021). https://arxiv.org/abs/2107.02151
27. Groenewold, H.: On the principles of elementary quantum mechanics. Physica **12**(7), 405–460 (1946).https://doi.org/10.1016/S0031-8914(46)80059-4
28. Moyal, J.E.: Quantum mechanics as a statistical theory. Math. Proc. Cambridge Philos. Soc. **45**(1), 99–124 (1949). https://doi.org/10.1017/S0305004100000487. https://www.cambridge.org/core/product/identifier/S0305004100000487/type/journal_article
29. Wigner, E.: On the quantum correction for thermodynamic equilibrium. Phys. Rev. **40**(5), 749–759 (1932). https://doi.org/10.1103/PhysRev.40.749
30. Schubring, G.: Biographie. Felix Hausdorff, Gesammelte Werke. Band IB. by Egbert Brieskorn and Walter Purkert. Math. Intell. **43**(4), 94–98 (2021). https://doi.org/10.1007/s00283-021-10083-9
31. Kobe, S.: Ernst Ising-Physicist and Teacher. J. Stat. Phys. **88**(3), 991–995 (1997). https://doi.org/10.1023/B:JOSS.0000015184.19421.03
32. Sakurai, J.J., Napolitano, J.: Modern Quantum Mechanics, 2nd edn. Dorling Kindersley, New Delhi (2014)
33. Salfi, J., Mol, J.A., Culcer, D., Rogge, S.: Charge-insensitive single-atom spin-orbit qubit in silicon. Phys. Rev. Lett. **116**(24), 246801 (2016)
34. Hendrickx, N.W., Massai, L., Mergenthaler, M., Schupp, F.J., Paredes, S., Bedell, S.W., Salis, G., Fuhrer, A.: Sweet-spot operation of a germanium hole spin qubit with highly anisotropic noise sensitivity. Nat. Mater. **23**(7), 920 (2024). https://doi.org/10.1038/s41563-024-01857-5. https://pmc.ncbi.nlm.nih.gov/articles/PMC11230914/
35. Wang, Z., Marcellina, E., Hamilton, A.R., Cullen, J.H., Rogge, S., Salfi, J., Culcer, D.: Optimal Operation Points for Ultrafast, Highly Coherent Ge Hole Spin-Orbit Qubits. NPJ Quantum Inf. **7**(1), 1–8 (2021). https://doi.org/10.1038/s41534-021-00386-2. https://www.nature.com/articles/s41534-021-00386-2. Number: 1 Publisher: Nature Publishing Group
36. Jeschke, G.: 3.2: Electron Zeeman Hamiltonian (2022). https://chem.libretexts.org/Bookshelves/Physical_and_Theoretical_Chemistry_Textbook_Maps/Electron_Paramagnetic_Resonance_(Jenschke)/03%3A_Electron_Zeeman_Interaction/3.02%3A_Electron_Zeeman_Hamiltonian
37. Winkler, R.: Spin—orbit coupling effects in two-dimensional electron and hole systems. In: Springer Tracts in Modern Physics, vol. 191. Springer, Berlin, Heidelberg (2003). https://doi.org/10.1007/b13586

38. Luttinger, J.M.: Quantum theory of cyclotron resonance in semiconductors: general theory. Phys. Rev. **102**(4), 1030–1041 (1956). https://doi.org/10.1103/PhysRev.102.1030
39. Nicolas Rotaru P. Del Vecchio, O.M.: Hole spin in germanium-tin quantum dot. In: 2024 Workshop on Quantum Simulation with Spin Qubits (2024)
40. Kramers, H.A.: Théorie générale de la rotation paramagnétique dans les cristaux. Proc. Royal Netherlands Acad. Arts Sci. **33**(6-10), 959–972 (1930). https://www.dwc.knaw.nl/DL/publications/PU00014621.pdf
41. Burkard, G., Ladd, T.D., Nichol, J.M., Pan, A., Petta, J.R.: Semiconductor spin qubits. Rev. Modern Phys. **95**(2), 025003 (2023). https://doi.org/10.1103/RevModPhys.95.025003. http://arxiv.org/abs/2112.08863. ArXiv:2112.08863 [cond-mat, physics:physics, physics:quant-ph]
42. Levy, J.: Universal quantum computation with spin-1/2 pairs and heisenberg exchange. Phys. Rev. Lett. **89**, 147902 (2002). https://doi.org/10.1103/PhysRevLett.89.147902
43. Petta, J.R., Johnson, A.C., Taylor, J.M., Laird, E.A., Yacoby, A., Lukin, M.D., Marcus, C.M., Hanson, M.P., Gossard, A.C.: Coherent manipulation of coupled electron spins in semiconductor quantum dots. Science **309**(5744), 2180–2184 (2005). https://doi.org/10.1126/science.1116955
44. DiVincenzo, D.P., Bacon, D., Kempe, J., Burkard, G., Whaley, K.B.: Universal quantum computation with the exchange interaction. Nature **408**(6810), 339–342 (2000). https://doi.org/10.1038/35042541. https://www.nature.com/articles/35042541
45. Bosco, S.: Hole spin qubits in si and ge quantum dots: ultrafast gates and noise resilient qubits (2021). https://youtu.be/cwbfjKA_03k. IQubits
46. Crastina, D.: Semiconductors for spin qubits (2024). https://online-learning.tudelft.nl/courses/development-and-applications-of-germanium-quantum-technology/. Development and Applications of Germanium Quantum Technologies Course
47. Turton, R.: The Quantum Dot: A Journey Into the Future of Microelectronics. Oxford University Press (1996)
48. Veldhorst, M.: Semiconductor Qubits: Vision and Challenges (2024). https://online-learning.tudelft.nl/courses/development-and-applications-of-germanium-quantum-technology/. Development and Applications of Germanium Quantum Technologies Course
49. Burkard, G., Ladd, T.D., Pan, A., Nichol, J.M., Petta, J.R.: Semiconductor spin qubits. Rev. Modern Phys. **95**(2) (2023). https://doi.org/10.1103/revmodphys.95.025003
50. McJunkin, T.W.: Heterostructure modifications, fabrication improvements, and measurement automation of si/sige quantum dots for quantum computation. Ph.D. thesis, Wisconsin University, Madison (2021)

Chapter 5
Taking Care of Your Qubits

Abstract Now that we know all about qubits of various types, this chapter covers the most mainstream techniques for maintaining a qubit in the real, noisy world, aside from the algorithmic techniques we covered in Chap. 3. Without the techniques in this chapter, we won't even have a qubit to perform error correction on! Read to learn about cooling with a dilution refrigerator, and how to build pieces of the mechanical assembly that supports your QPU within.

5.1 Learning Objectives

Different types of qubits require different techniques to maintain their quantum states. A common thread is however cooling. In this chapter, our learning goals include (Fig. 5.1):

- Understand Helium 3–Helium 4 dilution refrigerators and how they work.
- Gain practical skills working with SolidWorks to design mechanical components of the assembly.
- Gain practical skills with KiCAD to design the electrical surrounds of a QPU.

5.2 Maintaining Quantum States

How do we maintain quantum states for the lengths required to do quantum computing? Each type of quantum computer has different requirements.

In superconducting quantum computing, the quantization of the energy in our LC circuit is such that the system's energy can only exist at discretized levels spaced by $\hbar\omega_r$. Unless the ambient temperature of the environment T nears 50 mK, these energy levels will not be significant as to be distinct from the noise [1]. There are two considerations that must be had when we wish to maintain the quantum state of a superconducting circuit. The first is that KT must be much less than $\hbar\omega_r = K(50\,\text{mK})(\frac{V_r}{\text{GHz}})$.

© The Author(s), under exclusive license to Springer Nature Switzerland AG 2026

M. S. Edwards, *An Introduction to Quantum Computing for Computer Engineers*,
https://doi.org/10.1007/978-3-032-03650-6_5

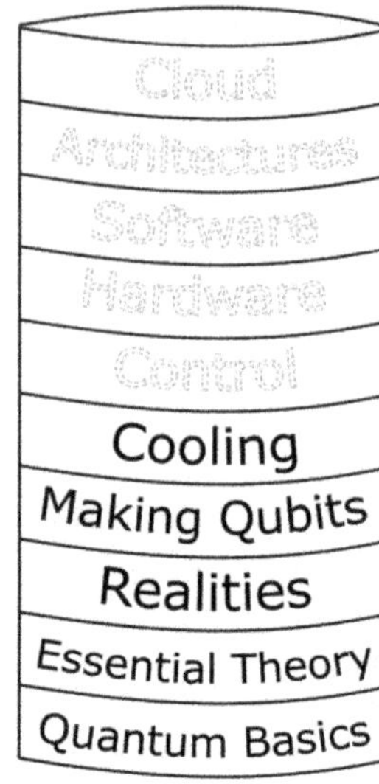

Fig. 5.1 Learning objective for this chapter: cooling

This gives us the requirement mentioned in [1], which we can make more specific by saying $T < 50\ \text{mK}(\frac{V_r}{\text{GHz}})$. This temperature requirement necessitates the use of a dilution refrigerator. These can typically bring us to around 20mK. The second consideration is that the energy level separation $\hbar\omega_r$ must be greater than the energy level widths $\hbar K$. This is related to the quality factor of the resonator $Q = \frac{\omega_r}{K} >> 1$. A value of 1 million is considered "high Q", while 50 is extremely low, for example. Coplanar waveguide resonators with Q between 10^3 and 10^8 are not difficult to fabricate [2].

A JJ has a critical current which decays significantly with magnetic fields at strengths such as 1 milliTesla. They also only operate at temperatures less than 100 mK. This also introduces some constraints into how we can operate our LC circuit.

For flux qubits, the sensitivity to flux noise is determined by $\frac{\partial E_n}{\partial \phi_x}$ and the sensitivity to critical current noise is determined by $-|\frac{\partial E_n}{\partial \beta}|$. When $\frac{E_J}{E_C} << 1$, we are less sensitive to flux noise.

For QDs, the energy level spacing is δ_s and we need $\delta_s > KT$ so that the qubit can have a well-defined state. To maintain the confinement of a charge carrier in a quantum dot, the energy barrier must also be larger than the thermal fluctuations. Otherwise the carrier cannot remain bound to the capacitive island. We need the tunnel resistance and the source resistance to be greater than the "quantum resistance" like $R_s, R_d >> h/e^2$. $h/e^2 = 25k\Omega$.

So far, among other things, these requirements commonly suggest that low temperatures are required. Cryogenic temperatures are also sometimes required in optical setups that involve superconducting photon detectors, for example. These temperature requirements can be met by a dilution refrigerator. This will be a main focus of the remainder of this chapter. We will first however note a few additional techniques to reduce noise and maintain your quantum state.

In ion trap quantum computers, we use a technique called trap cooling. We start with a "warm" state with a Lamb Dicke parameter[1] far greater than 1 denoted $|g, n\rangle$. We then drive the red sideband to $|e, n-1\rangle$. We let the ion relax via spontaneous emission to $|g, n-1\rangle$ and then repeat until $|g, 0\rangle$ is reached, the goal being to reach $k_B T << \hbar\omega_z$ [4, p. 312].

When it comes to superconducting and quantum dot setups, we will also use attenuators and circulators to limit noise into and out of a quantum device, respectively [5].

The goal of lowering noise is to have long coherence times. Trapped ion and neutral atoms have relatively long decoherence times, as compared to the shorter but gradually improving times associated with spin qubits, for example. While trapped ion coherence times are measured in seconds to minutes [6], and neutral atom coherence times up to 40 s have been observed [7], platforms like spin qubits are still limited by coherences in a range of around a hundred microseconds (Si QD: 88 us with Hahn echo [8], Ge QD: 100 us with dynamical decoupling [9]). The exact mechanisms for this decoherence are not always clear, and the physics of certain types of qubits are still open areas of research.

5.3 Helium 3–Helium 4 Dilution Refrigerators

How does a Helium 3–Helium 4 dilution refrigerator work?

When Helium 3 atoms are transferred to a diluted phase primarily containing Helium 4, a cooling process occurs which is what is taken advantage of in order to cool a Helium 3–Helium 4 dilution refrigerator [10, p.151].

The heat of mixing of Helium 3 and Helium 4 is the cooling capacity of the process.

The enthalpy of mixing is proportional to the following, where ΔC are the differences of the two phases' specific heats [10, p.152].

$$\int \Delta C dT$$

What is meant by the "enthalpy of mixing" is the change in energy that occurs when particles in the mixture interact [11]. This cooling actually comes from quantum mechanical effects, the difference in the zero-point motions of Helium 3 and Helium 4 and the different statistics of their properties at low temperatures [10, p. 163].

[1] Robert Dicke made scientific contributions to atomic physics, quantum optics, radar, cosmology and more. His contributions in cosmology led to the discovery of the cosmic microwave background [3].

A simple Helium 3–Helium 4 dilution refrigerator is composed of several chambers [10, p. 163–164]. The first and the coldest is the mixing chamber, which sits at around 0.01 K (much colder than the 77 K you would get from a Nitrogen dewar). In this chamber, Helium 3 atoms transfer from a concentrated to a dilute side. The Helium 3 then travels up to another chamber, the "still", which is kept at a temperature ($\doteq 0.7$ K) such that the vapor pressure of Helium 3 is much higher than that of Helium 4 and almost pure Helium 3 is evaporated. This causes the flow of Helium 3 from the mixing chamber to the still by osmotic pressure. Osmotic pressure is the difference in pressure across a phase boundary that can cause flow across the boundary [12]. The amount of Helium 3 in the dilute phase remains constant as new Helium 3 atoms constantly cross into the dilute side of the mixing chamber to replenish it. This continuous mixing creates a continuous cooling effect.

The evaporated Helium 3 goes to a room temperature Helium pump. After the pump, the Helium 3 is cooled by a liquid Helium 4 bath and condensed in a condenser before entering a heat exchanger in thermal contact with the still at around 0.7 K. This and potentially even more heat exchangers along the Helium's path back down the dilution refrigerator allow the Helium 3 to be cooled further prior to its re-entry into the concentrated side of the mixing chamber (Fig. 5.2).

Pulse tubes are used to pre-cool the fridge to temperatures where the Helium liquefies so the dilution process may begin. In a typical fridge, it might take 1 h for a warm sample to reach 1.4 K and 7 h to 300 mK. It may take around 30 min to re-condense the Helium.

A dilution refrigerator has stages that make it look a bit like a chandelier, corresponding to the different temperatures that lie on a gradient oriented vertically, with the coldest part (the mixing chamber) at the bottom and the warmest at the top of the fridge. The electronics and wiring mounted on these stages' cold plates need to be chosen for the specific temperature at each stage. Niobium titanium wire, for example, can be used after the 4K stage. Titanium and brass also don't have much magnetic moment (Ti the least), so they are good materials to use in the fridge. Longer wires help with thermal anchoring and so you will often see them loop on their way from one stage down to the next, on the way to the QPU at the bottom. We would also typically mount a quantum limited amplifier to the 4K stage plate in certain quantum computing setups, as we will cover in Chap. 7 on control electronics.

At the final stage, the cold plate is accompanied by a cold finger which includes a small box, a final assembly of metal, usually oxygen free copper. This will thermally anchor and house the QPU. Oxygen free copper has better thermal conductivity than other copper. In most quantum computing related setups, a magnetic field that may or may not be orientable is generated by a magnet kept somewhere near the bottom of the assembly. The cold finger therefore ideally positions the QPU near the center of this magnetic field.

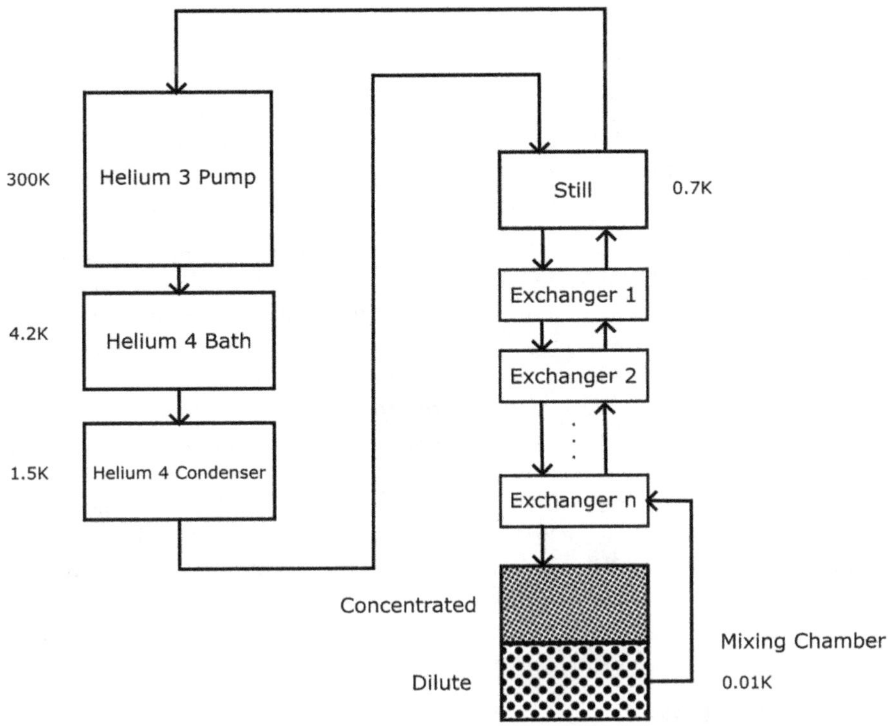

Fig. 5.2 He 3–He 4 dilution refrigerator

5.4 Tutorial: SolidWorks

When we are building a quantum computer, we usually do not build a dilution refrigerator from scratch. It is however common to design and machine specific parts in-house such as the cold finger and the box which houses the QPU. This box needs to thermally anchor the sample holder PCB which the QPU chip will be glued to, usually with PMMA. It needs to admit the DC and RF lines which have traveled the length of the fridge to the bottom where it is coldest. Various connectors can be used to connect these lines to the PCB such as a D-Sub or ZIF connector which the box must be designed to fit.

There are some further considerations which must go into the shape of the cold finger and the box. A solid piece connecting cold plate and cold finger will have better thermal anchoring, for examples, than a multi-piece solution that screws together. We should be able to open and close the box easily so that we may insert and remove samples (QPUs).

To design this assembly we would use a CAD software such as SolidWorks.

What does SolidWorks do?

At a high-level, SolidWorks allows us to work with three types of files. The first is the "part" file which allows us to design a single 2D drawing, called a "sketch", or 3D component called a "feature".

After we have designed the parts which will compose our assembly, we will import them into an assembly file and arrange them together. Finally, a 2D elevation diagram can be created which gives us an overview of the assembly we have designed.

At the beginning of a new project, we will always begin from the start window by creating a part. A new file will be created with no contents for us to edit. We can navigate this file by zooming in and out with the mouse wheel as well as holding control down to enable the arrow keys to move us around the canvas. One of the first things we should do is click to Options → Document → Units and set the units we wish to work with. Next we will take a look under Sketch at the top of the screen, where all the tools we will use at first to create a sketch are housed. In SolidWorks, 3D features are created from 2D cross-sections and paths by extruding them, revolving them and more. So, we always begin by drawing something in 2D.

5.4.1 2D Sketches

The first time we use a 2D drawing tool, SolidWorks will prompt us to select a plane to draw on. When we use the tool in the future, if we want to change the plane, we will have to select a different plane on the left side of the screen. Once we have selected a plane, we can begin to draw.

A few examples of drawing tools we can use are the line, arc, circle, ellipse, rectangle and polygon tools. Each is used slightly differently. For the line tool, we simply draw multiple line segments by click at each vertex. When we are done, we hit escape. To draw a spline curve, we click on each node in the curve and double-click on the last one to complete the curve. An arc is defined by three clicks: two to define the radius and another for the arc length. The circle, ellipse and rectangle tools and intuitive. To draw a polygon with n sides, you can enter n into the dialogue on the left of the screen after selecting the tool. The enter key can be used at any time to recall and begin using the most recently used tool again.

When we are drawing there are certain aids which appear on the canvas. For example, measurements and guiding lines appear to help us align our lines and curves with others that we have already drawn. Snapping is also enabled by default but can be disabled by holding down control.

When we are done drawing a sketch, we hit control and Q together (ctrl-Q) to create the sketch and exit from it. At this point the sketch becomes something we can move and resize as a whole, for example. To start a second sketch, we can simply start drawing on a plane again. To edit an existing one we can double-click on it to enter it again. When editing a sketch, we can drag on any node, line or curve to move it. Controls will also appear to edit the shapes of splines, for example, on the nodes of a spline curve.

To add measurements such as dimensions to a sketch, we can click the smart dimension tool and then click on whatever points, lines, curves or lengths we wish to add the measurements to. To better view a sketch, we can right click and select rotate view. We can also hit the space bar to select a specific viewpoint.

5.4.2 3D Features

When we create a 3D feature from a 2D sketch, the sketch becomes a part of the feature in question. It is therefore a good practice to dedicate a sketch to each feature that you know you want to create. One of the tools that you will use most heavily is the Extruded tool. Click Extruded and then click on a 2D sketch entity to extrude the sketch in 3D. Right click when the appropriate volume has been reached to lock in the extrusion. After you have extruded a shape in this way, any changes you make to the underlying sketch will update the extruded feature automatically.

Instead of the Extruded tool we can use the Swept tool to extrude a 2D section along a path instead of a straight line. A curve should be drawn which defines the path you will extrude along. A 2D section should be drawn near this path. Then, click on Swept. Next, click on the blue field in the dialogue on the left of the screen and the entity you wish to use as the cross-section. Click next on the pink field and the extrusion path. The extrusion will be performed.

A section can also be revolved about a line to create a 3D feature. Similarly to what we did for the Swept tool, a line should be drawn near your cross-section. Click Revolved and then on the line to use as the axis. Click on selected contours to choose the 2D entity to revolve.

One more tool you can use to turn 2D sketches into 3D features is Lofted. Lofted allows you to fill the space between two entities such as polygons. After clicking Lofted, click on the blue field in the left dialogue and then on the entities in order. You may also click on the red field to specify a guide curve which will shape the fill.

After a 3D feature has been created, any face or contour can be selected in order to move or adjust the extrusion/fill.

Display Style can be used to render the objects in different styles. Selected Scenes to add lights, shadows and environments to the picture.

Once we are done creating each of our parts, we should save them as part files.

5.4.3 Assembly

When we are ready to assemble our parts into a final assembly, we create an assembly file using New → Assembly. We can import the first part from the left dialogue, which will become our reference part. Other parts can be imported using Layout → Insert Component. The reference part cannot be moved or rotated but the others may (Fig. 5.3).

We can zoom in on any part in our assembly and edit it at any time, and even hide other parts to make it easy to see one or two at a time. To edit a particular part, we can right click and select Edit Part to modify its features and sketches. When we are done, we can right click on the first level in the hierarchy on the left side of the canvas and select Edit Assembly to do back to editing the whole assembly (Fig. 5.4).

Then it is important for us to arrange the parts so that we can confirm that they fit together, for example so that the sample holder box screws into the rest of the cold finger and so on.

We can render a part in the assembly with particular materials under appearances if desired.

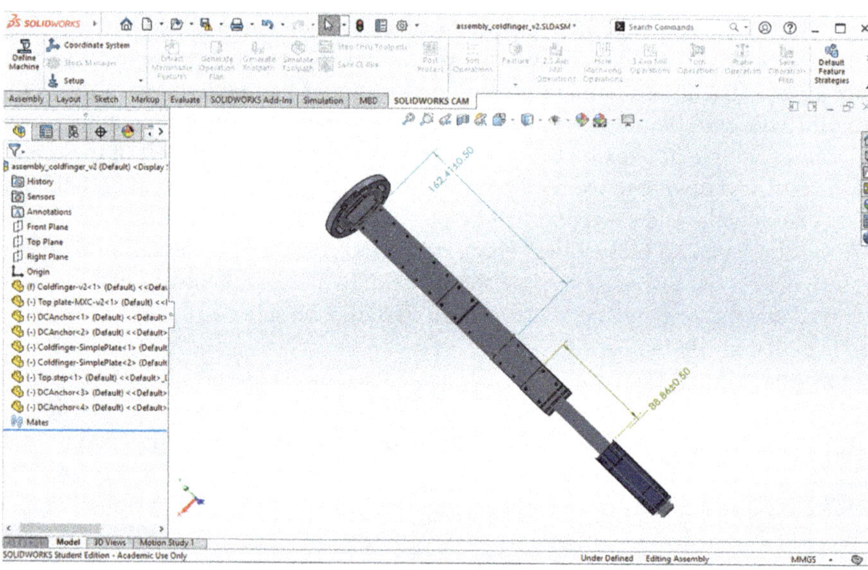

Fig. 5.3 Cold finger assembly

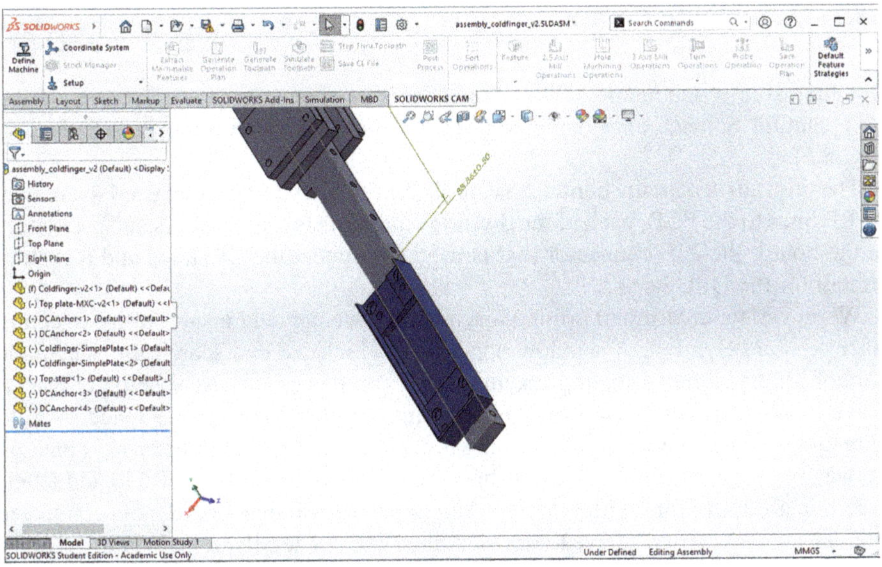

Fig. 5.4 Sample holder box

5.5 Tutorial: KiCAD

Another part of the assembly which it is common to build in-house is the sample carrier PCB which electrically connects the QPU to the outside world. As mentioned prior, the QPU chip is usually glued with PMMA to this PCB which is housed in the sample holder box. Wire bonding can be used to electrically connect a QPU to the traces on the PCB. We therefore design the PCB around a central (recessed) area where we glue the QPU chip, surrounded by bondpads for wire bonding. The PCB will route DC lines and more carefully designed RF lines (coplanar waveguides) to the connectors (D-Sub, ZIF, SMP, SMA, etc.) that connect the PCB to the wiring that has traveled the length of the fridge from the control electronics on the outside.

> **What does KiCAD do?**
>
> To design such a PCB we would use a different kind of CAD. A good open source and free software we can use for PCB design is KiCAD. KiCAD offers several editors that work together to provide a complete workflow for PCB design.

5.5.1 *Schematic Editor*

The first is the schematic editor. Consider the open-source PCB carrier from the Quantum Science and Technology Lab at the University of British Columbia (Fig. 5.5).

The relevant schematic contains symbols for the SMP connectors used to connect the RF lines to the PCB, each of the discrete components like resistors and capacitors on the board, the ZIF connector that is used to connect the DC lines and a custom symbol for the QPU itself.

When we are creating or editing a schematic, we can add a symbol by selecting Place → Add Symbol. A window appears in which we can search the schematic symbol library. After selecting a symbol, we can place it in the schematic by left clicking. We can then route the symbols' pins together by selecting Place → Add Wire and drawing the wire, left clicking to lock in each vertex in the wire's path on the way between terminals. If things become crowded, it can be useful to add labels to certain groups of pins with Place → Add Label. Matching labels denote connected pins. This way you don't have to draw another line. Generally, it is good to organize wires to that they are easy to follow individually and are grouped together in a way that shows their common functions; i.e. the DC lines from our QPU to the ZIF might all follow a similar path or paths.

What do we do if we need to modify or create a customized symbol like the one here for the QPU? We can start by selecting Tools → Symbol Editor, or we can double-click on the symbol we wish to edit and select Edit Symbol from the window

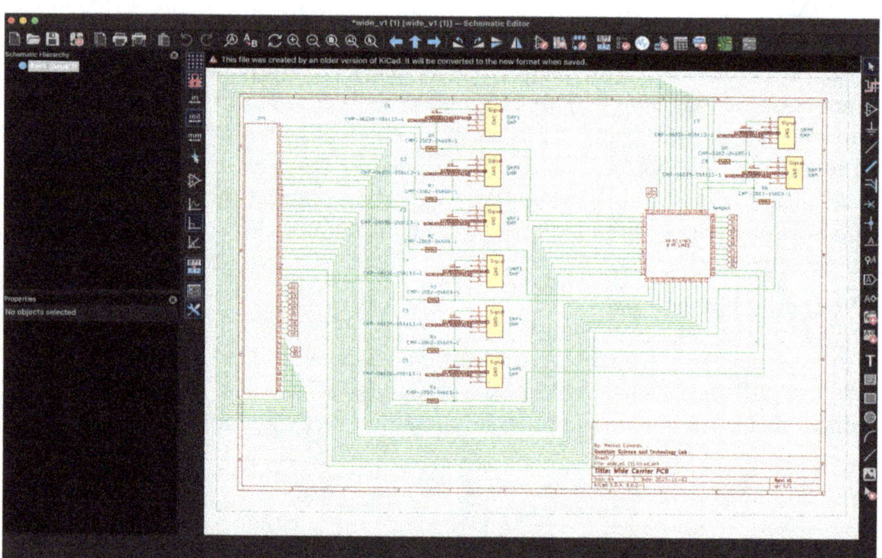

Fig. 5.5 Carrier PCB schematic

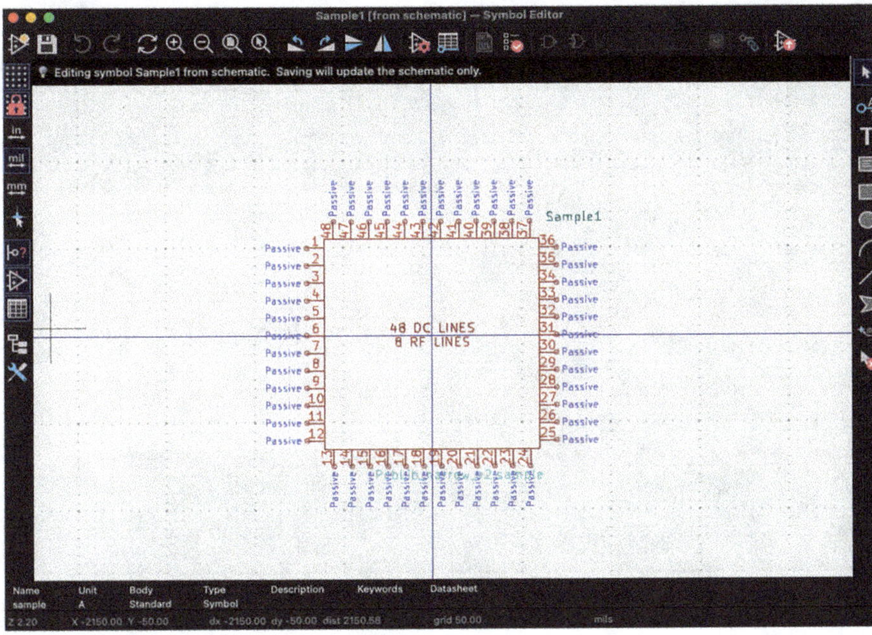

Fig. 5.6 Symbol editor

that appears (Fig. 5.6). To create a new symbol we must first select or create a symbol library, which can be global- or project-specific.

To draw a symbol we can use the Add Rectangle, Circle, Arc, Line and Polygon tools on the right-hand side of the window. It takes two clicks to define a rectangle by two of its corners. A circle is also drawn in two steps (center and radius). An arc simply has a defined start and end. Line segments can be drawn by clicking their vertices until the endpoint is double-clicked. A polygon is drawn similarly to a line but is a closed shape. Next we use the Add Pin tool to add pins to the symbol. A pin most importantly has a name, number and electrical type (Fig. 5.7). Since the bondpads are not connected to a QPU at this stage, we leave them all as passive pins, but once we have a QPU design we will know whether a line will be used as an input or output, for example.

Once we are done editing a symbol, we can add it to our schematic by clicking Add Symbol to Schematic.

When we think we are done editing a schematic, we can run the Electrical Rules Checker (ERC). This will check for electrical issues like any pins that we have accidentally left floating, and other errors such as whether any symbol libraries did not load which are referenced in the schematic. We should revise the schematic until the ERC passes.

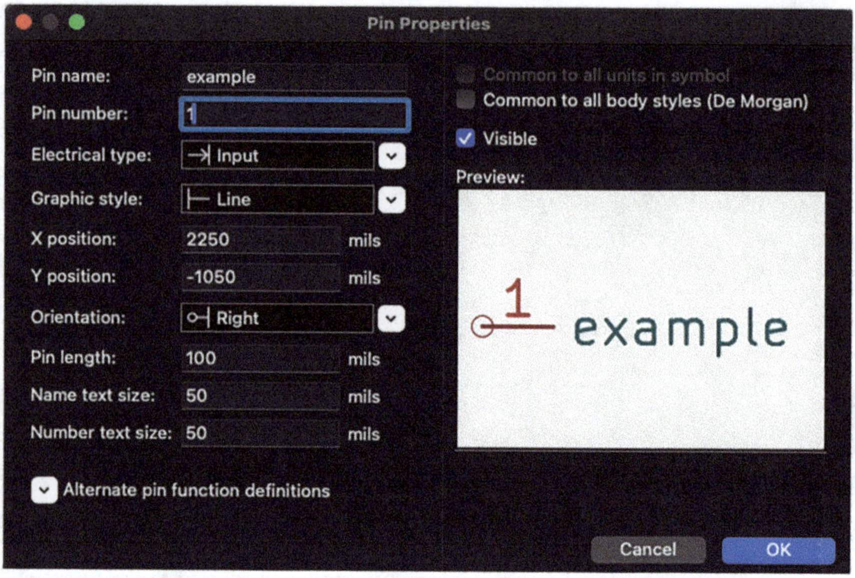

Fig. 5.7 Editing a pin

Then, we are ready to export a netlist with File → Export → Netlist. A netlist is a machine-readable file that specifies what pins are connected to what pins in our design. This will turn into what's called our ratsnest in the PCB editor we will use next.

5.5.2 PCB Editor

After closing the schematic editor and launching the PCB editor, we can go to File → Import → Netlist to import the netlist. The editor will populate with the parts we included in our schematic, and little lines will connect the pins that should be connected electrically according to the netlist. These aren't actual connections yet, these tangled, overlapping lines (the ratsnest) just tell us what needs to be connected for our PCB design to pass the next time we run ERC (Fig. 5.8).

To get a nice result in our editor, we need to arrange all the parts on our PCB and connect the pins that ought to be connected with traces. This is similar in a way to what we did in our schematic, but now the emphasis is on the arrangement of the components. Another aspect of the PCB editor that is different from the schematic editor is that it uses a footprint instead of a symbol to represent each component. This footprint should correspond to the actual physical footprint of the component it represents in the real world.

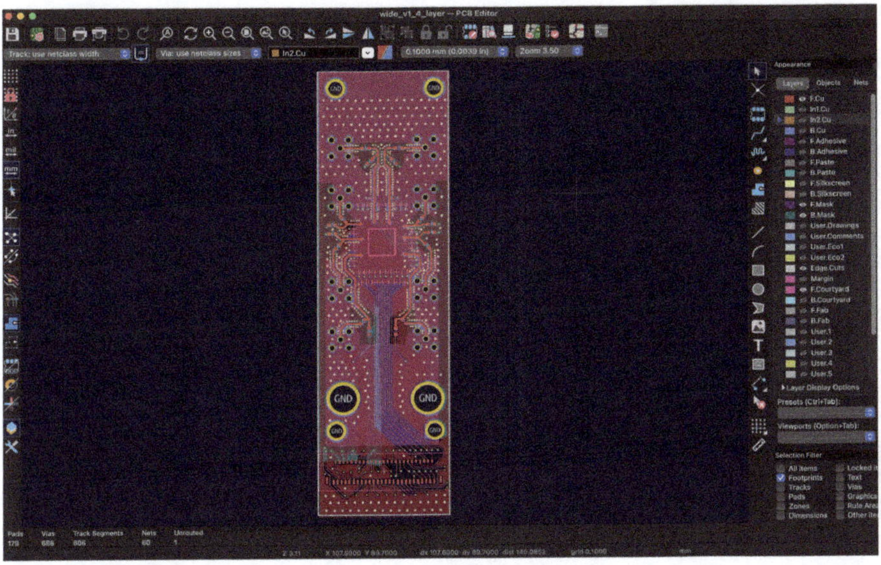

Fig. 5.8 PCB editor

To edit a footprint, we can launch the footprint editor directly, or we can double-click on a footprint and select Edit Footprint. A footprint should have pads which relate to the pins in the corresponding symbol. The pad number should be the same as the corresponding pin number in the symbol in order to relate them. A pad can be added to the footprint with the Add Pad tool in the right-hand toolbar.

Footprint libraries work similarly to symbol libraries (Fig. 5.9). If we have a number of footprints specific to a project, we should create a new project library and keep any new project-specific footprints we create under this library.

Some suppliers of electrical components like the ZIF connector, for example, provide downloadable libraries or footprints you can add to your project. To do this, you can select File → Add Library or File → Import → Footprint in the footprint editor. When we are done editing and organizing our footprints, we can add them to a PCB design with Place → Add Footprint.

Another step we should do before we begin routing our traces on the board is to setup our board stack. We can start this with File → Board Setup. Under Board Editor Layers we can edit the layers that we will work with. It is generally important to include a layer for solder paste, solder mask, and silkscreen for each of your copper layers on the outside of the board. These will make it possible for us to populate the PCB with physical parts once it has been fabricated.

Under Physical Stackup we can edit the physical layers of the board. We can include as many copper layers as we need to route our traces comfortably without any overlapping, separated by a dielectric like FR4. We need to select the thickness of the dielectric carefully when we are working with RF lines which should have around 50 Ω characteristic impedance z_0. The characteristic impedance of a coplanar waveguide depends on the trace width s, the width w of the gap between the trace and

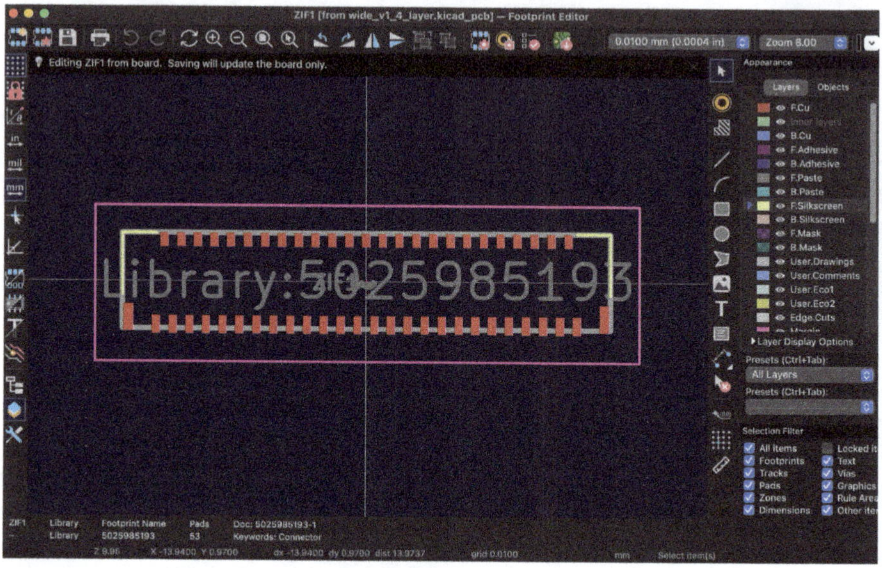

Fig. 5.9 Footprint editor

the ground plane on the face of the board, the dielectric thickness h and the relative dielectric constant ε_r [13]. The effective dielectric constant ε_{eff} is first calculated. We could use a track width of 0.36 mm, gap width of 0.1 mm and dielectric thickness of 0.22 between our first two layers of copper to achieve a z_0 of approximately 50 Ω with a ε_r of 4.29 for example. FR4 generally has a constant between 3.8 and 4.8 (Fig. 5.10).

$$
\begin{aligned}
z_0 &= \frac{60\pi}{\sqrt{\varepsilon_{\text{eff}}}} \frac{1}{\frac{K(k)}{K(k')} + \frac{K(k1)}{K(k1')}} \\
k &= \frac{s}{s+w} \\
k' &= \sqrt{1-k^2} \\
k1' &= \sqrt{1-k1^2} \\
k1 &= \frac{\tanh(\frac{\pi s}{4h})}{\tanh(\frac{\pi(s+w)}{4h})} \\
\varepsilon_{\text{eff}} &= \frac{1+\varepsilon_r \frac{K(k')}{K(k)} \frac{K(k1)}{K(k1')}}{1+\frac{K(k')}{K(k)} \frac{K(k1)}{K(k1')}}
\end{aligned}
$$

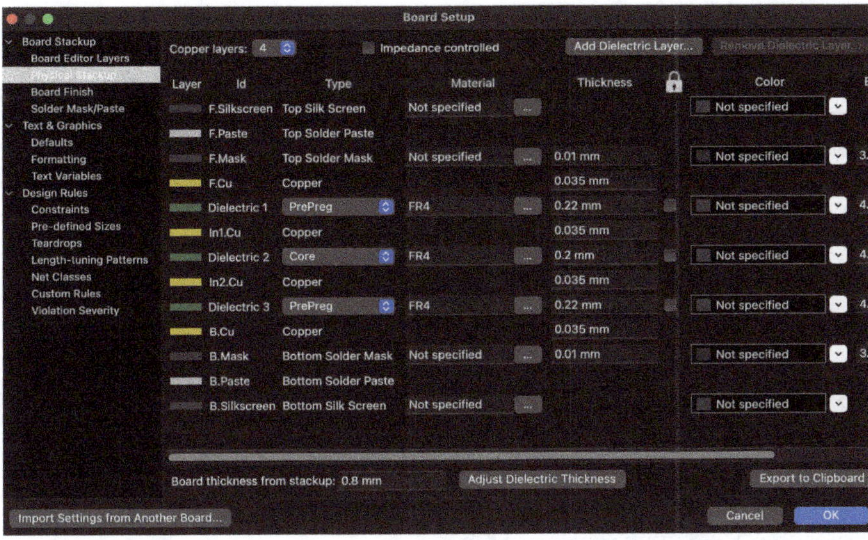

Fig. 5.10 Board stackup

Back in the primary window of the PCB editor, we can left click and drag to arrange our footprints within our PCB extents. The outline of the PCB can be drawn using the line, arc, rectangle tools, etc., that we mentioned before. A rectangle usually suffices. Then we can begin to draw our traces between them. Near the top left corner of the screen, we can enter the trace widths or use pre-defined widths. After selecting the desired width and Route Tracks from the right-hand toolbar we can left click to begin a trace and left click again to lock in a segment of the trace. The trace drawing will automatically conclude when we left click to end the trace at a pin. To change which layer we are drawing traces on, we can highlight the active layer under Layers on the right. In order to connect traces on two different layers, we can use the Add Vias tool.

Vias are also useful for ground plane stitching and shielding. We don't want any "dead copper" on our board which isn't connected to ground which could behave unpredictably in our magnetic field. We also want the ground planes on each copper layer to be a common ground. So, we should introduce vias spaced frequently throughout the available space of a multi-layer PCB to stitch the grounds together, including any copper "islands". To create the ground planes in the first place, we should use the Add Filled Zone tool. After selecting the tool, we will be prompted to select a net that it belongs to. If we don't have a ground net already, create one. Then you can click to outline the filled zone which should correspond to your PCB extents. Do this for each copper layer. You can press B to fill the zones, and you should see that some space is left around the traces you have drawn. To configure the amount of space around the traces, we can enter them under File → Board Setup → Design Rules → Net Classes under clearances. If more than one set of clearances, track widths, via sizes, etc., are needed then you may create multiple netclasses. Each net can be assigned under Board Setup to the appropriate netclass. Yes, these are the same nets from the netlist.

When we are done with our routing, we need to draw a solder mask area everywhere except for where we will solder our surface mount components. We need solder *paste* where we *will* solder our components.

When a board is finally done, it needs to pass ERC before you export Gerbers or other standardized design files for a PCB shop to fabricate from. You can change the ERC rules under File → Board Setup → Design Rules → Constraints but beware! These rules should be within the tolerances of your chosen PCB printing shop.

References

1. Bardin, J.C., Sank, D., Naaman, O., Jeffrey, E.: Quantum Computing: An Introduction for Microwave Engineers. IEEE Microwave Mag. **21**(8), 24–44 (2020). https://doi.org/10.1109/MMM.2020.2993475. https://ieeexplore.ieee.org/document/9134701/?arnumber=9134701. Conference Name: IEEE Microwave Magazine
2. de Sousa, R.: Quantum Devices 1 at University of Victoria (2024). Graduate Course
3. Peebles, P.J.E., Wilkinson, D.T.: Robert henry Dicke (1916–1997) physicist whose work led to the discovery of the cosmic microwave background. Nature **386**(6624), 448–448 (1997)https://doi.org/10.1038/386448a0. https://www.nature.com/articles/386448a0. Publisher: Nature Publishing Group
4. Nielsen, M.A., Chuang, I.L.: Quantum Computation and Quantum Information 10th, Anniversary Cambridge University Press, Cambridge; New York (2010)
5. Blais, A., Grimsmo, A.L., Girvin, S.M., Wallraff, A.: Circuit Quantum Electrodynamics. Rev. Mod. Phys. **93**(2), 025005 (2021). ArXiv:2005.12667 [quant-ph] 10.1103/RevModPhys.93.025005. http://arxiv.org/abs/2005.12667
6. Quantum Computing 101: Introduction, Evaluation, and Applications. https://ionq.com/resources/quantum-computing-101-introduction-evaluation-applications
7. Smith-Goodson, P.: Atom computing announces record-breaking 1225-qubit quantum computer. https://www.forbes.com/sites/moorinsights/2023/10/24/atom-computing-announces-record-breaking-1225-qubit-quantum-computer/. Section: Cloud
8. Piot, N., Brun, B., Schmitt, V., Zihlmann, S., Michal, V.P., Apra, A., Abadillo-Uriel, J.C., Jehl, X., Bertrand, B., Niebojewski, H., Hutin, L., Vinet, M., Urdampilleta, M., Meunier, T., Niquet, Y.M., Maurand, R., Franceschi, S.D.: A single hole spin with enhanced coherence in natural silicon. Nat. Nanotechnol. **17**(10), 1072–1077 (2022). https://doi.org/10.1038/s41565-022-01196-z. https://www.nature.com/articles/s41565-022-01196-z. Publisher: Nature Publishing Group
9. Hendrickx, N.W., Lawrie, W.I.L., Russ, M., Van riggelen, F., de Snoo, S.L., Schouten, R.N., Sammak, A., Scappucci, G., Veldhorst, M.: A four-qubit germanium quantum processor. Nature **591**(7851), 580–585,585A–585N (2021). https://doi.org/10.1038/s41586-021-03332-6. https://www.proquest.com/docview/2505728932/abstract/F030BA68FB854195PQ/1. Num Pages: 580-585,585A-585N Place: London, United States Publisher: Nature Publishing Group Section: Article
10. Pobell, F.: Matter and Methods at Low Temperatures. Springer, Berlin, Heidelberg (2007). https://doi.org/10.1007/978-3-540-46360-3
11. Schmitz, K.S.: Chapter 5—Thermodynamics of the Liquid State. In: K.S. Schmitz (ed.) Physical Chemistry, pp. 203–260. Elsevier, Boston (2017). https://doi.org/10.1016/B978-0-12-800514-9.00005-5. https://www.sciencedirect.com/science/article/pii/B9780128005149000055

12. 13.7: Osmotic Pressure - Chemistry LibreTexts. https://chem.libretexts.org/Bookshelves/General_Chemistry/Map%3A_General_Chemistry_(Petrucci_et_al.)/13%3A_Solutions_and_their_Physical_Properties/13.07%3A_Osmotic_Pressure
13. Wadell, B.C.: Transmission line design handbook. Artech House (1991)

Chapter 6
Talking to Your Qubit

Abstract We have understood how a qubit works, how to error correct it and keep it cool. Now we need to operate it. This chapter dives into the circuits and techniques for initializing, transforming and reading out your qubits' states. We cover superconducting, photonic, annealing, nuclear magnetic resonance (NMR), ion trap and spin qubit quantum computers' operation.

6.1 Learning Objectives

This is where the fun begins. Now we can begin to operate our quantum computer and perform real calculations. In this chapter we will aim to (Fig. 6.1):

- Learn about initializing the different types of quantum computers.
- Learn how to perform gates (and alternatives) on the appropriate types of quantum computers.
- Understand how to get information out of our quantum computers through readout.

6.2 Common Control Dynamics

From DiVincenzo's criteria [1] we know that we must be able to initialize our qubits, perform a universal set of gates on them, and be able to read them out. The question of coherence time belongs in another chapter and of scaling yet another. However, we will look at how the initialization, operation and readout of popular forms of qubits to compare. The approaches taken to controlling each type of qubit discussed in Chap. 4 differ. However, there are some overlaps. As we know, superconducting and spin qubits can both be housed in dilution refrigerators. Their operation also shares some elements.

A dilution refrigerator, for example, like the one discussed in the previous chapter does not only involve cooling. The compressor, gas handling system, stages, shields, chambers, plates, coldfinger and sample holder we discussed before are only a part

© The Author(s), under exclusive license to Springer Nature Switzerland AG 2026
M. S. Edwards, *An Introduction to Quantum Computing for Computer Engineers*,
https://doi.org/10.1007/978-3-032-03650-6_6

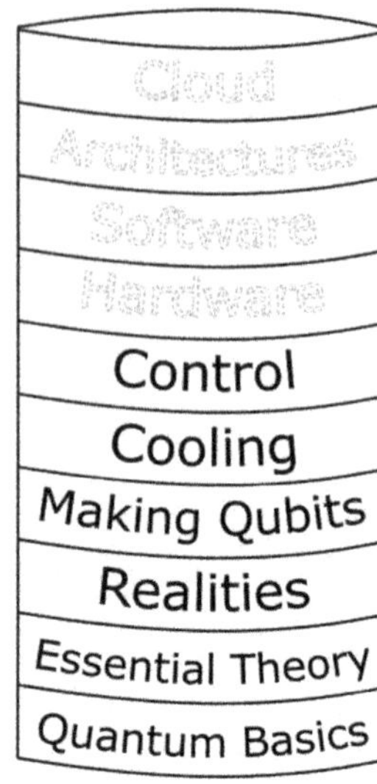

Fig. 6.1 Learning objective for this chapter: control

of the story. The refrigerator also needs to contain DC and microwave wiring, low-pass filters and amplifiers at the appropriate stages which lead electrical signals into and out from the device sitting at tens of milliKelvin at the bottom of the entire assembly. Connected to this internal wiring and electronics through Fischer, BNC and microwave cables will be room-temperature control electronics (which we will cover in Chap. 7), DACs, breakout box(es) and at least one control computer.

The electrical signals in question will include both DC offsets that allow us to tune our devices into operational regimes as well as microwave (or baseband as seen in low-frequency examples like [2]) control signals from AWGs and RF sources. A PCB will connect our chip out to these control lines. Rabi[1] driving with Rabi pulses allows us to control our qubits, whether they are quantum dot, superconducting, NMR based, etc. These control pulses have dynamics described by the Rabi Hamiltonian:

$$H = \frac{\hbar\omega_0}{2}\sigma_z + \hbar\omega_R \cos(\omega t + \phi)\sigma_x$$

Here $\omega_0 = g_N B_z$ is our Larmor frequency, $\omega_R = |g_N| B_x$ is the control pulse strength and $\omega_0 >> \omega_R$.

[1] Isidor Isaac Rabi worked on the atomic bomb and won the Nobel prize for physics in 1944 [3].

6.3 Initialization

Also common to many implementations of quantum information processing devices is the use of measurement to either indicate or cause initialization to a known state such as the ground state. One scheme, known as active reset, involves measuring into an energy eigenstate and then using the measurement outcome to perform Pauli X gates until the ground state is reached. For example, for a two-qubit reset, we would perform (Fig. 6.2):

Note that this scheme works even for mixed states, as the outcome of a computational basis measurement will be pure. Measurement itself has its own section in this chapter and it looks a little different between implementations, but we will go into the technical details of the various apparatuses in a moment.

6.3.1 Superconducting

One easy way to initialize a superconducting qubit is to simply wait for the length of the energy relaxation time T_1 [4]. Then the state will have relaxed to $|0\rangle$. You may think that an energy eigenstate like $|1\rangle$ is supposed to be stationary and therefore not relax to $|0\rangle$. However, this is an ideality which doesn't consider the coupling of our qubit to the environment which causes energy relaxation. Alternative approaches are more involved [5]. It is possible to do an all-microwave superconducting qubit reset, meaning that the same type of control electronics as used in control and readout are also used for initialization. This has its benefits. One approach to all-microwave reset [6] takes advantage of the response of the superconducting resonator, which is state-dependent. This gets around some issues that we will see with approaches using flux tuning to reset our qubit.

In particular, we can couple our qubit to a decaying external quantum system and this sometimes involves flux tuning. In one embodiment of this idea [7] which is highlighted in [5] we can flux-tune a qubit into resonance with a large-bandwidth, Purcell-filtered resonator. In readout applications, a Purcell filter is used to impede the propagation of emitted qubit-frequency photons [8]. However, in [7], the qubit thermalizes to the ground state due to Purcell decay. Purcell decay refers to the enhancement of the spontaneous emission rate of a quantum system that is coupled

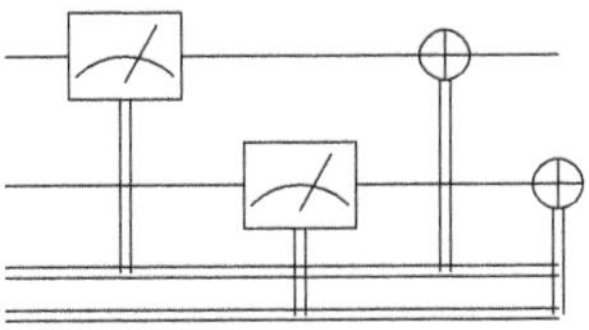

Fig. 6.2 Two-qubit active reset

to a resonator. The interaction of our qubit with the external resonator increases the density of photons emitted by the qubit. When a photon is emitted, of course, the qubit loses energy and relaxes. An issue with this approach is that the flux pulses involved can effect nearby qubits through cross-talk as well as subsequent gates [5].

A projective measurement can be used to herald the ground state [5]. When we herald a state it means that we learn indirectly about it, so that we can be sure we have our qubit in a particular state before we continue with our computation, for example. In [9] a second measurement pulse is introduced to an existing qubit control and measurement sequence. This second pulse allows the authors to confirm when the qubit is in the ground state. The measurement (readout) pulse is reflected off of the resonator. It gains a phase shift that is dependent on the state of the qubit. This signal is amplified by a Josephson parametric amplifier and indicates the state of the qubit.

A projective measurement can also be used to prepare a ground state using feedback [10]. This is similar to the active-reset scheme we will also see used in spin-qubits. In [10] the authors proposed a readout-based measurement scheme that basically involves using quantum-non-demolition (QND) readout to measure qubits. As the reader is aware, when qubits are measured they collapse into energy eigenstates. Calling a measurement "non-demolition" does not mean that it does not collapse the state. It only means that the qubit may still be subsequently used as a qubit. In the readout section of this chapter, we will encounter destructive measurements where this is not the case. After the QND measurement of the qubits, the measurement outcome tells us whether the qubits are in the ground state. We simply post-select the desired state preparations. In this kind of scheme, if a qubit is instead found in the $|1\rangle$ state, we can go further by performing an "active reset" and applying a Pauli X gate.

There are many more initialization schemes for superconducting qubits, such as microwave sideband cooling, but the ones we've discussed are representative of a few classes and we are here limited in space.

6.3.2 *Photonic*

There are several schemes for generating single photons for use in quantum computing. One such scheme involves attenuating a laser to a weak coherent state of just one photon [11, p.288]. A laser outputs the states $|\alpha\rangle$ of the following coherent form, in the Fock basis $|n\rangle$.

$$|\alpha\rangle = \exp(|\alpha|^2/2) \sum_{n=0}^{\infty} \frac{\alpha^n}{\sqrt{n!}} |n\rangle$$

$$|n\rangle = \frac{(a^\dagger)^n |0\rangle}{\sqrt{n!}}$$

Attenuating the laser to this regime involves tuning α such that the leading order terms in the sum dominate. This creates the effect that whenever light is generated, it is most likely a single photon.

The probability of detecting n photons is Poisson, with parameter $|\alpha|^2$.

$$p(n) = |\langle n|\alpha\rangle|^2 = \exp(-|\alpha|^2)\frac{|\alpha|^{2^n}}{n!}$$

The average number of photons is $\langle\alpha| a^\dagger a |\alpha\rangle = |\alpha|^2$.

$$\begin{aligned}(\Delta n)^2 &= \langle\alpha| n^2 |\alpha\rangle - \langle\alpha| n |\alpha\rangle^2 \\ &= \langle\alpha| a^\dagger a a^\dagger a |\alpha\rangle - |\alpha|^4 \\ &= \langle\alpha| a^{\dagger^2} a^2 |\alpha\rangle + |\alpha|^2 - |\alpha|^4 \\ &= |\alpha|^2\end{aligned}$$

However, the leading term $|0\rangle$ will be the most dominant and therefore light will not be generated the majority of the time. We must introduce a method for detecting when a photon is generated that does not destroy the photon.

A second approach to generating single photons involving nonlinear optical media is to make use of parametric down-conversion [11, p. 288]. We still can't predict beforehand exactly when photons will be generated, but parametric down-conversion transforms single photons into pairs of photons with correlated frequencies and momenta such that one photon heralds the other. This means that detecting one photon lets us know that we also have the other.

6.3.3 Annealers

The initialization procedure for annealing-based quantum computers is unique and a key step in the annealing process [12]. If you will recall, this Hamiltonian describes D-Wave's annealers:

$$H = -\frac{1}{2}\sum_{i=1}^{N}(\Delta\sigma_x + \varepsilon_i\sigma_z) + \sum_{i<j} J_{ij}\sigma_{iz}\sigma_{jz}$$

Initializing a D-Wave system involves making Δ large. Δ is controlled by the transverse magnetic field. We get an initial Hamiltonian that looks like the following:

$$H_{\text{init}} = -\frac{1}{2}\Delta\sum_i \sigma_{ix}.$$

We then prepare the qubits each in state $|+\rangle = \frac{|\uparrow\rangle+|\downarrow\rangle}{\sqrt{2}}$. $|+\rangle$ is the eigenstate of σ_x where $\sigma_x |+\rangle = |+\rangle$. The system is now prepared to follow an annealing schedule which will cause the energy of the system to minimize to a ground state representative of a solution to a quadratic unconstrained Boolean optimization (QUBO) problem.

6.3.4 NMR

NMR systems are typically in equilibrium at room temperature [11, p. 324], and spin energy $\hbar\omega$ is less than KT. The initial state of the spins in the molecular liquid are therefore random. Experiments typically begin with several minutes in which we wait for the nuclei to reach thermal equilibrium [11, p. 325]. The initial thermal equilibrium state is then the following [11, p. 328]. Here, $\beta = 1/KT$ and $Z = tr(\exp(-\beta H))$.

$$\varrho = \frac{\exp(-\beta H)}{Z}$$

$\varrho \doteq 2^{-n}(1 - \beta H)$ at high temperatures, where n is the number of qubits and $\beta \doteq 10^{-4}$ is typical [11, p. 329]. Off-diagonal elements in ϱ are very small since spin-spin couplings are much smaller than precession frequencies. Therefore, ϱ effectively looks like a mixture of pure states.

We can also polarize our nuclear spins with a strong external magnetic field and use what is called "effective pure state" preparation [11, p. 343]. However, it is challenging to prepare a specific state like $|0\ldots0\rangle$ or $|1\ldots1\rangle$ with modern technology [11, p. 328].

6.3.5 Spin Qubits

When we discuss the initialization of spin qubits, we discuss it together with readout, following the example of other sources such as [13]. This is because readout is used in initialization schemes, similarly to what we saw in approaches with superconducting qubits. Single-spin readout in GaAs of a Loss–DiVincenzo style quantum dot qubit was first demonstrated in 2004 by Elzerman [14]. Initialization and readout are both accomplished through the use of single-carrier tunneling to and/or from the quantum dot and a reservoir. We saw in chapter four how to design a reservoir like those that appear in [15]. These reservoirs are large traces of metal which are home to many charge carriers, as opposed to the single carrier that will be trapped in our dot.

To create the conditions for tunneling, a gate voltage pulse $V_g(t)$ of $\doteq 100\ ps$ is used to lower the energy levels of the quantum dot with respect to the Fermi level of the reservoir [13]. The particular voltage used is device-specific and depends on the lever arm of the gate in question, which is determined experimentally in the lab. The lever arm α expresses the relationship of V_g with the electrochemical potential μ of the dot such that $\mu = \mu_0 - e\alpha V_g$. Here μ_0 is the electrochemical potential when $V_g = 0$.

Elzerman's initialization scheme involved first lowering the energy levels E_0 and E_1 of $|0\rangle$ and $|1\rangle$, respectively, in the QD both below the Fermi level E_F of the reservoir so that the dot would be loaded in either of the $|0\rangle$ or $|1\rangle$ state randomly. Then, the QD levels were raised so that $E_0 < E_F < E_1$, allowing for a charge carrier to tunnel once more. If a tunneling is detected by a nearby Quantum Point Contact (QPC), it indicates a $|1\rangle$ state measurement outcome. No current in the QPC indicates $|0\rangle$. A simpler initialization scheme is to just initially lower one state below E_F, so that this state will be populated by tunneling from the reservoir.

With singlet-triplet qubits, a Pauli Spin Blockade (PSB) is used to initialize the double quantum dots. The idea of the PSB is simple. It is based on the idea, already discussed in this book, of the Pauli exclusion principle which says that two identical fermions cannot occupy the same state and must differ, for example, by a quantum number such as spin. If a first quantum dot is already loaded with a charge carrier of known spin, we can use this to our advantage to learn the spin of a carrier in a nearby dot. We would adjust the detuning (the difference in relative energy levels between the two dots) so that tunneling is possible from the second to the first dot. Then we would detect via a nearby charge sensor, for example, any movement of charge between the dots. If current is induced in the sensor by the movement of a charge, then we know that the carrier originally in the second dot has a spin opposite the spin of the carrier that started (and remains) in the first dot. Otherwise, the spins are the same, and this explains why tunneling could not occur. The carriers are identical besides their spatial distributions and spins. The conditions for tunneling are otherwise met. So, in our case, a current in the sensor would indicate a singlet state and no current would indicate a triplet (Fig. 6.3).

In GaAs and Si, respectively, the energy splittings between the singlet and triplet states are on the order of meV and μeV [13]. In Ge, the energy splitting between $|S\rangle$ and $|T_{-}\rangle$, for example, can depend on the tunnel coupling t_c, g-factor sum $\sum g$, detuning ε and more [16]. One can immediately see this when considering the singlet and triplet energies presented in [16].

Fig. 6.3 Pauli spin blockade

$$E_S \doteq \frac{\varepsilon}{2} - \sqrt{\frac{\varepsilon^2}{4} + 2t_c^2}$$
$$E_{T_\pm} \doteq \pm \frac{\sum g}{2} \mu_B B$$

Charge carriers are separated into the two different quantum dots as a second step in the initialization of a singlet-triplet qubit. When we have initialized a singlet in a single dot, for example, we wish to tunnel one carrier to the other dot without losing its spin state. This is possible when the tunneling is fast with respect to magnetic field gradients [13]. Otherwise, the singlet state relaxes. Using latched readout gets around this limitation while allowing for higher gradient magnetic fields, which are helpful for driving faster single-qubit gates.

Consider next the initialization schemes for Triple Quantum Dot (TQD) exchange-only qubits [13], for which the $|1\rangle$ state is encoded in a superposition of the two triplet states. Remember that PSB allows tunneling when spins differ, determining parity. PSB cannot however distinguish between the cases when spins are both up or both down. Therefore, it will not distinguish between all of the triplet projections. The same PSB protocol used to initialize singlet-triplet qubits can be used to initialize TQD exchange-only qubits. We will use two carriers in two dots for this protocol, leaving the third uninitialized.

6.3.6 *Ion Trap*

State preparation is done by driving transitions with a laser. We cool the trapped ions using optical pumping into the hyperfine and motional ground state [11, p. 324].

Consider the motional Hamiltonian $\hbar\omega(a^\dagger a + \frac{1}{2})$. A resonant laser pulse can drive an optical dipole transition. This involves the dissipation of the mechanical energy of the motional mode through an interaction with the laser pulse.

The driven dipole Hamiltonian is the following [17]:

$$H = \hbar\omega(a^\dagger a + \frac{1}{2}) - \hbar\Delta\,|e\rangle\,\langle e| + \frac{\hbar\omega}{2} \times (|e\rangle\,\langle g|\exp(ik_z z) + h.c.).$$

Here Δ is the detuning $\omega_L - \omega_0$ of the laser light frequency ω_L from the resonance frequency ω_0. Ω is the Rabi frequency. In the case of optical pumping, $\Omega = \mathbf{d} \cdot \boldsymbol{\varepsilon}\mathcal{E}_0/\hbar$ where $\mathbf{d}$ is the dipole moment, $\mathcal{E}_0$ the laser field amplitude and $\boldsymbol{\varepsilon}$ the polarization. $k_z = \mathbf{k}$ the wavevector and $z = |\mathbf{k}|cos\theta$. Here θ is the difference between the laser field direction and the direction of the ion's motion.

With probability $\langle n'|\exp(ik_z z)|n\rangle$, the state will transition from $|n\rangle$ to $|n'\rangle$. Energy will be lost in packet(s) of $\hbar k_z$ as photon(s) are emitted into the laser mode. Laser parameters may be chosen so that transitions to lower energy states are most likely to occur.

6.4 Gates and Alternatives

Another requirement for useful quantum computing is a universal set of quantum gates. Gates are implemented in different ways depending on the specific type of quantum computer one is building, but the high-level theory is the same across the board. Mathematically, a gate is a $2^N \times 2^N$ matrix that multiplies our N-qubit state in an eigenvalue equation.

> **What kinds of gates do we use in quantum computing?**
> A universal gateset must be able to create superposition, create entanglement, have real and imaginary gates, and not be contained in the stabilizer set [18] Furthermore, a universal gateset should be able to approximate any unitary transformation.

There are many gatesets that are considered universal, including the often used Clifford + T and Toffoli + H + P, gatesets. More universal gatesets include ZZ interaction + fixed axis rotations and Molmer–Sorenson + fixed axis rotations. In this chapter we will see universal gatesets including CZ + fixed axis rotations, beamsplitters + phase shifters + entangled states and more.

6.4.1 Superconducting

In superconducting qubits, we make use of microwave drive signals and flux biases [4] to implement quantum gates.

We can model a driving microwave signal of frequency ω_d and phase ϕ_d in the following way. We begin with an equation of motion for the resonator field $a(t)$ which tells us how the resonator dynamics are affected by a drive signal, the input field $b_{\text{in}}(t)$ [19].

$$\dot{a}(t) = i[H_s, a(t)] - \frac{\kappa}{2}a(t) + \sqrt{\kappa}(b_{in}(t) + \beta(t))$$

Here $\beta t = A(t)\exp(-i\omega_d t - i\phi_d)$ the coherent classical component of the input field. Its time-varying amplitude is $A(t)$. This $\beta(t)$ corresponds to a term H_d in the system Hamiltonian H_s which would be $H_s = \hbar\omega_q a^\dagger a + \frac{\alpha}{2}a^\dagger a^\dagger aa + H_d$ in the case of an anharmonic oscillator like a Transmon, for example.

$$H_d = \hbar[\varepsilon(t)a^\dagger\exp(-i\omega_d t - i\phi_d) + \varepsilon^*(t)a\exp(i\omega_d t + i\phi_d)]$$

When we add a drive source to a superconducting qubit circuit, the effect is to add off-diagonal terms to the Hamiltonian so that adjacent levels, or quantum basis states, undergo mixing [4]. Schematically, a drive source can be modeled by an AC voltage source and a capacitor, as the drive is capacitively coupled to the device. This source and this capacitor are in series with one another, and in parallel with the nonlinear inductor and the capacitor of the LC circuit.

Z-axis rotation is controlled in superconducting qubit devices by controlling the qubit frequency with the flux tuning line [4]. This flux tuning line can be inductively coupled to the Transmon's SQUID [20, p. 199] (note here we consider transmons that have a SQUID in the place of the single JJ). Z gates may also be implemented using virtual Z rotations, or changes of the reference frame when we allow the spin to precess. This technique simply involves waiting and keeping track of our reference frames in a way sometimes called "book-keeping" [20, p. 57]. X and Y rotations are achieved by coupling a microwave generator to the transmon [4]. This can be done inductively or capacitively. The rotation that is caused of the state vector is determined by the carrier phase of the microwave drive signal. The difference between an X rotation and a Y rotation is a shift of $\frac{\pi}{2}$. The amount of rotation is determined by the integrated envelope amplitude. The speed of the rotation depends on the signal's amplitude.

In order to interact superconducting qubits, we can capacitively couple them and they may undergo a periodic energy exchange [4]. The interaction Hamiltonian for two capacitively coupled Transmons is [20, p. 200]:

$$H = \hbar g_C(\sigma_+^{(1)}\sigma_-^{(2)} + \sigma_-^{(1)}\sigma_+^{(2)})$$

$$\hbar g_C = \frac{\sqrt{2}e^2 C_c}{C_1 C_2}\left(\frac{E_{J1}E_{J2}}{E_{C1}E_{C2}}\right)^{1/4}$$

Here C_c, C_1, C_2 are the capacitances of the coupling capacitor, and the shunting capacitors of the two transmons. The E_{Jn}, E_{Cm} are the energies of respective components. An alternative to direct capacitive coupling is to use a tunable coupling similar to what we saw in the section on annealers, which can be done either inductively or capacitively.

The rate of the exchange will depend on the coupling strength which is determined by the coupling capacitance. The exchange does not take place if the qubits have the same frequency and state. Therefore, the interaction can be turned on and off by tuning the relative qubit frequencies with the Z flux tuning control line.

A native gate to capacitively coupled flux tunable transmons is the iSWAP gate [4] [20, p. 201]. Here ω_s gives the frequency of the energy exchange.

$$\begin{bmatrix} 1 & 0 & 0 & 0 \\ 0 & \cos(\omega_s \Delta t/2) & -j\sin(\omega_s \Delta t/2) & 0 \\ 0 & -j\sin(\omega_s \Delta t/2) & \cos(\omega_s \Delta t/2) & 0 \\ 0 & 0 & 0 & 1 \end{bmatrix}$$

As we will also see with spin qubits, where the SWAP gate is native, implementing the gate amounts to turning the exchange on for the right amount of time. The iSWAP

Fig. 6.4 Phase shifter

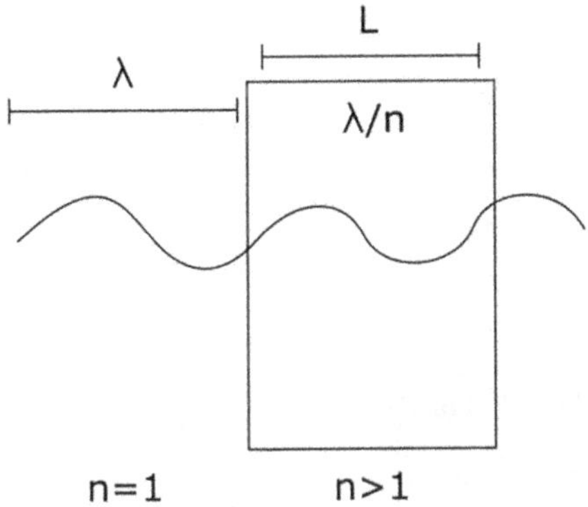

is achieved by turning on the exchange for a time Δt such that $\Delta t \omega_s = \pi$. Then our cosine entries disappear, and the sine entries become equal to 1.

Alternative methods to implement two-qubit gates including microwave-based parametric and cross-resonance gates exist as well, but for the sake of brevity we refer the reader to [20, p. 202] for details regarding these alternative approaches.

6.4.2 *Photonic*

There are several components which allow us to operate on our photonic qubits. Gates for dual rail photonic qubits are passive optical elements. A ϕ-phase shifter shifts the phase of an incoming photon by phase ϕ. The phase accumulated in a phase shifter is (Fig. 6.4):

$$\phi = \frac{2\pi}{\lambda_n} L = \frac{2\pi}{\lambda} Ln$$

The transformation operator is the exponential $\exp(i\delta\phi)$, where $\delta\phi = \frac{2\pi}{\lambda} L(n-1)$ [11, p. 291]. Applied to one rail, this implements a single-qubit rotation.

$$\begin{bmatrix} \exp(i\delta\phi) & 0 \\ 0 & 1 \end{bmatrix}$$

Polarizers are another kind of passive optical device. They can be implemented using small strips of conductive polymers that are separated by a spacing that is much less than the wavelength of incident light. The energy associated with the electromagnetic wave's oscillation, in a dimension determined by the polarizer's orientation, is dissipated as the current induced in the strips is converted to heat. The result is that the light which passes through the polarizer becomes polarized in a particular dimension (Fig. 6.5).

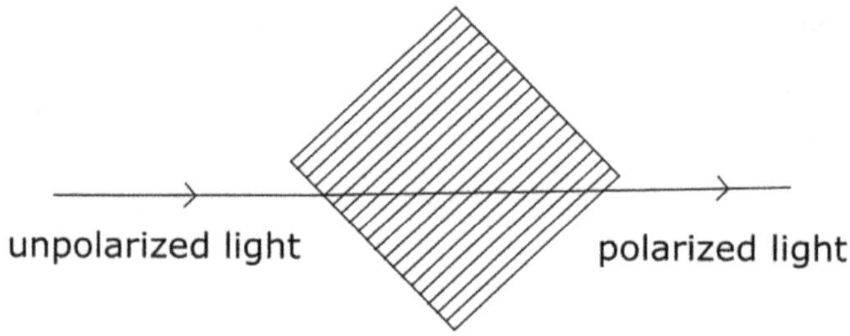

Fig. 6.5 Polarizer

In order to have a useful gateset, we need to be able to create superposition. To this end we use a half-silvered mirror which will reflect and transmit 50% of the incident light. The 50–50 beamsplitter with a phase shifter together act as a Hadamard up to an overall phase [11, p. 292].

$$H = \frac{1}{\sqrt{2}}\begin{bmatrix}1 & 1\\ 1 & -1\end{bmatrix}$$

$$|b_1, b_0\rangle = H\,|a_1, a_0\rangle$$

Consider a single photon incident on the 0th mode.

$$|b_1, b_0\rangle = H a_0^\dagger\,|0, 0\rangle$$
$$|b_1, b_0\rangle = H\,|1, 0\rangle$$

Remember that a $|0\rangle$ state is encoded as a photon on the 0th rail and a $|1\rangle$ state as a photon on the 1st rail.

$$|b_1, b_0\rangle = \frac{1}{\sqrt{2}}(a_0^\dagger + a_1^\dagger)\,|0, 0\rangle\,|b_1, b_0\rangle = \frac{1}{\sqrt{2}}(|1, 0\rangle + |0, 1\rangle)$$
$$= \frac{1}{\sqrt{2}}(|0\rangle + |1\rangle)$$

If we are working in the Fock basis, the last line reads differently. Given one incident photon on one input mode, the output becomes a Bell state. The particular Bell state depends on the particular input mode.

Fig. 6.6 Beamsplitter

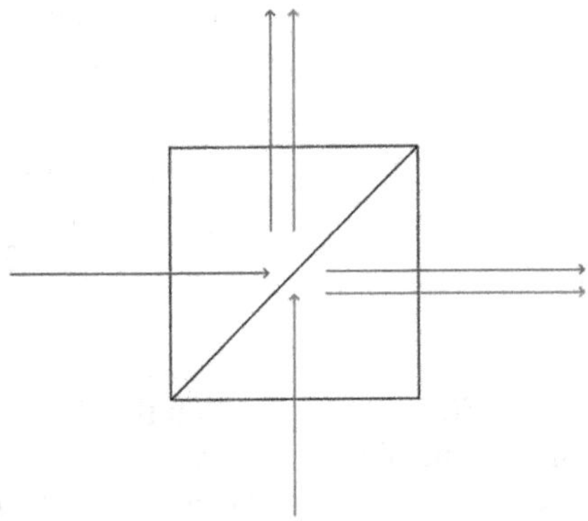

$$
\begin{aligned}
|b_1, b_0\rangle &= Ha_0^\dagger |0,0\rangle \\
|b_1, b_0\rangle &= H|1,0\rangle \\
|b_1, b_0\rangle &= \frac{1}{\sqrt{2}}(|1,0\rangle + |0,1\rangle) \\
&= \frac{1}{\sqrt{2}}(|10\rangle + |01\rangle) \\
|b_1, b_0\rangle &= Ha_1^\dagger |0,0\rangle \\
|b_1, b_0\rangle &= H|0,1\rangle \\
|b_1, b_0\rangle &= \frac{1}{\sqrt{2}}(|1,0\rangle - |0,1\rangle) \\
&= \frac{1}{\sqrt{2}}(|10\rangle - |01\rangle)
\end{aligned}
$$

The 50–50 beamsplitter can be used in such a way that it has two input modes and two output modes. This allows us to operate on more than one incident photon, for example (Fig. 6.6).

We can make use of Xanadu's package for photonic quantum computing called StrawberryFields to do a beamsplitter experiment. The *BSGate()* operator takes two parameters, the first (θ) determines the transmittivity of the splitter. A value of $\pi/4$ gives the 50–50 beamsplitter. The second parameter ϕ gives the reflection amplitude.

In general, the beamsplitter has the following action:

$$B(\theta, \phi) = \exp(\theta(\exp(i\phi)a_0 a_1^\dagger - \exp(-i\phi)a_0^\dagger a_1)).$$

What happens if we have more than one photon in the system and a 50–50 beamsplitter?

$$|b_1, b_0\rangle = Ha_0^\dagger a_1^\dagger |0,0\rangle$$

See that a photon incident on the 2nd mode yields superposition like before, but with a different sign since the negative term in the Hamiltonian comes into play.

$$|b_1, b_0\rangle = \frac{1}{2}(a_0^\dagger - a_1^\dagger)(a_0^\dagger + a_1^\dagger)\,|0, 0\rangle$$
$$\frac{1}{\sqrt{2}}(|2, 0\rangle - |0, 2\rangle)$$

This state expresses that whenever we measure the optical signal coming out of our 50–50 beamsplitter with these photons having been incident in both modes, we will find that the photons have "bunched" together into one or the other of the output modes. This is called the Hong-Ou-Mandel effect.

Example: Photon Bunching with StrawberryFields
We can check this using StrawberryFields, roughly following a demo provided by Xanadu [21]. We will use *S2gate(0.1)* to create a Gaussian two-mode squeezed vacuum state which has correlated pairs of photons in the two optical modes. Then, we will detect photons coming out of our 50–50 beamsplitter. For this we will make use of a provided function *threshold_detection_prob*. When we pass a list to this function, it represents the configuration of photon detections in the output modes. So, *[1, 1]* indicates that we detect a photon in each output mode, *[1, 0]* indicates our detector (which can only click or not click) detects photons in the first mode but not the second, and so on.

```
from thewalrus import threshold_detection_prob

# create a 2-mode quantum program
prog = sf.Program(2)

with prog.context as q:
    ops.S2gate(0.1) | q
    ops.BSgate(np.pi / 4, np.pi / 2) | q

eng = sf.Engine("gaussian")
results = eng.run(prog)
state = results.state

p_click = threshold_detection_prob(state.means(), state.cov(), [1, 1])
print(f"Coincidence probability: {np.round(p_click.real, 4)}")

p_first = threshold_detection_prob(state.means(), state.cov(), [1, 0])
print(f"First bunched photon configuration probability: {np.round(p_first.real, 4)}")

p_second = threshold_detection_prob(state.means(), state.cov(), [0, 1])
print(f"Second bunched photon configuration probability: {np.round(p_second.real, 4)}")
```

```
Coincidence probability: 0.0
First bunched photon configuration probability: 0.005
Second bunched photon configuration probability: 0.005
```

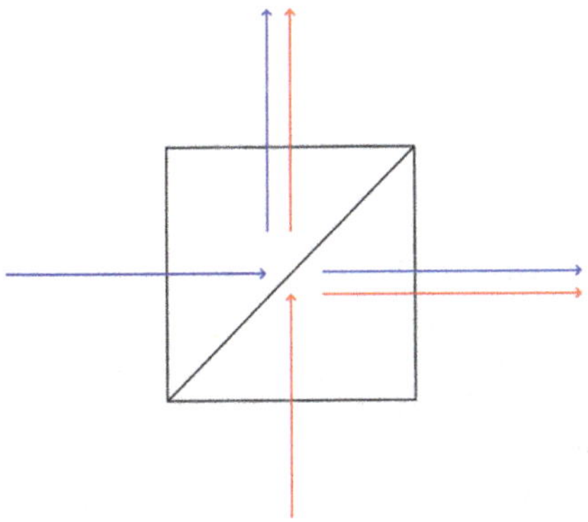

Fig. 6.7 Polarizing beamsplitter

Our experiment confirms that while we don't always see photons coming out of our circuit, when we do, they are most likely bunched! If we did not round these probabilities to 4 decimal places we would see a small coincidence probability instead of this 0.000 coincidence probability. The reason is that threshold detectors can't perfectly count photons. We will discuss more on the detection of photons and measurement of optical quantum states shortly.

Now onto another type of beamsplitter. A polarizing beamsplitter splits incident photons of orthogonal polarizations into different paths. We can implement this in StrawberryFields using the *Interferometer(np.array([[0, 1], [1, 0]]))* gate (Fig. 6.7).

Waveplates rotate between polarizations. These can be made from birefringent materials, which are crystals that have different indexes (n_o, n_e) of refraction along different crystal axes so that $|V\rangle$ and $|H\rangle$ pick up different phases. Birefringent crystals are naturally occurring.

$$\Delta\phi = \frac{2\pi}{\lambda_0}(n_o - n_e)L$$

Here λ_0 is the wavelength of the light in air, L is the length of the waveplate and $\Delta\phi$ is the resulting phase difference.

Entanglement in photonic quantum computing comes from how our photons are generated. We use a process called spontaneous parametric down-conversion to generate photons. Heralded down-conversion generates a pair of photons and can result in entangled photon pairs. One use of such a photon pair is to herald the arrival of the other.

An optical CZ operation can be implemented using a nonlinear Kerr medium [11, p. 294]. The medium applies a unitary transformation $K = \exp(i\chi La^\dagger ab^\dagger b)$, where L is the crystal length, χ relates to the index of refraction and a, b are optical modes. The medium's atoms mediate an interaction between the photons [11, p. 293].

$$
\begin{aligned}
K|00\rangle &= |00\rangle \\
K|01\rangle &= |01\rangle \\
K|10\rangle &= |10\rangle \\
K|11\rangle &= \exp(i\chi L)|11\rangle
\end{aligned}
$$

We can choose a Kerr medium[2] with length L and index of refraction such that $\chi L = \pi$. Then $K|11\rangle = -|11\rangle$, and this implements a CZ gate. It is also possible to implement a CNOT gate with the added use of a beamsplitter [11, p. 294] but either is sufficient to satisfy the requirement for interaction between modes in a universal gateset.

We saw in Chap. 4 how displacement and squeezing work, which are the operators we mainly use in CVQC. In StrawberryFields we can make use of the *Squeezed(r=magnitude, p=angle)* to specify a squeezed state, and similarly, we can use *DGate(r=magnitude, phi=angle)* to specify a displacement, etc.

6.4.3 Annealers

Instead of using gates, annealing [12] involves lowering the parameter Δ of $-\frac{1}{2}\sum_{i=1}^{N}(\Delta\sigma_x + \varepsilon_i\sigma_z) + \sum_{i<j} J_{ij}\sigma_{iz}\sigma_{jz}$ slowly, almost adiabatically, until the system settles into a low-energy state that represents a solution to a QUBO problem. Many NP-hard problems map without a large overhead onto the Ising model which this approach solves. However, it is difficult to prove that the approach is or is not more efficient than a classical one since it is impossibly complex to simulate what occurs during an anneal.

An anneal involves lowering Δ according to an annealing schedule $S(t)$ which defines a linear slope for $\Delta(t)$ s.t. $\Delta(0) = 1$ and $\Delta(t_A) = 0$. When we reach the ground state at $t = t_A$, $H = H_{Ising}$ and we can sample a solution to our problem.

Theoretically, quantum tunneling occurs during this anneal which causes the system to reach the ground state quickly. This also means we can tunnel out of local minima where a classical annealer might get stuck. This tunneling has been shown to happen if the qubits are coherent [12]. However, the question generally remains of whether we are taking advantage of it effectively.

To choose an annealing schedule, we can make use of the adiabatic theorem (or we can try trial and error). According to the adiabatic theorem, we can stay in the ground state E_0 if we vary Δ slowly enough and there is a certain energy difference between our states $E_j(S(t))$ and $E_{j+1}(S(t))$. If we vary too quickly, we can cause adiabatic errors. An adiabatic error is an excitement of the state into a higher energy eigenstate which is not the best solution to our QUBO in the end.

[2] An alternative to using a nonlinear crystal is to implement the KLM protocol [22] which stands for Knill, Laflamme, Milburn (the author of this book was instructed by Dr. Laflamme at IQC) which gave a groundbreaking approach to implementing a Z gate using only linear optical elements.

Then if we start in the ground state at $t = 0$ s.t. $|\psi(0)\rangle = |E_j(0)\rangle$, and assuming $S(0) = 0$, $S(t_f) = 1$, we can bound our annealing time. Here $E_G = E_1 - E_0$ and $E_C > E_1$.

$$t_f \gg Max_{s\in[0,1]} \frac{\hbar| \langle E_i(s)| \frac{\partial H(s)}{\partial s} |E_j(s)\rangle |}{|E_i(s) - E_j(s)|^2} \text{ for } i \neq j$$

The result is that $t_A >> \frac{\hbar E_C}{E_G^2}$. The smaller the gap between our first two energy eigenstates, the more annealing time is required.

6.4.4 NMR

Nuclear spin qubits are controlled by Rabi pulses with the following form. Here $\omega_R >> \omega_0$ is the Rabi frequency or RF strength, and ω_0 the Larmor frequency.

$$H = \frac{\hbar\omega_0}{2} S_z + \hbar\omega_R \cos(\omega t + \phi) S_x$$
$$\omega_0 = g_N B_z$$
$$\omega_R = |g_N| b_x$$

In the rotating frame, the Hamiltonian is:

$$H_{\text{RWA}} = \frac{\hbar}{2}(\omega - \omega_0) S_z + \frac{\hbar\omega}{2}(S_x \cos(\gamma) + S_y \sin(\gamma))$$

X and Y rotations are implemented by resonant $\omega = \omega_0$ driving. These driving tones are RF pulses with controlled phase. The rotation angle is determined by the length of the pulse. Rabi driving with in-phase pulses can drive rotations about the x axis and out-of-phase pulses can drive rotations about the y axis.

We implement Z rotations by different methods. We can achieve a Z rotation by applications of timed Y and X rotations. Or, we can basically change our reference frame with the application of gates at a fraction of a period of a rotation in the rotating frame. Such reference frame changes must be kept track of as they accumulate.

As long as $U^2 = I$, we can write $\exp(-itU)$ as $\cos(t)I - i\sin(t)U$ instead of having to expand $\exp(-itU) = \sum_{k=0}^{\infty} \frac{1}{k!}(-itU)^k$.

$$
\begin{aligned}
\exp(-itU) &= \sum_{k=0}^{\infty} \frac{(-it)^k}{k!} U^k \\
&= \sum_{k=0}^{\infty} \frac{(-it)^{2k}}{2k!} I + \sum_{k=0}^{\infty} \frac{(-it)^{2k+1}}{(2k+1)!} U \\
&= \cos(t)I - i\sin(t)U
\end{aligned}
$$

So, for example,

$$
\begin{aligned}
R_x(\theta) &= \exp(-i\theta/2X) = \cos(\theta/2)I - i\sin(\theta/2)X \\
R_y(\theta) &= \exp(-i\theta/2Y) = \cos(\theta/2)I - i\sin(\theta/2)Y \\
R_z(\theta) &= \exp(-i\theta/2Z) = \cos(\theta/2)I - i\sin(\theta/2)Z
\end{aligned}
$$

And any single-qubit gate can be written as $U = \exp(i\alpha)R_n(\beta)R_m(\gamma)R_n(\delta)$ for non-parallel n, m unit vectors and real numbers $\alpha, \beta, \gamma, \delta$. This is called an Euler angle decomposition. Therefore,

$$
\begin{aligned}
R_z &= R_x(\pi/2)R_y(\theta)R_x(\pi/2) \\
&= \exp(-\pi/4\sigma_x)(\cos(\theta/2)I + \sin(\theta/2)\sigma_y)\exp(i\pi/4\sigma_x) \\
&= \cos(\theta/2)I - i\sin(\theta/2)\sigma_z \\
&= \exp(-i\theta/2\sigma_z)
\end{aligned}
$$

The J coupling means the state of one qubit effects the precession of the other. This makes it difficult to implement single-qubit gates. We can however apply timed pulses to synchronize the precessions of the qubits. For example, we can apply a π pulse on one qubit at $T/2$ to reverse the direction of the evolution of the a qubit coupled to it. Then at time T, the coupled qubit will have returned to its original state and a second pulse can restore the state of the first qubit. This achieves an identity operation.

$$
\begin{aligned}
U_{evolve} &= \exp(-i\frac{HT}{\hbar}) \\
U_{tot} &= U_{-x}(\pi)U_{evolve}(T)U_x(\pi)U_{evolve}(T) \\
&= \exp(i\pi/2JT\sigma_z^{(1)}\sigma_z^{(2)})\exp(-i\pi/2\,JT\sigma_z^{(1)}\sigma_z^{(2)}) \\
&= I
\end{aligned}
$$

In terms of interaction between spins, we can implement a CNOT using refocusing pulses and single-qubit pulses [11, p. 332]. First, we can implement a CZ using

$$\sqrt{i}\exp(iZ_1Z_2\pi/4)\exp(-iZ_1\pi/4)\exp(-iZ_2\pi/4) = \begin{bmatrix} 1 & 0 & 0 & 0 \\ 0 & 1 & 0 & 0 \\ 0 & 0 & 1 & 0 \\ 0 & 0 & 0 & -1 \end{bmatrix}$$

Then, we can achieve the CNOT by the sequence

$$U_{CNOT} = \begin{bmatrix} 1 & 0 & 0 & 0 \\ 0 & 1 & 0 & 0 \\ 0 & 0 & 0 & 1 \\ 0 & 0 & 1 & 0 \end{bmatrix} = \frac{1}{\sqrt{2}} \begin{bmatrix} 1 & 1 & 0 & 0 \\ 1 & -1 & 0 & 0 \\ 0 & 0 & 1 & 1 \\ 0 & 0 & 1 & -1 \end{bmatrix} \begin{bmatrix} 1 & 0 & 0 & 0 \\ 0 & 1 & 0 & 0 \\ 0 & 0 & 1 & 0 \\ 0 & 0 & 0 & -1 \end{bmatrix} \frac{1}{\sqrt{2}} \begin{bmatrix} 1 & 1 & 0 & 0 \\ 1 & -1 & 0 & 0 \\ 0 & 0 & 1 & 1 \\ 0 & 0 & 1 & -1 \end{bmatrix} = (I \otimes H)U_{CZ}(I \otimes H)$$

6.4.5 *Spin Qubits*

In spin qubit devices we can make use of microwave drives to implement quantum gates, similarly to how we use them with superconducting qubit devices [23]. RF pulses at the qubits' resonant frequencies in the few GHz range are used to control them. These pulses are sent in the form of electrical voltage signals on top gates or barriers which are in close proximity to the physical quantum dots, so that the resulting electric field can penetrate into the heterostructure in which the charge carrier of the qubit is trapped. The interaction of the spin with this electromagnetic field is the Spin Orbit Interaction (SOI). A stronger SOI allows for faster qubit driving, but also means that the qubit will be more sensitive to electromagnetic noise. Specific qubits can be targeted by driving at their unique qubit resonance frequencies. This technique is called resonant driving, or Electric Dipole Spin Resonance (EDSR).

The electric spin resonance model treats the AC field as a point $B_{ac}\cos(\omega t + \phi)x$ [13]. The effective rotating frame Hamiltonian is $H = (g\mu_B B_0 - \hbar\omega)S_z + g\mu_B(B_{ac}/2)S_x$. When the electron or hole spin is driven by a resonant pulse, the S_z term becomes 0 since $\hbar\omega = g\mu_B B_0$. The spin will oscillate at the Rabi frequency $\frac{g\mu_B B_{ac}}{2h}$.

An alternative approach to spin qubit gates using "hopping" between dots was originally proposed by Loss and DiVincenzo [24]. This approach has recently been demonstrated, for example, in a ten qubit hole-spin processor [2]. Instead of involving microwave pulses, single-qubit gates involve shuttling of charge carriers between quantum dots with different local g-factors. Due to the different g-factors between two dots, the spin precession in each site will be about a different quantization axis. For example, holes in planar Ge heterostructures have out-of-plane g-factors on the order of 100 times larger than the in-plane component [2]. This means even a small tilt of B from the in-plane g-tensor strongly re-orients the quantization axis in the out-of-plane direction. By simply allowing the spin to precess in each site for a calculated amount of time, we gain two-axis control of the qubit. As we saw earlier, this is sufficient to implement any single-qubit rotation. This remains feasible at small magnetic fields which are desirable because in this regime quantum coherence is improved.

The authors of [2] point out that this approach "vastly improves" on prior demonstrations of universal qubit control in terms of power dissipation. They see a 20x reduction from the dipole spin resonance technique. Increased orthogonality of the quantization axes would further reduce dissipated power. However, g-factors are hard to control and are sensitive to the electrostatic and strain environments. So, it is conceivable that finer control of the g-factor will be an area worth exploring.

Two-qubit qubit gates that we can implement simply by allowing the Heisenberg exchange Hamiltonian to evolve are called native gates. Consider the following realization of the SWAP gate.

We begin with the Heisenberg exchange Hamiltonian which describes the interaction of two neighboring spin qubits. This is a more specific case of the more general one-dimensional Heisenberg chain model. This model assumes that interactions only occur between adjacent sites, for example on a one-dimensional ring of spins. The coupling causes these nearest neighbors to have their energies minimized when they are aligned [25].

$$H = J(\sigma_x^{(2)} \otimes \sigma_x^{(1)} + \sigma_y^{(2)} \otimes \sigma_y^{(1)} + \sigma_z^{(2)} \otimes \sigma_z^{(1)})$$

As we learned before, the propagator of the state is found by exponentiating the Hamiltonian.

$$U(t_2 - t_1) = \exp(-iH(t_2 - t_1)/\hbar)$$

Example: SWAP Gate

We can solve for the parameters t_2 and t_1 needed to make this propagator equal the SWAP gate by comparing it to the SWAP matrix, which we know is:

$$SWAP = \begin{bmatrix} 1\,0\,0\,0 \\ 0\,0\,1\,0 \\ 0\,1\,0\,0 \\ 0\,0\,0\,1 \end{bmatrix}$$

$$\exp(-iH(t_2 - t_1)/\hbar) = \exp(-iJ(t_2 - t_1)\sigma_x^{(2)} \otimes \sigma_x^{(1)}/\hbar)\exp(-iJ(t_2 - t_1)\sigma_y^{(2)} \otimes \sigma_y^{(1)}/\hbar)$$

$$\exp(-iJ(t_2 - t_1)\sigma_z^{(2)} \otimes \sigma_z^{(1)}/\hbar)$$

$$= (\cos(J(t_2 - t_1)/\hbar)I - i\sin(J(t_2 - t_1)/\hbar)\sigma_x^{(2)} \otimes \sigma_x^{(1)})$$

$$(\cos(J(t_2 - t_1)/\hbar)I - i\sin(J(t_2 - t_1)/\hbar)\sigma_y^{(2)} \otimes \sigma_y^{(1)})$$

$$(\cos(J(t_2 - t_1)/\hbar)I - i\sin(J(t_2 - t_1)/\hbar)\sigma_z^{(2)} \otimes \sigma_z^{(1)})$$

$$= (\cos(J(t_2 - t_1)/\hbar) - i\sin(J(t_2 - t_1)/\hbar))(|00\rangle\langle 00| + |11\rangle\langle 11|)$$

$$+ (\cos(J(t_2 - t_1)/\hbar) + i\sin(J(t_2 - t_1)/\hbar))$$

$$(\cos^2(J(t_2 - t_1)/\hbar) - sin^2(J(t_2 - t_1)/\hbar))(|01\rangle\langle 01| + |10\rangle\langle 10|)$$

$$- 2i\sin(J(t_2 - t_1)/\hbar)\cos(J(t_2 - t_1)/\hbar)$$

$$(\cos(J(t_2 - t_1)/\hbar) + i\sin(J(t_2 - t_1)/\hbar))(|01\rangle\langle 10| + |10\rangle\langle 01|)$$

For this operator to equal the SWAP operator, we set:

$$\cos^2(J(t_2 - t_1)/\hbar) - sin^2(J(t_2 - t_1)/\hbar) = 0$$

By the double angle formula:

$$\begin{aligned}
\cos^2(J(t_2 - t_1)/\hbar) - sin^2(J(t_2 - t_1)/\hbar) &= \cos(2\,J(t_2 - t_1)/\hbar) \\
\cos(2\,J(t_2 - t_1)/\hbar) &= 0 \\
2\,J(t_2 - t_1)/\hbar &= \pi n + \pi/2 \\
(t_2 - t_1) &= \frac{\hbar\pi n}{2J} + \frac{\pi\hbar}{4J}
\end{aligned}$$

If we let $n = 0$ then we can see that our operator implements a SWAP up to a global phase by inspecting the other nonzero terms of our operator matrix.

$$- 2i\sin(J(t_2 - t_1)/\hbar)\cos(J(t_2 - t_1)/\hbar)(\cos(J(t_2 - t_1)/\hbar) + i\sin(J(t_2 - t_1)/\hbar))$$

$$= -2i\sin(\pi/4)\cos(\pi/4)(\cos(\pi/4) + i\sin(\pi/4))$$

$$= \frac{1-i}{\sqrt{2}}$$

$$(\cos(J(t_2 - t_1)/\hbar) - i\sin(J(t_2 - t_1)/\hbar))$$

$$= (\cos(\pi/4 + n\pi/2) - i\sin(\pi/4 + n\pi/2))$$

$$= (\cos(\pi/4) - i\sin(\pi/4))$$

$$= \frac{1-i}{\sqrt{2}}$$

Arbitrary single-qubit rotations along with at least one two-qubit gate are sufficient for universal quantum control, and the native SWAP gate is a natural choice for the two-qubit gate. However, we can also implement other two-qubit gates by driving and detuning.

When it comes to two-qubit gates that require more control than simply turning the exchange on and off again, we drive transitions in the level diagram of the many-body quantum system. This is as opposed to resonantly driving individual qubits' state's transitions. We are interested for example in how the exchange interaction between dots creates splittings of our resonance lines in a many-body energy diagram. Hendrick et al. [26] showed that J/h is tunable from $35-65$ MHz for neighboring Germanium QDs in their hole-spin processor with one charge loaded in each site. [16] reported a similar range of $8-54$ [27].

A general controlled rotation CROT gate is enabled by this kind of exchange interaction J between neighboring quantum dots which causes resonance lines to split into two [28]. The resonance frequency of the target qubit depends on the state of the control qubit. The CNOT, CZ, etc., are special cases of the more general CROT gate.

We can define rotation about the x axis by $R_x(\theta)$.

$$\begin{aligned} R_x(\theta) &= \exp(-i\frac{\theta}{2}\sigma_z) \\ &= \begin{bmatrix} \cos(\frac{\theta}{2}) & -j\sin\left(\frac{\theta}{2}\right) \\ -j\sin\left(\frac{\theta}{2}\right) & \cos\left(frac\theta 2\right) \end{bmatrix} \end{aligned}$$

A controlled rotation about an axis w is given as a block matrix.

$$CROT = \begin{bmatrix} I & \mathbf{0} \\ \mathbf{0} & R_w \end{bmatrix} \tag{6.1}$$

The CNOT gate is a particular instance of this CROT where $R_w = R_x(\pi/2) = \sigma_x$.

$$\text{CNOT} = \begin{bmatrix} I & \mathbf{0} \\ \mathbf{0} & \sigma_x \end{bmatrix} \tag{6.2}$$

Hendrick et al. [26] for example implements a two-qubit CNOT gate for two hole-spin Ge QDs. The Pauli spin blockade suppressing transport current for a particular bias voltage $V = 1mV$ is used to initialize the two-qubit system to the $|\downarrow\downarrow\rangle$ state.

The authors solve for an exchange coupling $J = 4Ut_{12}^2/(U^2-(\alpha\varepsilon-U_0)^2)$ where U is the charging energy of the QDs, $\alpha = 0.23$ is the lever arm of the two gates and $t_{12}/h = 1.8$ GHz is the interdot tunnel coupling which can be tuned via a central barrier.

By an applied voltage to the central barrier between the two QDs, the exchange interaction is set to $J/h = 39$ MHz at the symmetry point (1, 1) with one equal charge in each site. This regime enables fast selective driving.

Full control of the two-qubit Hilbert space is attained by applying microwave pulses at the four resonant frequencies to drive transitions between the $|\uparrow\uparrow\rangle$, $|\uparrow\downarrow\rangle$, $|\downarrow\uparrow\rangle$ and $|\downarrow\downarrow\rangle$ states. Because the splitting depends on the states of both qubits, a pulse at a single resonant frequency can cause a conditional rotation of the target qubit. The problem of accidentally driving transitions besides the desired transition slightly off-resonantly is combated by choosing driving speeds so that the time of the resonant π pulse is equal to the time of an off-resonant 4π pulse (two complete rotations). The authors find that they can drive a fast CX operation with times $T_{CX,QD1} = 55$ ns and $T_{CX,QD2} = 75$ ns targeting each qubit respectively.

The authors find that there is a small shift in the resonant frequencies of the qubits in response to the pulse control. In order to rectify this, they apply phase corrections to all following pulses. They observe that the phase rotations have a 4π periodicity and suggest an echoing pulse correcting residual environment coupling.

Other gates such as CPHASE [28], CX [26, 28], CZ [28], CS^{-1} [29], SWAP [29], i-Toffoli [28] and Toffoli-like [29] gates have also been realized using the same hole-spin Ge-based platform as this CROT, and we invite the reader to investigate these. Even more (too many to list) demonstrations have been made on Silicon and Gallium Arsenide platforms but a good review article can be found in [13].

When working with a qubit encoding other than the standard Loss–DiVincenzo qubit, we find slight variations on the approaches to single- and two-qubit gates. For example, when operating a singlet-triplet qubit, we use the exchange to perform single-qubit gates, and an Ising interaction between coupled Singlet-Triplet qubits can be used to implement a CZ gate; one can use the exchange interaction to drive single-qubit operations for EO qubits, and capacitive charge-coupling in the high detuning regime can be used to perform two-qubit gates on EO qubits [13].

6.4.6 *Ion Trap*

In ion trap QC, we use Raman[3] transitions involving two lasers with frequencies ω_1, ω_2 where $\omega_1 - \omega_2 = \omega_q$ to drive our qubits. Applying a field tuned to ω_q activates the following Hamiltonian term [11, p. 319].

$$H_I = \frac{\hbar\Omega}{2}(S_+\exp(i\phi) + S_-\exp(-i\phi))$$

This helps us to achieve x rotations $R_x(\theta) = \exp(-i\theta S_x)$ and y rotations $R_y(\theta) = \exp(-i\theta S_y)$ by timing the pulse durations and selecting the appropriate value of ϕ. Any two-axis control is sufficient for universal quantum computation. However, it is relatively inefficient to control a qubit by two axes that have a small angular offset from one another. Instead, qubit operation is most efficient when the two axes of

[3] C.V. Raman was awarded the Nobel prize in physics in 1930 for his discovery of Raman scattering [30].

rotation are orthogonal, so that we spend the minimum amount of time waiting for rotation during each gate.

The first proposal for an ion trap gate was the Cirac–Zoller gate in 1995 [31], which was largely responsible for beginning the field of ion trap quantum computing. The Cirac–Zoller gate implements the controlled phase CZ gate. It's just a fun coincidence that CZ looks like an abbreviation for Cirac–Zoller!

The first step in the implementation of the Cirac–Zoller (CZ) gate is to apply a π pulse on the red sideband of the first ion in order to map its state to the ions' motion. The π pulse on the red sideband maps the excited state of the first ion $|e_0\rangle$ to the motional state by introducing a phase, sign and a bit flip on the motional state. The e_0 state also relaxes to g_0.

$$\begin{aligned}
|g_0g_10\rangle &\xrightarrow{\pi_0} |g_0g_10\rangle \\
|e_0g_10\rangle &\xrightarrow{\pi_0} -i\,|g_0g_11\rangle \\
|g_0e_10\rangle &\xrightarrow{\pi_0} |g_0e_10\rangle \\
|e_0e_10\rangle &\xrightarrow{\pi_0} -i\,|g_0e_11\rangle
\end{aligned}$$

A 2π pulse on the red sideband of a second ion then yields a conditional phase. The Cirac–Zoller gate uses an auxiliary non-qubit state. The $2\pi_1$ pulse changes the sign of the $-i\,|g_0g_11\rangle$ state without affecting any change on the other three states by rotating through the auxiliary state. A control laser is tuned to $\omega_{aux} + \omega_z$ where ω_z is the frequency one would use to drive a transition between $|e_1\rangle$ and $|g_1\rangle$, and ω_{aux} is an additional offset that uniquely enable driving between the target state and the auxiliary state $|a\rangle$ [31] [11, p. 320]. In the space spanned by the auxiliary state and the target state, the $2\pi_1$ pulse is a $R_x(2\pi)$ rotation which implements a phase flip.

$$\begin{aligned}
|g_0g_10\rangle &\xrightarrow{2\pi_1} |g_0g_10\rangle \\
-i\,|g_0g_11\rangle &\xrightarrow{2\pi_1} i\,|g_0g_11\rangle \\
|g_0e_10\rangle &\xrightarrow{2\pi_1} |g_0e_10\rangle \\
-i\,|g_0e_11\rangle &\xrightarrow{2\pi_1} -i\,|g_0e_11\rangle
\end{aligned}$$

Another π pulse on the red sideband of the first ion restores the ion and disentangles the motional bus.

$$\begin{aligned}
|g_0g_10\rangle &\xrightarrow{\pi_0} |g_0g_10\rangle \\
-i\,|g_0g_11\rangle &\xrightarrow{\pi_0} |e_0g_10\rangle \\
|g_0e_10\rangle &\xrightarrow{\pi_0} |g_0e_10\rangle \\
-i\,|g_0e_11\rangle &\xrightarrow{\pi_0} -\,|e_0e_10\rangle
\end{aligned}$$

You can see how the phase introduced on the $|e_0e_10\rangle$ state is exactly what we would expect from applying a CZ gate to a two-qubit system in theory. Consider the following CZ applied to an equal superposition of four states.

$$\begin{aligned}
&CZ\frac{1}{\sqrt{4}}(|00\rangle + |01\rangle + |10\rangle + |11\rangle) \\
&= \frac{1}{\sqrt{4}}\begin{bmatrix} 1 & 0 & 0 & 0 \\ 0 & 1 & 0 & 0 \\ 0 & 0 & 1 & 0 \\ 0 & 0 & 0 & -1 \end{bmatrix}\begin{bmatrix} 1 \\ 1 \\ 1 \\ 1 \end{bmatrix} \\
&= \frac{1}{\sqrt{4}}\begin{bmatrix} 1 \\ 1 \\ 1 \\ -1 \end{bmatrix} \\
&= \frac{1}{\sqrt{4}}(|00\rangle + |01\rangle + |10\rangle - |11\rangle)
\end{aligned}$$

We can also implement a SWAP-like operation using a laser at frequency $\omega_q - \omega_z$. We choose the phase such that $R_y(\pi)$ is performed on the subspace spanned by the ion and motional states $|10\rangle$, $|01\rangle$ [11, p. 320]. A CNOT is achieved by $CNOT_{ij} = H_j\overline{SWAP}_jC_i(Z)SWAP_jH_j$ [11, p. 321].

6.5 Readout

In the previous chapter we discussed the great lengths to which we go to keep our qubit from interacting with the environment and decohering. However, if the qubit is forever completely isolated, we cannot learn its state, or the outcome of any quantum computation. This clearly won't do!

> **What is readout in quantum computing?**
> Each quantum computing implementation involves a readout mechanism. This readout mechanism couples the qubit to a part of the environment, for a limited time, so that we may learn about it.

There are strategies to keep the signals by which we couple our qubits to the outside world from causing too much mayhem, such as the attenuators which absorb the room-temperature thermal noise before it can get to our sample in the case of superconducting and spin qubits. These strategies have been learned over decades and continue to be improved upon.

When performing microwave field detection, it is useful to think of the quantum state in phase-space in a similar way to what we saw with regards to photonic quantum computing and continuous variables. The Wigner function is related, as we know, to the quadratures' measurement outcome probability distribution. The better separated two phase-space signals are (with their uncertainties due to vacuum noise, etc.), the less overlap there will be between them and the less error we will encounter in distinguishing them.

A typical setup for microwave field detection is described in [19]. It involves circuit components at several different temperature stages which include a stage at base temperature (10mK), a stage at 4K as well as one at room temperature (300K). A microwave source provides an RF signal that is attenuated before interacting with the resonator. The output from the resonator is amplified by a quantum limited amplifier such as [32] and then by a High Electron Mobility Transistor (HEMT) amplifier. The signal is down-converted using a mixer and local oscillator before being finally digitized. A circulator is also often introduced before the amplification of the signal to combat noise.

After the two-stage amplification, we down-convert the signal into a frequency range that can be readily sampled by a digitizer.

This type of microwave field detection is relevant for qubits of multiple types (superconducting, spin qubit) that are operated in the microwave regime. Other types of quantum computers' readout mechanisms involve other, non-overlapping details which we will go into in the following sections.

6.5.1 Superconducting

There are several approaches to superconducting qubit readout but the most common is dispersive qubit readout [19] [20, p. 202]. In the dispersive regime, we have a Transmon Hamiltonian:

$$H_{\text{disp}} \doteq \hbar(\omega_r + \chi\sigma_z)a^\dagger a + \frac{\hbar\omega_q}{2}\sigma_z$$

Here, $\chi = \frac{g^2}{\Delta}$.

The dispersive regime describes a Transmon capacitively coupled to another (cavity) resonator. The detuning between the two is large with respect to their coupling $\frac{|\Delta|}{|g|} >> 1$ [20, p. 195]. This model neglects nonlinearity in the cavity that is introduced by the Transmon [19]. The resonator frequency is dependent on the qubit state [19]. Specifically, the qubit state corresponds to the following expectations and resonator frequencies (Table 6.1).

The (coherent) state of the cavity resonator can therefore be used to determine the spin of the qubit. Moreover, measuring the qubit state this way is a quantum-non-demolition (QND) measurement. We can interpret the cavity resonator as a qubit measurement apparatus. The Hamiltonian of the apparatus is $\hbar(\omega_r + \chi\sigma_z)a^\dagger a$ while the qubit Hamiltonian is $\frac{\hbar\omega_q}{2}\sigma_z$ and we note that $[\hbar(\omega_r + \chi\sigma_z)a^\dagger a, \frac{\hbar\omega_q}{2}\sigma_z] = 0$. Since

Table 6.1 State-dependent resonator frequency and expectation

Qubit state	$\langle \sigma_z \rangle$	Resonator frequency
$\vert g\rangle$	-1	$\omega_r - \chi$
$\vert e\rangle$	1	$\omega_r + \chi$

these commute, we can find an eigenbasis for both the qubit and the measurement apparatus. We can therefore readout the qubit in a QND way.

Practically, we can measure the qubit state in the following ways [19]. A $\omega_r \pm \chi$ frequency tone on the resonator is transmitted if the qubit is in state $|g\rangle$ but is reflected if it is in state $|e\rangle$. We can alternatively drive the resonator at ω_r in order to observe an accumulation of phase on the transmitted signal that is dependent on the state of the qubit.

For other (less prevalent) approaches to superconducting qubit readout we refer the reader to [19, p. 33].

6.5.2 *Photonic*

In photonic quantum computers, we can make use of Single Photon Detectors (SPDs) to perform readout. These can be used in different measurement schemes such as homodyne and heterodyne detection (Fig. 6.8).

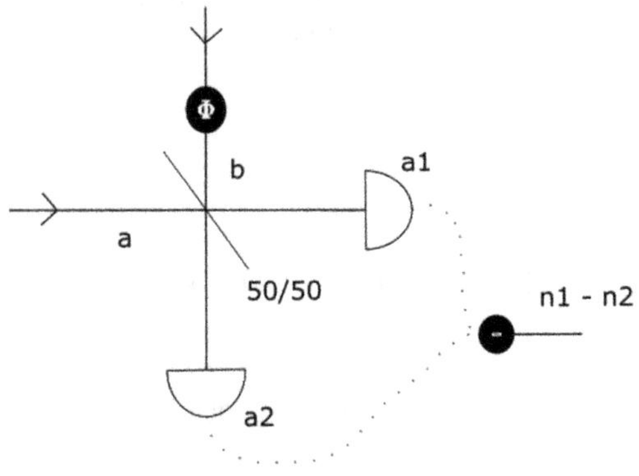

Fig. 6.8 Homodyne detection

$$x_\phi = \frac{1}{\sqrt{2}}(\exp(-i\phi)a + \exp(i\phi)a^\dagger)$$

$$x_0 = \frac{1}{\sqrt{2}}(a + a^\dagger) = x$$

$$x_{\pi/2} = \frac{1}{\sqrt{2}}(a - a^\dagger) = p$$

$$[x, p] = 1$$

In this picture of homodyne detection, the two photodiodes do photon counting. These gives us two photon numbers $n_1 = a_1^\dagger a_1$ and $n_2 = a_2^\dagger a_2$. By taking the difference of the voltages from the two SPDs, we learn about the state.

$$\begin{aligned}
n_1 - n_2 &= a_1^\dagger a_1 - a_2^\dagger a_2 \\
&= \frac{1}{\sqrt{2}}(a^\dagger + \exp(-i\phi^2)b^\dagger)\frac{1}{\sqrt{2}}(a + \exp(i\phi)b) \\
&\quad - \frac{1}{\sqrt{2}}(a^\dagger - \exp(-i\phi)b^\dagger)\frac{1}{\sqrt{2}}(a - \exp(i\phi)b) \\
&= b^\dagger \exp(-i\phi)a + b\exp(i\phi)a^\dagger
\end{aligned}$$

To the phase shifter we can input a local oscillator (LO) field, and then the setup helps us to measure the signal field a. The LO field should be more intense than the signal. The more perfect the mode overlap between the LO and the signal input, the better in terms of noise. We can measure a vacuum state with the setup. The signal field may be a squeezed vacuum $S(\varepsilon)|0\rangle$ instead of vacuum $|0\rangle$ in which case we could expect to see as much as 6 dBm lower noise power. We might have noise power in the range of -60 to -75 dBm, for example [33].

Say we had a signal field $|\psi\rangle$ and LO $|\beta\rangle$. Then,

$$\begin{aligned}
&b|\beta\rangle = \beta|\beta\rangle \\
&\langle\beta| b^\dagger = \beta^* |\beta\rangle \\
&\text{Let } \beta \in \mathcal{R}. \\
&\langle\psi|\langle\beta|(n_1 - n_2)|\beta\rangle|\psi\rangle \\
&= \exp(-i\phi)\langle\beta| b^\dagger |\beta\rangle \langle\psi| a |\psi\rangle + \exp(i\phi)\langle\beta| b |\beta\rangle \langle\psi| a^\dagger |\psi\rangle \\
&= \sqrt{2}\beta \langle\psi| \frac{1}{\sqrt{2}}(\exp(-i\phi)a + \exp(i\phi)a^\dagger)|\psi\rangle \\
&= \sqrt{2}\beta \langle\psi| x_\phi |\psi\rangle
\end{aligned}$$

The Positive Operator Valued Measure (POVM) corresponding to a heterodyne measurement is a projection onto coherent states (Fig 6.9).

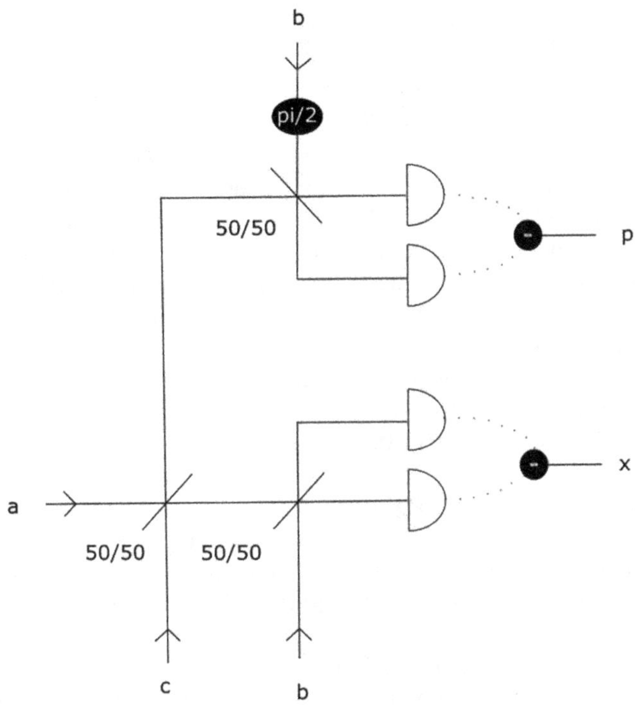

Fig. 6.9 Heterodyne detection

$$F_{x,p} = \frac{1}{\pi} |\alpha\rangle \langle\alpha|$$
$$\alpha = \sqrt{2}(x + ip)$$
$$\frac{1}{\pi} \int |\alpha\rangle \langle\alpha| \, d^2\alpha = 1$$
$$Tr[\varrho F_x \varrho] = Q(\alpha)$$

Here $Q(\alpha)$ is a quasiprobability. It can be mapped reversibly from $Tr[\varrho F_x \varrho]$. To operate our setup, we input $|\beta\rangle$ as b and $|\psi\rangle$ as field a. In mode c we input the vacuum state.

6.5.3 *Annealers*

D-Wave's fifth-generation annealers have 5000 qubits. Needless to say, the interconnect problem would be a very real roadblock for the company if there were not ways to readout these qubits with few lines. The more lines running in and out of a cryostat, the greater the heat load and noise. D-Wave uses a specific technique so that they can readout their qubits with only two lines which we will briefly cover here [34].

The qubits themselves sit at around 10 milliKelvin while the measurement electronics will sit at room temperature. The qubit needs to be coupled via some sensor to a transmission line that will carry the signal out to the room-temperature electronics. To achieve this, the authors of [34] show a sort of quantum analogue of a shift register. Recall that when we have a flux qubit, the height of the barrier between our two wells is controllable by a parameter $\beta_L = \frac{E_l}{E_J}$ the ratio of the energies of our potential energy terms. The asymmetry of the wells is controllable by the external flux ϕ_x. In a similar way to how a classical shift register is composed of an array of connected latches which latch their state for a time before it is propagated to the next latch, a SQUID can be made to "latch" a flux state by increasing the height of the barrier between the wells. Recall that a qubit is usually formed by treating a low-energy state in the left well as $|0\rangle$ and a low-energy state in the right well as $|1\rangle$ or vice versa.

A Quantum Flux Parametron (QFP) circuit, which is a SQUID with controllable well asymmetry and barrier height, can be designed such that the flux from a qubit induces a current in it. We can then latch the state induced by the qubit's flux programmatically in the QFP. Coupling this QFP to another, and another, ...in a chain allows us to propagate the state across a chip.

When it comes to the sensor itself, a naive approach would be to measure the voltage across a SQUID that experiences a flux from the qubit and a bias near the critical current. The SQUID's critical current would depend in this case on the flux state. The downside of this is that it requires a readout line for every sensor / qubit. Instead, we can measure inductance using an LC resonator or "tank circuit" at the end of a chain of QFPs. The tank is coupled to the QFP chain by a pair of SQUIDs called the tune and sense SQUIDs which allow us to get the right inductance setpoint.

6.5.4 NMR

NMR differs from the other quantum information processor implementation methodologies in that we deal with a bulk of qubits which exist in the multitude of molecules in our molecular bath. Instead of working with one electron or one photon, we are working with many nuclei. However, we treat each molecule as an identical system and what we effectively achieve is many instances of our experiment, identically implemented in each molecule. This neither helps or hurts us when it comes to qubit control. However, when it comes to readout it actually helps us.

The signal from a single spin is far too minuscule to measure. We measure instead the expectation value of a larger sample. In particular, around 10^8 molecules are usually required to create a measurable signal [11, p. 324] Considering the Hamiltonian H for a qubit, which will be the same for each corresponding qubit in all of our identical molecules, we have that $\varrho(t) = \exp(-iHt)\varrho(0)\exp(iHt)$. The signal is then an expectation such as $< M_+(t) >= Tr(\varrho(t)(\sigma_+^{(1)}\sigma_+^{(2)}\dots))$.

We measure this by resonantly tuning a coil in the frequency of the qubit(s) we want to measure. Spin that is precessing in XY will produce an oscillating field in the corresponding planar coil.

6.5.5 Spin Qubits

When it comes to quantum dots, we generally use spin-to-charge conversion to measure our qubits. Two approaches can be taken to spin-to-charge conversion. Either we can tunnel our charge carrier to a reservoir, in which case our readout mechanism is destructive, or we can tunnel between dots and use a Pauli Spin Blockade effect. Furthermore, we can make use of different circuits to detect this charge movement. A Single Electron Transistor (SET) or a Quantum Point Contact (QPC) can be used to perform resistive or capacitive measurements.

A QPC measures the change in current due to a charge transition induced via the capacitive coupling between the QPC and the dot. A SET can be used similarly but in the case of the SET it is the varying resistance across the SET that is seen by our readout signal.

We can also use a plunger or barrier gate to do gate-based dispersive reflectometry. Dispersive readout refers to the mechanism of induction used to sense a charge transition. Gate-based dispersive readout uses a small gate to do this.

The case where we tunnel to a reservoir is called Elzerman's readout since it was first demonstrated by Elzerman in 2004 [14]. In this scheme, the energy levels of the spin up and down are aligned with the reservoir such that only one state can tunnel out of the dot.

Pauli spin blockade measurement requires two quantum dots and charge carriers as it is based on the Pauli exclusion principle between two particles. At low magnetic field, Pauli spin blockade measurement is most relevant [35]. We achieve a spin-selective charge transition by slowly varying the detuning between dots. The S(1,1) state will become the S(0,2) state but the T $|\uparrow\uparrow\rangle$ and $|\downarrow\downarrow\rangle$ states will not undergo charge transitions. The ancilla qubit needs to be in a known state and therefore needs to be reset between measurements. It should also be reset to combat decoherence.

Latched readout is a technique that requires a reservoir but results in more signal (time, magnitude).

Whether we are doing gate-based dispersive readout or using an SET to detect charge configuration, we will use RF reflectometry. When using a SET, for example, this involves creating an RF measurement pulse which is reflected off of a circuit that contains our SET [36]. The circuit has three main components. The first is a bias tee, which is a resistor in parallel with a capacitor. This allows us to mix AC and DC signals. The DC component is used to bias our SET into a regime where it is maximally sensitive to the charge motion between dots. After the bias tee, our signal encounters a "tank" LC circuit. This tank circuit is an impedance matching circuit which gives us a matching condition.

$$\Gamma = \frac{Z_L - Z_0}{Z_L + Z_0}$$

$$Z_L = 2\pi f L i + \frac{1}{\frac{1}{R_{SD}} + 2\pi f C i}$$

We have 100% transmission when $\Gamma = 0$, i.e. when $Z_L = Z_0$. Here Z_0 is the input impedance (or output impedance of the RF signal generator). R_{SD} is the resistance of the SET. We obtain the matching condition when $R_{SD} = \frac{L}{CZ_0}$. We have a resonant frequency $f_{res} = \frac{1}{2\pi\sqrt{LC}}$.

The intuitive idea here is that if we tune the SET onto the rising edge of a Coulomb peak, where the resistance has a large rate of change, then the change in charge configuration due to a tunneling event will change the charge occupation of the SET as well. The resistance of the SET comes from the Coulomb blockade effect which scales in n^2 with n charge carriers occupying the island. Therefore a tunneling will be affect the matching condition in the RFR setup. We can see a resonance in the phase and separately in the amplitude of the AC signal reflected off of the bias tee + tank circuit + SET as a dip in the power near the resonance frequency and as a change in the period of the phase with respect to the frequency, respectively. The height of the dip and/or the period of the phase will vary with the occupation dependent R_{SD}. Often, one of these signals has a higher Signal-to-Noise Ratio (SNR) than the other. It is sufficient to use one of these signals to perform RF Reflectometry.

6.5.6 Ion Trap

Readout is done by what is called "shelving" and "fluorescent readout". When a laser is shone on an ion, the state may be excited. This will depend on whether the laser beam is tuned to a transition from the current state to a higher energy state. This means that the excitation is dependent on the initial state. If the initial state is conducive and an excitation does occur, this excited state only has a short lifetime. When the state relaxes again, the gained energy will be released and a photon will be emitted. A photomultiplier tube with around 30% quantum efficiency can collect around 30 of these photons per second. A $|1\rangle$ state is labeled "dark" and a $|0\rangle$ state is "bright", for example, where cyclic fluorescence would indicate repeated excitation to state $|s\rangle$ and relaxation to the initial state $|0\rangle$ [20, p. 153–154]. Shelving refers to the technique of transferring the population of one (qubit) state to another (ancillary) state, which is more conducive to fluorescent readout due to its more favorable energy separation from the other computational qubit state. Shelving is sometimes performed prior to a cyclic fluorescent readout.

References

1. DiVincenzo, D.P.: The physical implementation of quantum computation. Fortschr. Phys. **48**(9–11), 771–783 (2000). https://doi.org/10.1002/1521-3978(200009)48:9/11<771::aid-prop771>3.0.co;2-e
2. Wang, C.A., John, V., Tidjani, H., Yu, C.X., Ivlev, A., Déprez, C., van Riggelen-Doelman, F., Woods, B.D., Hendrickx, N.W., Lawrie, W.I.L., Stehouwer, L.E.A., Oosterhout, S., Sammak, A., Friesen, M., Scappucci, G., de Snoo, S.L., Rimbach-Russ, M., Borsoi, F., Veldhorst, M.: Operating semiconductor quantum processors with hopping spins (2024). http://arxiv.org/abs/2402.18382. ArXiv:2402.18382 [cond-mat]
3. Isidor Isaac Rabi—Facts. URL https://www.nobelprize.org/prizes/physics/1944/rabi/facts/
4. Bardin, J.C., Sank, D., Naaman, O., Jeffrey, E.: Quantum computing: an introduction for microwave engineers. IEEE Microwave Magaz. **21**(8), 24–44 (2020)
5. Magnard, P., Kurpiers, P., Royer, B., Walter, T., Besse, J.C., Gasparinetti, S., Pechal, M., Heinsoo, J., Storz, S., Blais, A., Wallraff, A.: Fast and unconditional all-microwave reset of a superconducting qubit. Phys. Rev. Lett. **121**(6), 060502 (2018). https://doi.org/10.1103/PhysRevLett.121.060502
6. Geerlings, K., Leghtas, Z., Pop, I.M., Shankar, S., Frunzio, L., Schoelkopf, R.J., Mirrahimi, M., Devoret, M.H.: Demonstrating a driven reset protocol for a superconducting qubit. Phys. Rev. Lett. **110**(12), 120501 (2013). https://doi.org/10.1103/PhysRevLett.110.120501
7. Reed, M.D., Johnson, B.R., Houck, A.A., DiCarlo, L., Chow, J.M., Schuster, D.I., Frunzio, L., Schoelkopf, R.J.: Fast reset and suppressing spontaneous emission of a superconducting qubit. Appl. Phys. Lett. **96**(20), 203110 (2010). https://doi.org/10.1063/1.3435463
8. Sete, E.A., Martinis, J.M., Korotkov, A.N.: Quantum theory of a bandpass Purcell filter for qubit readout. Phys. Rev. A **92**(1), 012325 (2015). https://doi.org/10.1103/PhysRevA.92.012325
9. Johnson, J.E., Macklin, C., Slichter, D.H., Vijay, R., Weingarten, E.B., Clarke, J., Siddiqi, I.: Heralded state preparation in a superconducting qubit. Phys. Rev. Lett. **109**(5), 050506 (2012). https://doi.org/10.1103/PhysRevLett.109.050506
10. Ristè, D., van Leeuwen, J.G., Ku, H.S., Lehnert, K.W., DiCarlo, L.: Initialization by measurement of a superconducting quantum bit circuit. Phys. Rev. Lett. **109**(5), 050507 (2012). https://doi.org/10.1103/PhysRevLett.109.050507
11. Nielsen, M.A., Chuang, I.L.: Quantum Computation and Quantum Information 10th, Anniversary Cambridge University Press, Cambridge; New York
12. de Sousa, R.: Quantum devices 1 at university of Victoria (2024). Graduate course
13. Burkard, G., Ladd, T.D., Nichol, J.M., Pan, A., Petta, J.R.: Semiconductor spin qubits. Rev. Mod. Phys. **95**(2), 025003 (2023). https://doi.org/10.1103/RevModPhys.95.025003. ArXiv:2112.08863 [cond-mat, physics:physics, physics:quant-ph]
14. Elzerman, J.M., Hanson, R., Willems van Beveren, L.H., Witkamp, B., Vandersypen, L.M.K., Kouwenhoven, L.P.: Single-shot read-out of an individual electron spin in a quantum dot. Nature **430**(6998), 431–435 (2004). https://doi.org/10.1038/nature02693
15. McJunkin, T.W.: Heterostructure modifications, fabrication improvements, and measurement automation of si/sige quantum dots for quantum computation. Ph.D. thesis, Wisconsin University, Madison (2021)
16. Wang, C.A., Déprez, C., Tidjani, H., Lawrie, W.I.L., Hendrickx, N.W., Sammak, A., Scappucci, G., Veldhorst, M.: Probing resonating valence bonds on a programmable germanium quantum simulator. NPJ Quantum Inf. **9**, 58 (2023). https://doi.org/10.1038/s41534-023-00727-3
17. Eschner, J., Morigi, G., Schmidt-Kaler, F., Blatt, R.: Laser cooling of trappedions. JOSA B **20**(5), 1003–1015 (2003). https://doi.org/10.1364/JOSAB.20.001003
18. Wong, T.: Quantum computing, universal gate sets. https://www.scottaaronson.com/qclec/16.pdf
19. Blais, A., Grimsmo, A.L., Girvin, S.M., Wallraff, A.: Circuit quantum electrodynamics. Rev. Mod. Phys. **93**(2), 025005 (2021). https://doi.org/10.1103/RevModPhys.93.025005. ArXiv:2005.12667 [quant-ph]

20. Majidy, S., Wilson, C., Laflamme, R.: Building Quantum Computers: A Practical Introduction. Cambridge University Press (2024)
21. Bulmer, J.: Threshold detector probabilities of Gaussian states—strawberry fields. https://strawberryfields.ai/photonics/demos/run_threshold_detector_tutorial.html
22. Knill, E., Laflamme, R., Milburn, G.J.: A scheme for efficient quantum computation with linear optics. Nature **409**(6816), 46–52 (2001). https://doi.org/10.1038/35051009
23. Veldhorst, M., Hwang, J.C.C., Yang, C.H., Leenstra, A.W., de Ronde, B., Dehollain, J.P., Muhonen, J.T., Hudson, F.E., Itoh, K.M., Morello, A., Dzurak, A.S.: An addressable quantum dot qubit with fault-tolerant control-fidelity. Nat. Nanotech. **9**(12), 981–985 (2014). https://doi.org/10.1038/nnano.2014.216
24. Loss, D., DiVincenzo, D.P.: Quantum computation with quantum dots. Phys. Rev. A **57**, 120–126 (1998). https://doi.org/10.1103/PhysRevA.57.120
25. Baxter, R.J.: One-dimensional anisotropic Heisenberg chain. Ann. Phys. **70**(2), 323–337 (1972)
26. Hendrickx, N.W., Franke, D.P., Sammak, A., Scappucci, G., Veldhorst, M.: Fast two-qubit logic with holes in germanium. Nature **577**(7791), 487–491,491A–491K (2020). https://doi.org/10.1038/s41586-019-1919-3. https://www.proquest.com/docview/2350114222/abstract/6E1FF2C0C06B489BPQ/1
27. Fang, Y., Philippopoulos, P., Culcer, D., Coish, W.A., Chesi, S.: Recent advances in hole-spin qubits. Mater. Quantum Technol. **3**(1), 012003 (2023). https://doi.org/10.1088/2633-4356/acb87e
28. Hendrickx, N.W., Lawrie, W.I.L., Russ, M., Van riggelen, F., de Snoo, S.L., Schouten, R.N., Sammak, A., Scappucci, G., Veldhorst, M.: A four-qubit germanium quantum processor. Nature **591**(7851), 580–585,585A–585N (2021). https://doi.org/10.1038/s41586-021-03332-6. https://www.proquest.com/docview/2505728932/abstract/F030BA68FB854195PQ/1
29. van Riggelen, F., Lawrie, W.I.L., Russ, M., Hendrickx, N.W., Sammak, A., Rispler, M., Terhal, B.M., Scappucci, G., Veldhorst, M.: Phase flip code with semiconductor spin qubits. NPJ Quantum Inf. **8**(1), 1–7 (2022). https://doi.org/10.1038/s41534-022-00639-8
30. Raman, C.V.Z: Life, Career, Nobel Prize, Biography, and Facts. Britannica (2025). https://www.britannica.com/biography/C-V-Raman
31. Cirac, J.I., Zoller, P.: Quantum computations with cold trapped ions. Phys. Rev. Lett. **74**, 4091–4094 (1995). https://doi.org/10.1103/PhysRevLett.74.4091
32. Khalifa, M., Feldmann, P., Salfi, J.: Kinetic inductance parametric converter. Phys. Rev. Appl. **22**(2), 024025 (2024). https://doi.org/10.1103/PhysRevApplied.22.024025
33. Chelkowski, S.: Squeezed light and laser interferometric gravitational wave detectors. Ph.D. thesis, Hannover U. (2007)
34. Whittaker, J.D., Swenson, L.J., Volkmann, M.H., Spear, P., Altomare, F., Berkley, A.J., Bumble, B., Bunyk, P., Day, P.K., Eom, B.H., Harris, R., Hilton, J.P., Hoskinson, E., Johnson, M.W., Kleinsasser, A., Ladizinsky, E., Lanting, T., Oh, T., Perminov, I., Tolkacheva, E., Yao, J.: A frequency and sensitivity tunable microresonator array for high-speed quantum processor readout. J. Appl. Phys. **119**(1), 014506 (2016). https://doi.org/10.1063/1.4939161
35. Scppucci, G., Veldhorst, M., Greplova, E., Andresen, S., Bermejo, P., Chatterjee, A., Crastina, D., Katsaros, G., Li, R., Palyi, A.: Development and applications of germanium quantum technologies at tu delft (2024). Online course
36. Liu, Y.Y., Philips, S., Orona, L., Samkharadze, N., McJunkin, T., MacQuarrie, E., Eriksson, M., Vandersypen, L., Yacoby, A.: Radio-frequency reflectometry in silicon-based quantum dots. Phys. Rev. Appl. **16**(1), 014057 (2021). https://doi.org/10.1103/PhysRevApplied.16.014057

Chapter 7
Control Hardware

Abstract The previous chapter covered the techniques for operating a quantum processor. However, these techniques must employ certain scientific instruments such as Analog to Digital Converters (ADCs), Digital to Analog Converters (DACs), voltage sources, lock-ins, digitizers, high-frequency and baseband Arbitrary Waveform Generators (AWGs), laser sources and more. A working understanding of these instruments is crucial for any quantum computer engineer. We provide a case study of IonQ's quantum computer control systems near the end of the chapter.

7.1 Learning Objectives

As we delve into the instrumentation that is used to control quantum processors, we will aim to do the following (Fig. 7.1):

- Understand the roles of specialized modules that handle timing and control flow
- Understand the roles of instruments that are used in electronic quantum computing setups
- Understand the roles of instruments that are used in photonic quantum computing setups
- See how these instruments are used in many settings, including in the case of IonQ's ion trap-based quantum computer.

The considerations that one would normally consider a part of a processor architecture bleed into this category of instrumentation or control electronics when we are thinking about quantum computers. This is because a Quantum Processing Unit (QPU) as they've come to be called is actually less analogous to a classical CPU and more analogous to an Arithmetic Logic Unit (ALU) or Graphics Processing Unit (GPU) in its function. It is responsible only for the execution of certain logical operations that amount basically to matrix multiplication.

© The Author(s), under exclusive license to Springer Nature Switzerland AG 2026

M. S. Edwards, *An Introduction to Quantum Computing for Computer Engineers*,
https://doi.org/10.1007/978-3-032-03650-6_7

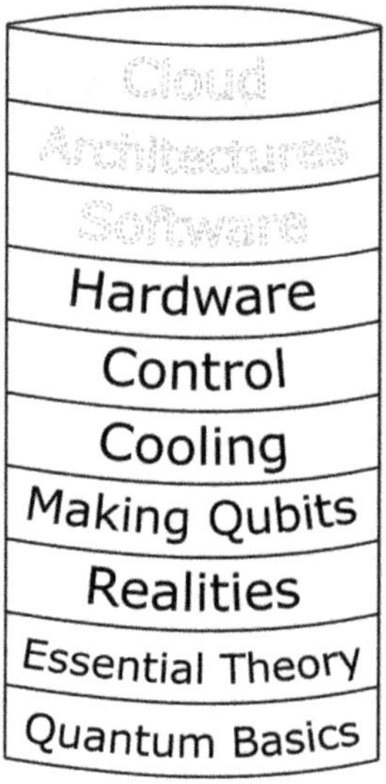

Fig. 7.1 Learning objective for this chapter: hardware

> **What are the roles of specialized QC modules?**
>
> Whereas in a classical computer processor we would have an ALU, registers, caches, buses, a clock, a control unit and memory, in a quantum computer all of these functions are performed outside of our QPU by a combination of specialized instrumentation and software.

There are interesting analogies to be made since elements of the quantum computing stack are directly comparable to those of the classical computer, like the assembly languages such as Quantum Assembly (QASM) [1], Quantum Macro Assembler (QMASM) [2] and Blackbird [3] that have been developed by companies and labs including IBM, Los Alamos National Lab and Xanadu. However, the systems that run these quantum assembly languages are very different from the systems that run X86, ARM or MIPS.

Most experimental quantum computing setups are still driven by the needs of specific experiments which may aim, for example, to probe certain physics. The choice of control hardware therefore does not *yet* conform in general to a standard (though there is some early work toward this kind of standard being done), but instead is flexibly determined by experimental requirements. Experiments are designed so that theories may be tested. Theories can only be tested if their parameters can be varied in isolation. For example, a study on the coherence time of a certain new type of qubit might refer to the results of Hahn echo and dynamical decoupling experiments with similar types of qubits done in the past as controls. The variable changed, then, is something that affects the coupling of the new—perhaps more isolated qubit—

to the environment. To keep everything else the same, we need to replicate the control experiment(s) which will require us to use control pulses from AWGs and high-frequency sources to implement dynamical decoupling sequences, and carefully timed measurements using signal amplifiers, down-converters and digitizers. This is one example of how we might determine what goes into our setup. In another example, maybe we are motivated specifically to improve the setup (and not the qubit) by introducing a new type of amplifier to amplify the readout signal. In this case we might be looking to compare the noise in our setup to the noise in other setups which are otherwise identical. You get the idea.

7.2 The Clock

When it comes to quantum computing timing is extremely important. We want to make the best use of the coherence time. With gate operation times in spin qubits, for example, around 10 nanoseconds [4] it is important that we can generate pulses in less than this.

As we learned earlier, logical gate implementations in multiple types of quantum processors involve allowing a two-qubit Hamiltonian to evolve for a particular amount of time. This interaction can be turned on and turned off, for example, in spin qubits by pulsing a barrier with a voltage signal. At the same time, we will often pulse one or the other gate with a resonant tone to cause a rotation of the qubit state. We use the term "gate" here to refer to a metal top-gate on the QPU itself and not to refer to a unitary. Confusing, I know!

We need to turn the interaction on for just the right amount of time so that the desired operation occurs. We also need to synchronize our pulses to the barrier and to the gate. There are several techniques for doing this, but the CPU clock on a classical control computer just won't cut it.

We need to share timing information between instruments such as the high-frequency RF source and the AWG that might generate our control signals. One approach is to share a reference clock signal between the instruments and another is to send triggers to the instruments. A trigger is a precisely timed digital signal that tells the instrument to begin playing its waveform. It is also possible to share more information between instruments in advanced setups that use, for example, transceiver links. A dedicated piece of hardware called a synchronization module will often be responsible for providing these timing related signals to the various instruments in your setup. Picosecond accuracy (signal jitter fractions of ps) and nanosecond signal resolution are achieved in modern setups, and the feedback loop of measurement to next signal generation required for operations like active qubit reset happen in around a hundred nanoseconds [5].

7.3 Control Flow

In some setups, the readout instrument (digitizer) and the signal generator (AWG / RF source) can be programmed to work together so that as soon as one qubit state is measured, a particular signal will be played. However, this kind of readout $\rightarrow$ signal decision map does not yet address arbitrary control flow.

As we have mentioned, control flow is not handled in any way by a QPU, which is more analogous to a GPU than a CPU. It has no program counter or instruction register. Instead, control flow (the decision of what operation to execute next, etc.) must be handled outside of the QPU. To this end, exciting dedicated hardware has been developed such as the Pulse Processing Unit (PPU) [6], which can execute programs in real-time that include classical conditionals and logic as well as pulse generation and readout sequences. Since the PPU is implemented on an FPGA and sits next to the control electronics, the lengthy and non-deterministic delay of communicating with a CPU is not incurred. We will take a look at how to program such a PPU in the next section which focuses on control software.

7.4 Arbitrary Waveform Generators

An Arbitrary Waveform Generator or AWG is useful for generating pulses used to control your qubit. The key characteristics for an AWG are the bandwidth, usually in MHz, and the sample rate which is usually in Gigasamples per second (GSa/s) [7]. The sample rate determines the bandwidth of the waveform output by the AWG and is limited by the rate at which the Digital to Analog Converter (DAC) can take in samples. The bandwidth refers to the maximum frequency of the output signal from the AWG. It is roughly half the sample rate since the DAC needs to generate at least two data points per period. Due to other inefficiencies, such as the use of a filter to convert the voltage-stepped output of the DAC to a smooth signal, another 10% or so is taken off of the sample rate.

An AWG also has a dynamic range which gives the ratio of highest to lowest voltage signals it can generate. The range of voltages that can be represented is determined by the resolution of its internal binary representation of voltage pre-DAC. The more bits, the higher resolution.

Let's go into a few more characteristics of AWGs before we see how to use one to drive a qubit [8]. AWGs can be programmed to repeat a user-defined waveform multiple times, or to play it only once. In both cases, we can time our waveform using an external or internal trigger. An AWG can change or sweep the output wavelength through a voltage controlled oscillator. Onboard memory with GSa per channel can usually be programmed to contain a waveform so that the waveforms are specified fully prior to being played. Having a dedicated memory onboard for this speeds up playback as no communication is required with a computer, for example, during playback. Though, in order to originally program the memory contents, a connection to a computer over GPIB (an interface bus standard developed in the 1970s) is helpful. An AWG also includes an attenuator. The outputs are usually BNC connectors and should be terminated with $\doteq 50\,\Omega$. Some AWGs are combined with digitizers into one module especially for use cases in quantum computing where, as we already saw, active reset is a necessity. This helps reduce communication and response time.

It should be possible to decompose an arbitrary gateset to AWG pulses using the following relationship [9].

$$U_n \dots U_1 U_0 = T \prod_{n=0}^{n} \exp(-\frac{i}{2}\Theta_n(t)(I_n\sigma_x + Q_n\sigma_y))$$

$$\Theta(t) = -CV \int_0^t s(t')\mathrm{d}t'$$

Here s is a waveform envelope and T ensures that pulses are ordered in time according to the order of the U_n. Θ gives the angle by which a state is rotated. C is a constant that encodes device characteristics and V is the waveform amplitude.

A microwave source provides a high-frequency ω_{LO} that is enveloped by an AWG signal $s(t)$. A combined signal with a frequency $\omega_{\mathrm{LO}} \pm \omega_{\mathrm{AWG}}$ is generally produced by an IQ (in-phase / quadrature) mixer. This combined frequency is resonant with the target qubit. The driving Hamiltonian in the rotating frame is given by the following.

$$H_d = -\frac{C}{2} V s(t) \begin{bmatrix} 0 & \exp(i(\delta\omega t + \phi)) \\ \exp(-i(\delta\omega t + \phi)) & 0 \end{bmatrix}$$

t and V determine the polar angle of the qubit's rotation. The phase offset between sequential RF pulses defines the azimuthal angle between the axes of rotation [10].

Some common pulses used for qubit control include the π pulse which, as the name suggests, rotates the quantum state in the Bloch sphere by π radians. The π pulse is a specific instance of a rotation operator about the z axis R_z.

Example: Rotation about Z Axis

For example,

$$\begin{aligned} R_z(\pi)\,|+\rangle &= \exp\left(-i\frac{\pi}{2}\sigma_z\right)|+\rangle \\ &= \left(\cos\left(\frac{\pi}{2}\right) I - i\ \sin\left(\frac{\pi}{2}\right)\sigma_z\right)|+\rangle \\ &= \sigma_z\,|+\rangle \\ &= |-\rangle \end{aligned}$$

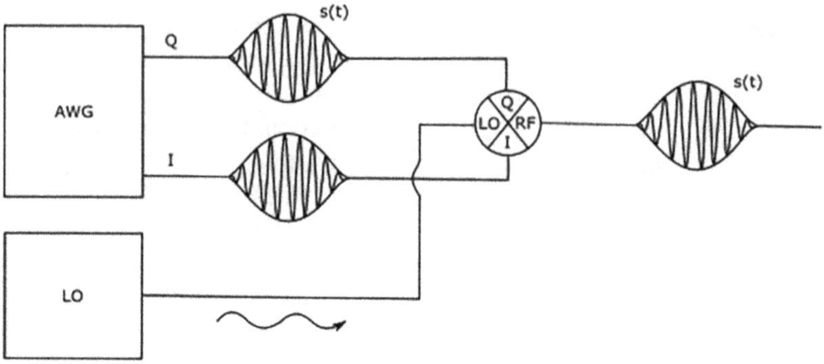

Fig. 7.2 Rabi driving

Similarly, a $\frac{\pi}{2}$ pulse will rotate the state $\frac{\pi}{2}$ radians in the Bloch sphere, and so on (Fig. 7.2).

7.5 Mixers

In the previous section we discussed a setup involving an IQ mixer. Let's briefly talk about IQ mixers and how they work. The basic idea with how we use an IQ mixer is that we can achieve the modulation of our final output signal's amplitude and phase indirectly by modulating just the amplitudes of two input signals, I and Q, which is easier.

The relationship of the output's amplitude to the inputs is given by $\sqrt{I^2 + Q^2}$. The phase of the output is related by $\tan^{-1}(Q/I)$.

An IQ mixer is composed of two hybrid and two balanced mixers [11] and can be used in several configurations. We are mainly interested in the configuration where an I and a Q signal both at frequency ω_{AWG} are mixed with a local oscillator with a high frequency ω_{LO} to produce an output on the RF port. IQ mixers can however be used to demodulate as well, in a reverse configuration to this, converting an RF signal to baseband (Fig. 7.3).

A mixer can also be used to down-convert or up-convert signals to lower or higher frequencies respectively, in exactly the way we saw in the AWG section when two frequencies $\omega_{\text{LO}} \pm \omega_{\text{AWG}}$ were produced by the IQ mixer.

Fig. 7.3 IQ mixer

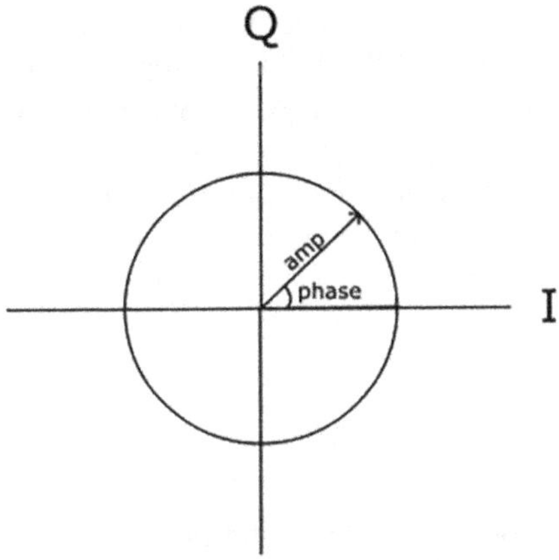

7.6 RF Sources

In the previous sections, we discussed how an AWG and a mixer may be used in conjunction with a local oscillator, a high-frequency RF source, to achieve qubit control. Let's look briefly at the key specifications of RF sources.

RF sources are lower frequency generators than, say, microwave generators. We typically operate a spin qubit at a few GHz, which is at the upper end of the radiofrequency range of a few kHz to a few GHz.

Some key characteristics to consider when selecting an RF source are the following [12]. The output power of an RF source is typically provided in dBm. It is also important to consider the frequency range and modulation bandwidth. We will get into the ways in which RF sources may produce modulated signals in a moment. Phase noise is given in dBc/Hz and specifies the ratio of noise power to carrier power, over a single Hertz bandwidth and at an offset. Spectral purity gives the stability of a signal and is also given in dBc, the ratio of the signal's harmonics with the power of the carrier signal. The sample rate of the generator's DAC will also limit the output frequency and is worth considering.

RF sources come in two varieties [12]. The first is vector signal generators and the second is analog signal generators. A vector signal generator uses digital to analog conversion techniques to generate the output waveform. The "vector" referred to is the vector in the IQ plane that would represent the signal if it were generated from I and Q signals by an IQ mixer. An analog signal generator does not use digital techniques to generate waveforms. Both types of generators are widely available.

For example, the Rhode and Schwarz SGS100A is available as either a vector or a continuous wave source. The SGS100A does not just generate a fixed tone, it can take I and Q signals and convert them to a modulated output RF signal, performing the function of multiple components in the diagram of a qubit control circuit we provided in the AWG section. This was done in a quantum computing setup by Hendrickx et al. in 2021 [13].

However, Direct Digital Synthesis (DDS) is a new option that further eliminates the need for this kind of setup involving multiple modules. A high-frequency signal can be generated by DDS that is precisely modulated by digital electronics and digital to analog conversion, right within an AWG. So, we eliminate most of the elements of our circuit which are subsumed in the AWG with DDS capabilities. Direct digital synthesis involves an *internal* reference oscillator, frequency control register, a numerically controlled oscillator, DAC and finally filtering.

7.7 Digitizers

As opposed to the instruments we have discussed so far which used techniques such as Digital to Analog Conversion (DAC) to generate waveforms, digitizers use Analog to Digital Conversion (ADC) to digitize waveforms.

A key characteristic of a digitizer is its bandwidth. The bandwidth refers to the frequency range of analog inputs that the digitizer can sample. It is good for the upper end of the bandwidth to exceed the frequency of the signal you wish to sample, by as much as two times [14].

The sample rate is another key characteristic. The sample rate is the rate at which samples are taken from the analog input. The Nyquist[1] frequency is half of the sampling rate. The Nyquist frequency is the highest frequency that can be represented by the discrete signal created by sampling. This makes sense because you need at least two points per period to represent a wave. This is called the Nyquist-Shannon sampling theorem.

Dynamic range is the final characteristic we will discuss. The dynamic range of a digitizer is the ratio of highest and lowest signal amplitudes that the digitizer can read [14]. The dynamic range is determined by how many bits are used to encode a voltage. The more bits, the higher voltage resolution, just like we saw with AWGs.

7.8 Lock-In Amplifiers

Lock-in amplifiers are another tool we can use to measure signals, and particularly noisy ones [16]. With a lock-in we measure the amplitude and phase of a signal with respect to a reference. Most lock-ins can provide results in two formats. The first is the signal demodulated by a cosine to yield an in-phase component X and demodulated by a sine to yield a quadrature component Y. The second format adds a step by calculating $R = \sqrt{X^2 + Y^2}, \theta = \tan^{-1}(X/Y)$ much as we have seen before.

How does this work? It involves a mixing stage and a filtering stage which both combine to perform demodulation. The noisy signal is first sent with the cosine

[1] Nyquist was an electrical engineer whose work on the Nyquist Sampling Frequency was cited by Claude Shannon in 1948 [15].

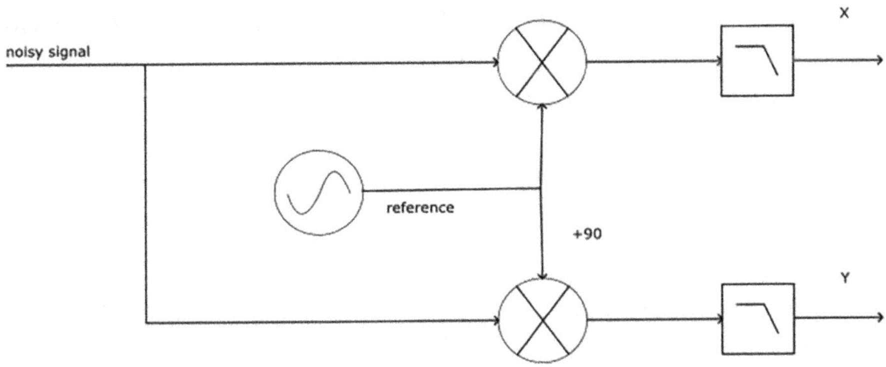

Fig. 7.4 Lock-in Amp

reference to a frequency mixer that multiplies them. The result is a combined signal with frequency $\omega \pm \omega$ in the case where the frequencies of the noisy signal and the reference are the same. A low pass filter then is used to remove the DC component of this combined signal. The filter bandwidth is a configurable cut-off frequency where signal power is attenuated by half. It is related to the so-called "time constant" T, which a user can usually set, by $\frac{1}{2\pi T}$. A larger filter bandwidth means lower SNR, but faster measurement and higher time resolution. The noisy signal is similarly sent to a frequency mixer with the sine reference signal of the same frequency, and this mixer's output is also filtered. Finally we calculate $R = \sqrt{X^2 + Y^2}, \theta = \tan^{-1}(X/Y)$ which can be done using an IQ mixer.

Some key characteristics to consider when choosing a lock-in are the frequency range and measurement bandwidth. A lock-in should also have a low input noise. Its dynamic reserve, also important, is its capability of rejecting unwanted signal components while still providing accurate results (Fig. 7.4).

7.9 Digital to Analog Converters

Digital to Analog Converters (DACs) are a part of many of the modules which we have already discussed, so it is worth briefly discussing their own characteristics. A DAC converts an n-bit digital value to a phyiscal output like a voltage. If different n-bit values are provided in succession then a signal like an analog waveform can be generated. For example, a 6-bit DAC can generate $2^6 = 64$ distinct voltages. This is the DACs resolution. A DAC also has a maximum sampling rate, which is how much time is required between each n-bit input for the correct voltages all to be generated correctly in sequence. The dynamic range of a DAC is the ratio of highest to lowest signals it can generate in dB. The noise, such as quantization noise, introduced to the ideal generated signal by the DAC is also important to consider. A final characteristic of DACs is monotonicty which is the ability of the generated signal to change only in the direction that the digital input changes.

7.10 Analog to Digital Converters

An Analog to Digital Converter (ADC) performs the reverse operation of a DAC, converting an analog signal like a continuously varying voltage signal to a digital one. Per the Nyquist-Shannon sampling theorem, an ADC can represent a signal with a frequency up to half its sampling rate. An ADC converts its input to an n-bit digital output, with a resolution of 2^n. Since voltages will either be rounded down or up to the nearest bitstring, an error called quantization error is introduced. Ideally, this error is uniformly distributed. ADCs are also characterized by jitter, which is a small variance Δt in its sampling clock which results in some additional noise.

7.11 Amplifiers

Current and voltage amplifiers are often used to amplify signals in a quantum dot or superconducting setup. A current amplifier outputs a current proportional to its input, and will often have a variable gain. Current sources like this are usually designed so that they drive an impedance that approaches 0 [17]. A voltage amplifier outputs a voltage proportional to its input.

In quantum computing setups, amplification is typically done at different temperature stages within the dilution refrigerator. For example, in the following setup we show an amplifier at the 4K stage and at room temperature. In the depicted setup, the first amplifier on the 4K stage is typically a quantum-limited amplifier. The second amplifier is a High Electron Mobility Transistor (HEMT) amplifier (Fig. 7.5).

The transformation corresponding to the quantum-limited amplifier involves a power gain G, signal mode with amplitude a_f and idler mode with amplitude h [18].

$$a_{\text{amp}} = \sqrt{G}a_f + \sqrt{G-1}h^\dagger$$

$[a_{\text{amp}}, a^\dagger] = 1$ and the noise introduced by the amp is accounted for in $h^\dagger$. It is minimized as $\langle h^\dagger h \rangle \to 0$. What makes a quantum-limited amplifier "quantum-limited" is that it only adds quantum noise. The state-of-the-art closely approximates this ideality.

The transformation that corresponds to the HEMT amplifier follows the same equation for a_{HEMT} as a_{amp} above, only the noise is greater $(5 < \doteq \langle h^\dagger h \rangle < \doteq 10)$ [18].

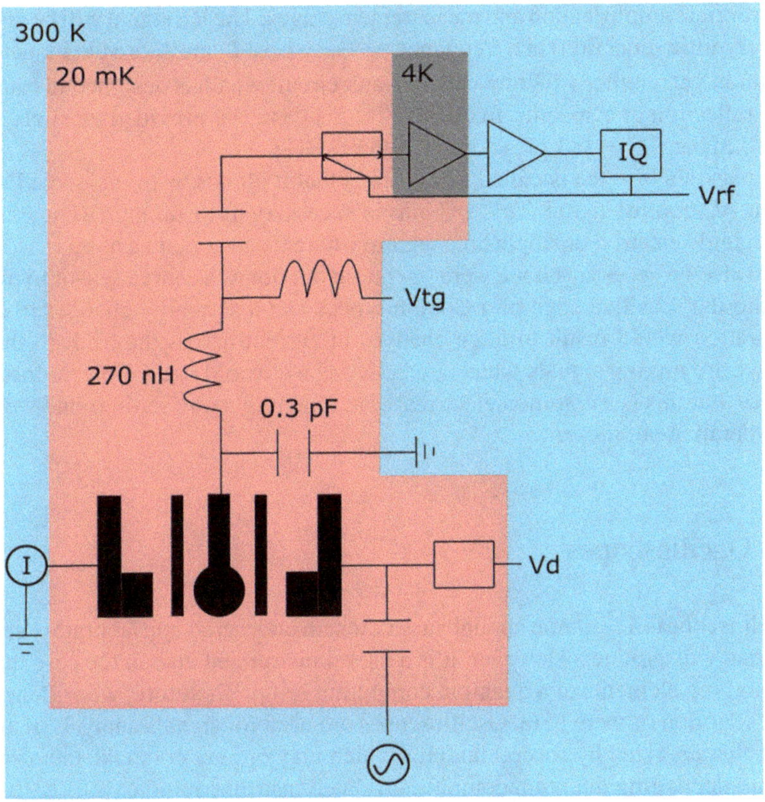

Fig. 7.5 Quantum dot setup

7.12 Miscellaneous Electrical Components

In the previous section, we depicted a setup for controlling a qubit defined in a quantum dot. This involved amplifiers as well as several other components including inductors, capacitors, resistors and a directional coupler. We briefly discuss the roles of these here. In the diagram, V_d denotes a voltage on the drain of the quantum dot which is essentially a Single Electron Transistor (SET). V_d is connected to the drain by a resistor and a capacitor also connects an AWG to the drain. This structure of resistor and capacitor which allows the AC and DC signals mentioned to mix is called a *bias tee*. Of course, the capacitor allows AC signals to pass while the resistor allows DC to pass. We have a similar setup when it comes to the RF signal V_{rf} and DC signal V_{tg} applied to the top gate. This time the bias tee uses an inductor and a capacitor which allows DC and AC to pass respectively. In this case, the RF signal passes through a directional coupler prior to the capacitor of the bias tee. This is so that a signal may be applied the dot through the bias tee and then the reflection of that signal may be routed away from the RF signal generator and into the appropriate amplification and measurement stages. The RF signal will be resonant (have very little reflection) at a frequency $\frac{1}{2\pi\sqrt{LC}}$ where L and C are the inductor and capacitor in yet another structure called a tank circuit which is designed to cause this lack of reflection at a specific frequency. In this case, we provided example values $L = 270$ nH and $C = 0.3$ pF, so $f_{\text{res}} = 0.5592\,\text{GHz}$.

The reason a bias tee is called a bias tee is that it allows us to include a DC bias with our AC control signal. This DC bias is necessary, for example, to tune our dot into the single electron or single hole regime where it can contain a qubit in isolation. It would also be necessary if we were operating our dot as a charge sensor to tune the SET onto the sensitive edge of a Coulomb peak[2] so that nearby changes in charge configuration would result in large changes in current across the SET. In the case where we are mixing signals which are both AC we would use a diplexer instead of a bias tee, but this is a common electrical component in many settings and we won't go into detail about these.

7.13 Oscilloscopes

An oscilloscope is a common enough instrument in undergraduate laboratory sections across many disciplines. However, it is a useful instrument also in the case that you are debugging elements of a quantum computing setup. Therefore, we will include a brief description of them here. Oscilloscopes are used for signals analysis of all kind and are characterized by several features which may or may not make them suitable for use in debugging quantum computing-related instrumentation.

An important characteristic of an oscilloscope is its bandwidth. While you need to be careful before you think about connecting instruments like scopes, voltmeters and ammeters to sensitive nanoelectronics (don't do it!), an oscilloscope can certainly safely be used to verify that an AWG is outputting what it is programmed to before it is connected to the rest of a control setup, for example. This would mean that the frequency of the signal produced by the AWG, or other isolated instrument, would need to be within the bandwidth of the oscilloscope in order to be faithfully measured. This might mean that it makes sense to include the oscilloscope after down-conversion with a mixer. Generally, oscilloscopes can have bandwidths between 20 MHz and 1.5 GHz [20] but higher end oscilloscopes can be very expensive. The bandwidth is determined by the sampling rate, which is the number of measurements that the oscilloscope takes per unit time.

An oscilloscope internally uses an ADC. Therefore, a higher bit depth in the internal digital representation of the measured signal will result in a greater measurement accuracy.

Other features that are important to consider are whether the oscilloscope has triggering capabilities, so that its measurement of a signal can be aligned digitally with the production of that signal by another instrument. The speed at which the

[2] Charles-Augustin de Coulomb was a military engineer from 1764 to 1772 in the West Indies [19].

scope can measure and process a signal before displaying it on its screen may also matter for certain use cases. A larger memory depth is also helpful to display longer waveforms. An oscilloscope should have high input impedance to avoid loading the signal [20]. The scope's rise time should also be fast so that a quick change in the input signal can be faithfully reflected.

7.14 Vector Network Analyzers and Spectrum Analyzers

A Vector Network Analyzer (VNA) is a very useful tool for superconducting and quantum dot-based quantum computing, especially in the early phases of building a setup [21]. A VNA can sometimes double as a spectrum analyzer. A 2-port VNA provides an input signal to an electrical network and measures an output signal. It is possible to measure the reflected signal on the input line as well as the transmitted signal on the output line. A VNA would normally sweep a frequency range during a measurement to determine the frequency response of the Device Under Test (DUT). A spectrum analyzer is a 1-port instrument that is suited to measure unknown signals rather than the frequency response of a DUT to a known stimulus. The spectrum analyzer does not usually contain an input signal source.

The measurements that can be taken using a VNA are usually denoted S_{11}, S_{12}, S_{21}, S_{22} where the subscript refers to the port. $S_{21} = \frac{transmitted}{incident}$ and $S_{11} = \frac{reflected}{incident}$. To measure S_{12}, S_{22} requires reversing the orientation of the DUT with respect to the VNA ports. Remember that only one port doubles as source and receiver. $S_{12} = \frac{\text{transmitted}}{\text{incident}}$ and $S_{22} = \frac{\text{reflected}}{\text{incident}}$.

Some key characteristics of a VNA are its dynamic range, frequency range, trace noise and measurement speed. The frequency range gives the range of frequencies that can be generated and measured. A VNA usually displays DUT response on a frequency / dB plot in which case the frequency range limits the extent of the domain. The dynamic range gives the VNA's measurable attenuation range. Trace noise is the noise that is introduced into the measurement by the VNA itself. This is typically in milli-dB [21]. Measurement speed is self explanatory.

7.15 Laser Sources

The instruments we have covered thus far are about all you need for superconducting and quantum dot-based setups. They are also important parts of ion trap-based quantum computing setups, for example. However, when it comes to setups involving optics like ion trap and photonic quantum computers, there are a few more instruments that we need. Now we will shift gears to instrumentation that would be useful specifically for setups involving optics such as photonic quantum computing and ion trap setups. The first such instrument is a laser source. A laser source is a pn junction with an extremely reflective and a partially reflective end (Fig. 7.6).

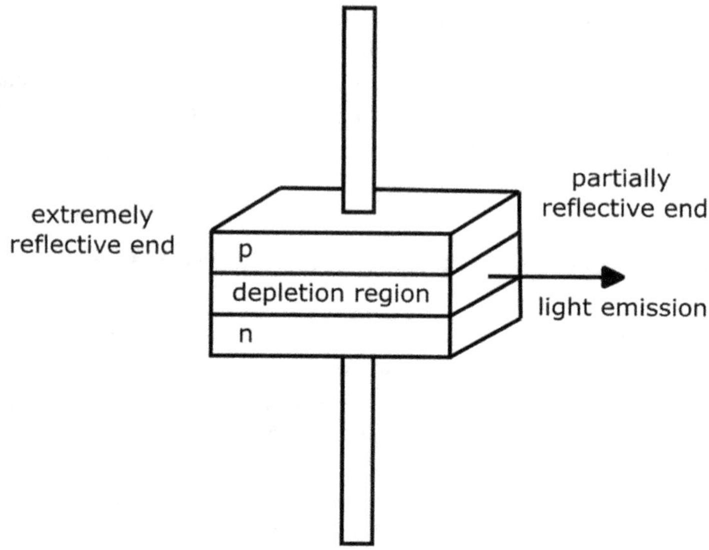

Fig. 7.6 Laser source

With the current through the device above a threshold, stimulated emission of photons will become dominant as opposed to spontaneous emission which will no longer be the prevalent phenomenon. We call this the lasing threshold. Light produced by a laser in the stimulated emission regime is coherent rather than chaotic. Laser pulses can be created by techniques such as gain switching, mode-locking, and gating (above and below threshold).

7.16 Single Photon Sources

More specific to photonic quantum computers are Single Photon Sources (SPS). Solid state single photon emitters include optically active quantum dots and color centers. Parametric down-conversion can be used to generate photons but is spontaneous and therefore we have the challenge that we do not know exactly when a photon will be emitted. While an ideal photon source would always generate a single photon with 100% probability, no such ideal source has ever been fabricated. Single photons are useful for their quantum mechanical properties which are explored in more detail in Chaps. 4 and 6.

A key characteristic of a single photon source is its efficiency. For example, a laser attenuated to reduce its mean photon number will generate photons in numbers that follow a Poisson distribution. The rate at which single photons are generated is usually much less than the rate at which no photons are, and this means a lower efficiency than an ideal single photon source.

7.17 Electro-Optic Modulators

Electro-optic modulators refer to devices which modulate either the phase, frequency, polarization or amplitude of a laser by an electro-optic effect. For example, an electric field can cause a change in the refractive index of Lithium Niobate (LiNbO3) which will in turn modulate light passing through it. Mach-Zender Modulators (MZMs) involve splitting light through two paths, one of which modulates the phase by an electro-optic effect so that it can be controlled whether the light interferes constructively or destructively when recombined [22]. The polarization of light passing through a nonlinear crystal can depend on the crystal orientation and the orientation of an applied electric field. With certain laser-controlled modulators, GHz range modulation bandwidths are achievable, though as we will see shortly RF control is often sufficient.

7.18 Acousto-Optic Modulators

Acousto-optic effects in a Bragg cell can be used to control the intensity, frequency, phase and polarization of transmitted light by RF. A Bragg cell is a material that transmits light and is vibrated by RF so that the light is reflected off the sound waves. One case we are particularly interested in due to its application in ion trap QC is intensity or amplitude modulation. The intensity of the RF signal controls the intensity of the beam diffracted through a Bragg cell. When the incident laser is at the Bragg angle (proportional to the wavelength of the light λ_{in} over twice the wavelength of the RF signal λ_{RF}), a diffraction pattern results in transmitted beams at angles that satisfy $2\lambda_{\text{RF}} \sin(\theta) = m\lambda$. The ratio of the intensity of the first order diffracted beam with the undeflected beam is denoted $\nu = \frac{I_1}{I_0}$ and has parallel and perpendicular components $\nu^{||} = \frac{|E_1^{||}|^2}{|E_0^{||}|^2}$ and $\nu^{\perp} = \frac{|E_1^{\perp}|^2}{|E_0^{\perp}|^2}$ [23]. Here $E_1^{||}$ denotes the electric field in the first-order beam polarized parallel to the RF wave, and so on.

The frequency of a diffracted beam of order m is also shifted by a multiple $m\omega$ of the frequency ω of the RF wave, and the phase is shifted similarly by the phase of the RF wave. Polarization control is also possible by engineering the RF wave.

7.19 Photon Detectors

In photonic quantum computing we will require Single Photon Detectors (SPDs) [24]. Characteristics of SPDs include the detection efficiency, dark-count rate, dead time and timing jitter. The detection efficiency is the probability that an incident photon is detected. The dark-count rate is the rate at which the detector gives “false positives”, seeming to detect a photon when none was incident. After a photon detection there is a period called the dead-time during which the detector cannot detect another incoming photon. The timing jitter refers to the timing mismatch between an incident photon and the electrical signal generated by the detector to indicate a photon detection.

Eisaman lists 36 types of SPDs [24]. Types of detectors include Photo-Multiplier Tubes (PMTs), single photon avalanche photodiodes, quantum dot field-effect transistor-based detectors, superconducting nanowire SPDs, up-conversion SPDs, superconducting tunnel junction based detectors, superconducting transition edge sensors, visible light photon counters and more. All of these convert photons to electrical signals. Some of them which rely on superconducting effects need to be operated at cryogenic temperatures.

Some photon detectors including PMTs are more useful for cases like ion trap-based QC where ions fluoresce multiple photons, as these cannot resolve photon number and instead indicate only when there are more than zero photons incident. In some photonic QC setups we are interested in resolving photon number, whereas in the case of detecting a bright or dark ion we are only interested in whether it is bright or dark.

7.20 Case Study: IonQ

In a 2024 Embedded FM podcast, Rick Altherr told all about IonQ's control setup for their ion trap-based quantum computers [25]. IonQ's quantum control systems are an excellent case study for us.

Starting with the clock as we did at the beginning of this chapter, IonQ quantum control systems run at 250 MHz or 4 ns using a dedicated timing module. The NTP 1588 precision time protocol is used. A single gate operation may take as long as 100 μs, but they must be aligned with the spin precession since the first initialization laser pulse with sub 10 ns accuracy. Also, when dealing with entangling gates, it is important that laser pulses to the two qubits are exactly simultaneous (within picoseconds). Phase-Locked Loops (PLLs) and a phase alignment clock distribution module are used to ensure that the AWG FPGAs therefore experience a clock edge within a few picoseconds of each other.

32 AWG channels across 8 AWGs are all synchronized by the clock distribution module. Off-the-shelf chips are used to figure out what the phase difference is from one module to the other due to length of wire between and the clock distribution module factors this in. This is necessary since an electrical signal can only travel so far along a length of wire in picoseconds.

These FPGA AWGs take into account phase adjustments that must be done to match the phase precession of the qubit. They also drive electro-optical modulators (EOMs), acousto-optic modulators (AOMs) and deflectors. These are controlled by RF. Generally, IQ waveforms from the AWGs are modulated into constant RF tones. An upconverter-like AOM then does modulation of the amplitude of a laser pulse. GPIOs are used to turn lasers on and off with a shutter. They do not use diode pumped lasers, rather they use constantly-on laser sources. These lasers finally drive Raman transitions between energy states of the 3–5 K Ytterbium or Barium ions.

Each AWG channel and GPIO dispatcher are all their own independent digital computers with specialized instruction sets. Each instruction includes a time-stamp of when to execute the instruction. An example instruction for an AWG includes:

- execution time
- memory location of samples defining waveform
- amplitude
- carrier frequency
- phase adjustment
- pulse length.

One master FPGA coordinates all the other FPGAs, another FPGA generates time-correlated GPIOs. The low-level control software is written in Rust.

Very slow running control systems run the cryostat, regulate liquid Helium, etc. These are standard control systems such as PID controllers. DACs and RF generators (DC voltages and RF) are used to generate the changing RF field for the ion trap. Servos also look at various waveforms output from the lasers before and after modulation for effects that need to be controlled for. Calibration must be done to aim the laser pulse to hit an ion correctly. Ion fluorescence and photon counting (bright and dark ions) varies with temp, pressure and also needs calibration via control loops. A lot of these calibrations happen slowly but then have real-time effects.

Users finally interact with the system through the cloud using QisKit, Pennylane or similar, as we will cover in the final chapter of this book.

References

1. Cross, A., Javadi-Abhari, A., Alexander, T., De Beaudrap, N., Bishop, L.S., Heidel, S., Ryan, C.A., Sivarajah, P., Smolin, J., Gambetta, J.M., Johnson, B.R.: Openqasm 3: A broader and deeper quantum assembly language. ACM Trans. Quant. Comput. **3**(3), 1–50 (2022) https://doi.org/10.1145/3505636
2. Pakin, S.: A quantum macro assembler. In: 2016 IEEE High Performance Extreme Computing Conference (HPEC), pp. 1–8 (2016). https://doi.org/10.1109/HPEC.2016.7761637
3. Blackbird Programming Language—Strawberry Fields. https://strawberryfields.ai/photonics/demos/run_blackbird.html
4. Fang, Y., Philippopoulos, P., Culcer, D., Coish, W.A., Chesi, S.: Recent advances in hole-spin qubits. Mater. Quant. Technol. **3**(1), 012003 (2023) https://iopscience.iop.org/article/10.1088/2633-4356/acb87e
5. Keysight: High-speed Signal Processing with Low Latency Feedback System for Quantum Bit Evaluation. https://www.keysight.com/us/en/assets/7018-05850/flyers/5992-2458.pdf. Section: Article Section
6. Pulse Processing Unit. https://www.quantum-machines.co/technology/pulse-processing-unit/
7. Know Your Top Arbitrary Waveform Generator Specifications—Technical Support Knowledge Center Open. https://docs.keysight.com/kkbopen/know-your-top-arbitrary-waveform-generator-specifications-598781340.html

8. Paul, S., Saikia, A., Majhi, V., Pandey, V.K.: Chapter 3—Transducers and amplifiers. In: Paul, S., Saikia, A., Majhi, V., Pandey, V.K. (eds.) Introduction to Biomedical Instrumentation and Its Applications, pp. 87–167. Academic Press (2022). https://doi.org/10.1016/B978-0-12-821674-3.00008-5. https://www.sciencedirect.com/science/article/pii/B9780128216743000085
9. Krantz, P., Kjaergaard, M., Yan, F., Orlando, T.P., Gustavsson, S., Oliver, W.D.: A quantum engineer's guide to superconducting qubits. Appl. Phys. Rev. **6**(2), 021318 (2019). https://doi.org/10.1063/1.5089550
10. Hendrickx, N.: Qubit arrays in germanium. Ph.D. thesis, Delft University of Technology (2021)
11. Sabah, S., Lorenz, R.: Design and calibration of IQ-mixers. In: 6th European Particle Accelerator Conference (EPAC 98), pp. 1589–1591 (1998)
12. Cartwright, T.J.: Signal Generators: The Different Types and How They Work. Keysight Blogs. https://www.keysight.com/blogs/en/tech/educ/2023/signal-generator
13. Hendrickx, N.W., Lawrie, W.I.L., Russ, M., Van riggelen, F., de Snoo, S.L., Schouten, R.N., Sammak, A., Scappucci, G., Veldhorst, M.: A four-qubit germanium quantum processor. Nature **591**(7851), 580–585, 585A, 585N (2021). https://doi.org/10.1038/s41586-021-03332-6. https://www.proquest.com/docview/2505728932/abstract/F030BA68FB854195PQ/1
14. Arthur Pini Greg Tate, O.R.: An Introduction to Modular Waveform Digitizers. https://spectrum-instrumentation.com/dl/an_waveformdigitizers_introduction.pdf
15. Markowsky, G.: Harry Nyquist. Pioneering Contributions to Information Theory, Britannica (2025). https://www.britannica.com/biography/Harry-Nyquist
16. Trbovic, J.: Principles of Lock-in Detection. Zurich Instruments (2019). https://www.zhinst.com/en/resources/principles-of-lock-in-detection
17. Terrell, D.L.: Amplifiers. In: Terrell, D.L. (ed.) Op Amps, 2nd edn., pp. 36–133. Newnes, Burlington (1996). https://doi.org/10.1016/B978-075069702-6/50003-2. https://www.sciencedirect.com/science/article/pii/B9780750697026500032
18. Blais, A., Grimsmo, A.L., Girvin, S.M., Wallraff, A.: Circuit quantum electrodynamics. Rev. Mod. Phys. **93**(2), 025005 (2021). ArXiv:2005.12667 [quant-ph] 10.1103/RevModPhys.93.025005. http://arxiv.org/abs/2005.12667
19. Charles-Augustin de Coulomb.: Biography, Discoveries, Law, & Facts, Britannica (2025). https://www.britannica.com/biography/Charles-Augustin-de-Coulomb
20. 15 Features You Should Look for Before Buying an Oscilloscope—Keysight Technologies. https://www.keysight.com/used/us/en/knowledge/guides/used-oscilloscope-buying-guide/15-features-you-should-look-for-before-buying-an-oscilloscope
21. Introduction to VNA Basics. https://download.tek.com/document/70W_60918_0_Tek_VNA_PR1.pdf
22. What is a Mach-Zehnder Modulator and How Does it Work? Synopsys. https://www.synopsys.com/glossary/what-is-a-mach-zehnder-modulator.html
23. Eklund, H., Roos, A., Eng, S.T.: Rotation of laser beam polarization in acousto-optic devices. Opt. Quant. Electron. **7**(2), 73–79 (1975) https://doi.org/10.1007/BF00631587
24. Eisaman, M.D., Fan, J., Migdall, A., Polyakov, S.V.: Invited review article: single-photon sources and detectors. Rev. Sci. Instrum. **82**(7), 071101 (2011). https://doi.org/10.1063/1.3610677
25. Altherr, R., White, E., White, C.: An ion of the highest fidelity. https://embedded.fm/transcripts/483

Chapter 8
Control Software

Abstract Now that we have familiarized ourselves with the control hardware that is used in various laboratory settings, we'll dive into the software side of things. In this chapter we will look at the software that sits directly on top of instrumentation as opposed to the software that might run simulations on your laptop or in the cloud, since we will cover that side of things in a later chapter. Code is provided for programming various types of instruments as examples throughout the chapter.

8.1 Learning Objectives

This chapter's focus on control software will help us to understand what happens between the hardware and the cloud tools you might be more familiar with (Fig. 8.1). Learning objectives for this chapter include:

- Get a sense of the layered landscape of control software for quantum computers
- Understand how a quantum computer differs architecturally from a CPU and how this impacts how we program it
- Gain practical programming knowledge through several code examples
- Understand how different software packages work and how they work together in a fully functional software stack like Open Quantum Design's.

8.2 Software Tooling Overview

Before we dive into the particulars of software, it is important to realize that all of the existing tools support certain programming models. Programming models for interacting with quantum computers are still being proposed [1]. It is increasingly important that programming models for quantum computers be considered holistically and not simply at a component level, as indeed it is important that the development of a quantum computer from end to end is considered holistically, which is

© The Author(s), under exclusive license to Springer Nature Switzerland AG 2026

M. S. Edwards, *An Introduction to Quantum Computing for Computer Engineers*,
https://doi.org/10.1007/978-3-032-03650-6_8

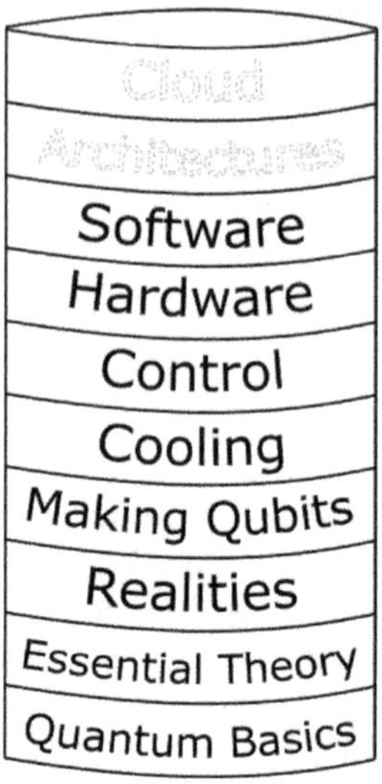

Fig. 8.1 Learning objective for this chapter: software

a main motivator of this book's publication. While the authors of [1] say that programmers generally need to work at all layers of the quantum programming stack including high-level algorithm descriptions, gates and hardware-adjacent expressions of algorithms, they do not specifically mention that there is still the world of quantum control systems, hardware and physics that need to be "engineered". We have addressed these to an introductory level in previous chapters, so now we can safely consider the software-only side of things. However, we will first put things into context. Consider the following model hierarchy proposed by Matteo et al.

Programming model	Quantum programs that map algorithms onto source code
Execution model	Abstractions related to execution such as compilation
Hardware model	Pulses and hardware-specific instructions

Matteo et. al. discuss the concept of an impedance mismatch between abstraction layers in a hierarchical programming model. This type of mismatch is associated with limitations in productivity and performance. In software we often discuss issues with abstraction hierarchies such as leaky abstractions, where the details that should be contained in one layer are referenced in another. There is generally an ideal separation of concerns between abstractions.

We generally like to hide functionality and implementation details in one layer from the next, until we have an API layer that is as expressive as possible. An expressive interface allows us to express our program concisely and without ambiguity at the same time. This Occam's razor approach can help us to decide what should be contained in an abstraction such as a class.

What are layers in quantum programming models?

Ideally a user-facing layer maps directly onto the user's mental model of their problem. Each layer beneath that will handle a detail such as memory management or pulse generation, for example.

We will see in the following examples several ways in which quantum software tools support (or do not support) the abstraction hierarchy discussed above.

A concrete example of how this abstraction hierarchy looks in practice is the following. We might have a quantum program for creating a Greenberger-Horne-Zelinger (GHZ) state defined in QisKit, for example.

Example: GHZ State in QisKit

```
circuit = QuantumCircuit(3)
circuit.h(0)
circuit.cx(0, 1)
circuit.cx(1, 2)
circuit.measure_all()
```

Before we can actually run this on a quantum backend we want to compile it to a hardware-aware circuit. To do this we need to provide details about the backend such as its specific gate durations, etc. We will also want to add error correction or dynamical decoupling to combat dephasing.

```
durations = InstructionDurations(
    [
        ("h", 0, 40),
        ("cx", [0, 1], 500),
        ("reset", None, 7),
        ("cx", [1, 2], 400),
        ("measure", None, 1000)
    ]
)
dynamical_decoupling_sequence = [XGate(), XGate()]
pass_manager = PassManager(
    [
        ALAPSchedule(durations),
        DynamicalDecoupling(durations, dynamical_decoupling_sequence)
    ]
)
circuit_with_decoupling = pass_manager.run(circuit)
```

This will finally generate a pulse sequence for us, which we can work with using Qiskit Pulse (soon to be Qiskit Dynamics). In this example, it is clear what information we need at each level in the abstraction hierarchy. But there are more elements to any working quantum software stack and some of the programming modalities do not have the same separation of concerns. When we consider lower level software, the waters may be a bit muddier. Here, when we say that software is low-level we mean that it directly interacts with quantum control systems.

The authors of [2] outline the four elements of the ecosystem of "open quantum hardware" that enable us to develop QCs. Prior to fabrication these include Hamiltonian simulation, which we can do with the Python package Qutip; QPU design and layout, which we would do (for example) with K-Layout as we saw in Chapter 4; and EM simulation which we could do with a package like QuDiPy. These don't really fit in this chapter as they are not control software, but they can aid with the pieces of the puzzle covered previously. We have by no means provided an exhaustive list of packages we may use for these pieces of the puzzle either.

Fabrication itself relies on another species of tools entirely which strictly do not fit in the category of control software, and so we reach the main topic of this chapter when we finally discuss the third of Shammah et. al's ecosystem elements. This they dub "Installation and Bring-Up" which is distinct from "Sustained Operation". What is the material difference between this third and fourth topic? They both involve control, calibration and data acquisition. What differs is that the third piece "Installation and Bring-Up" exclusively involves testing and characterization, while "Sustained Operation" has exclusive ownership of cloud infrastructure. Cloud quantum computing is the topic of a future chapter and therefore we will find ourselves discussing something like Shammah et. al's third element.

Many, if not most, research labs exist permanently in the "Installation and Bring-Up" phase of quantum computer operation. This is because of the nature of research. Experimental research, for example, aims to perform novel experiments that may be repeated perhaps 3 times before the group will move on to another topic which could result in a new publication. An experiment involving an experimental QPU, for example, will involve testing and characterization certainly, but not for the purpose of enabling sustained operations down the road. This testing and characterization is worthwhile in its own right and it is enabled by control, calibration and data acquisition software.

Control-oriented software in quantum computing enables the experimental community to carry out the physics experiments that allow quantum computing to advance at the most fundamental level. A variety of open and closed source solutions for controlling quantum computers exist. Models for quantum programming higher up in the stack have been developed involving programming, execution and hardware layers. We provide an overview and model of control-oriented software solutions (Labber, QCoDeS, ARTIQ, Qua, etc.) which are previously covered most effectively in hardware-related papers due to the divide between the low-level/hardware-oriented scientific community and the quantum computer science-oriented community. The latter prefers to work with more popularized frameworks (QisKit, Cirq, Pennylane, etc.) that are decoupled from underlying control systems historically, but a full-stack mentality is necessary to facilitate a productive quantum computer engineering effort.

The quantum programming model presented by Di Matteo et. al. [1] stops at the hardware-optimized gate model quantum circuit. This circuit is a set of unitaries that can be represented by a quantum channel, a directed acyclic graph, encoded in a graph Laplacian, or represented in a quantum assembly language, for example. You may have picked up on the fact that this is somewhat theoretical and disconnected from the underlying control system that must implement the circuit in the real world.

Even in the case of hardware-optimized quantum assembly, it is misleading that the code is termed "assembly" as the target is not a processor with an instruction register, program counter, etc. Rather, a complex control system involving multiple separate modules is involved and the particulars are individual to the type of quantum computer being used.

Many embodiments of such a control system will involve a control computer with a CPU as well as external modules connected through a high-speed bus which may have their own processors (such as Quantum Machines' Pulse Processing Unit (PPU)) or memories. A QPU is more akin to an ALU, not a CPU. What we would normally consider the elements of CPU architecture exist outside the QPU (Fig. 8.2).

Communication with the control computer can take ms, but communication between optimized modules on a high-speed bus can be on the order of μs. In many regimes, nanosecond resolution timing during an experiment is necessary. This level of accuracy is not possible when executing on a typical CPU, due to the communication latency, not to mention the indeterminacy of the timing of programs executing on a CPU which shares time between threads, programs, etc. according to a scheduling strategy. Therefore, in order to achieve both nanosecond resolution and programmability, data is often loaded into an external module by the control computer prior to an experiment as a precursor to the control system generating a pulse train, for example. A PPU also can be used for its completely pre-determined and dedicated scheduling which is observable to the user/programmer.

In some cases, this one-way programming model is not enough, as we wish to achieve a complete control pulse, measurement, logical processing, feedback loop with hundreds of ns latency for algorithms such as active qubit reset. This is a key ingredient for mid-circuit measurement. Instead, our model must involve a sort of distributed computation between processors (quantum and classical) and application specific circuits such as FPGAs which locally process information, and the carefully timed communication between them.

The particulars of any one distributed computing architecture are reflected in the control software that sits on top of it. For example, when programming in Quan-

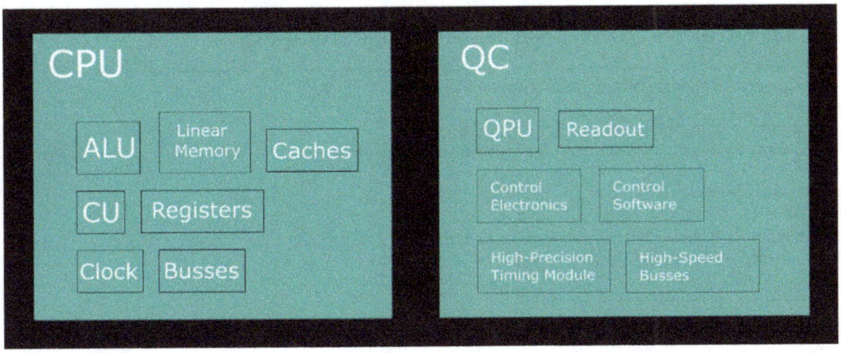

Fig. 8.2 Comparing CPU and QPU

tum Machines' Qua programming language, we are able to specify arbitrary logical conditions on measurement outcomes which, when met, determine that subsequent control pulses should be played. On the other hand, when using Keysight's Quantum Control System (QCS), we can only map measurement outcomes to control pulses without the intermediate logic. This is a reflection of the PPU's presence in the Quantum Machines architecture.

In Qua, active reset looks something like the following. Here we specify the condition $I > I\_threshold$ but any classical logic may be used to create a truthy or falsy expression here.

```
measure(readout_pulse, resonator, ...)
save(I, I1 + Q1)
with if_(I > I_threshold):
    play(pi_pulse, qubit)
```

With Keysight's Quantum Control System (QCS), we provide a list of ordered waveforms to play conditionally, corresponding to measurement outcomes 0 and 1.

```
on_zero = qcs.RFWaveform(10e−9, qcs.GaussianEnvelope(), 1, frequency)
on_one = qcs.RFWaveform(10e−9, qcs.GaussianEnvelope(), 0.75, frequency)

program.add_conditional_operation(
    operations=[on_zero, on_one], target=awg, register=classifier.register
)
```

In the lab (the Quantum Science and Technology Lab of the University of British Columbia) where the author cut his teeth on this types of software, the most used library was neither Qua or QCS but QCoDeS. QCoDeS is a modular data acquisition framework developed by Microsoft and available on GitHub. QCoDeS implements a variety of instrument drivers as Python classes. These instruments correspond to the ones we covered in the previous chapter including AWGs, sources, etc. A QCoDeS instrument driver wraps a specific instrument's programming interface with methods that allow a user to operate the instrument Pythonically.[1] For example, we could make use of the QDAC2 driver to coordinate some voltage ramping in the following fashion.

Example: QCoDeS QDAC2 Software Voltage Ramping

```
from qcodes_contrib_drivers.drivers.QDevil import QDAC2
import nidaqmx
from nidaqmx.constants import AcquisitionType as AcqType
from nidaqmx.constants import TerminalConfiguration as TermConfig

ramp_rate = 1 # V/sec
gain = 1
sample_rate_per_channel = 1e6
I_threshold = 2 # 1nA threshold current to stop the 1D measurement
```

[1] Saying something is "pythonic" means that it is in accordance with the PEPs or Python Enhancement Proposals accepted according to the Python governance model. The head of this model was Guido van Rossum, the creator of Python, until 2018 when he stepped down as "Benevolent Dictator for Life".

```
max_tries = 10 # number of tries before aborting a hardware sweep

V_LIMIT = 1.5

def read_simple(num_samples):
    with nidaqmx.Task('ai_task') as ai_task:
        ai_task.ai_channels.add_ai_voltage_chan(
            "Dev1/ai0",
            terminal_config=TermConfig.DIFF,
            min_val=-1,
            max_val=1
        )
        ai_task.timing.cfg_samp_clk_timing(
            sample_rate_per_channel,
            sample_mode=AcqType.FINITE,
            samps_per_chan=num_samples
        )
        ai_task.start()
        result = ai_task.read(number_of_samples_per_channel=num_samples)
        ai_task.stop()
    return result

def ramp_voltages(ch_nums, starts, ends, time_per_scan, with_read=True):
    # go to starts
    for i, ch in enumerate(ch_nums):
        attr = f'ch{ch.zfill(2)}'
        getattr(qdac, attr).dc_constant_V(starts[i])

    # setup slew rates
    for i, ch in enumerate(ch_nums):
        attr = f'ch{ch.zfill(2)}'
        getattr(qdac, attr).dc_slew_rate_V_per_s(abs(starts[i] - ends[i]) / time_per_scan)

    # start the sweeps together
    for i, ch in enumerate(ch_nums):
        attr = f'ch{ch.zfill(2)}'
        getattr(qdac, attr).dc_constant_V(ends[i])

    if with_read:
        return read_simple(int(time_per_scan * sample_rate_per_channel))

def ramp_together(channels, initial_voltages, final_voltages, steps, time_per_scan, with_read=True):

    volt_steps = []
    for k, channel in enumerate(channels):
        volt_steps.append(np.linspace(initial_voltages[k], final_voltages[k], steps))

    starts = []
    ends = []
    for step in range(len(volt_steps[0]) - 1):
        starts.append([])
        ends.append([])

    for c in range(len(channels)):
        for j in range(0, len(volt_steps[c])):

            if len(volt_steps[c]) > j+1:
                start = volt_steps[c][j]
                stop = volt_steps[c][j+1]

                starts[j].append(start)
                ends[j].append(stop)
            else:
                break

    if with_read:
        reads = []
```

```
        for start_slice, end_slice in zip(starts, ends):
            reads.append(ramp_voltages(channels, start_slice, end_slice, time_per_scan, with_read))
        return reads

    else:
        for start_slice, end_slice in zip(starts, ends):
            ramp_voltages(channels, start_slice, end_slice, time_per_scan, with_read)
```

The first function here, *read_simple*, sets up a National Instruments DAQ to take a voltage reading at a sample rate of 1e6 samples per second. The second function *ramp_voltages* calls the provided methods *dc_slew_rate_V_per_s* and *dc_constant_V* to set the slew rates and target voltages of the ramps respectively. *ramp_together* simply calls *ramp_voltages* several times, each time causing a slice of the ramp of equal length on each channel to be executed together.

These functions would be used in the following way to ramp voltages from a starting voltage to an ending voltage.

```
import numpy as np
import matplotlib.pyplot as plt

reads = ramp_together(['1'], [0.2], [0], 4, 0.5, with_read=True) # 4 steps works
plt.plot(reads[0])
plt.plot(reads[1])
plt.plot(reads[2])
```

It should be clear when you read this code that we require different information about the control system than was provided by our "hardware model" in the chapter's first Qiskit example. Indeed, this is lower level code than anything in the abstraction hierarchy discussed previously in that we need to know about sample rates, terminal configurations, gains, current thresholds, port numbers and more. The existing model is clearly of a "top-down" design and leaves much work to the designers of the control systems which should execute hardware-aware circuits, so that even more software is required to bridge the gap (Fig. 8.3).

You may also think this is a lot to be done on our CPU, and it is! So, this code is not so useful for experiments, maybe it is only useful for tasks like tuning the device into an operational regime. If we want to perform quick ramps during an experiment, we need to use the provided *dc_sweep* method which allows us to pre-program our QDAC2 to perform a ramp from its own memory during an experiment. This code

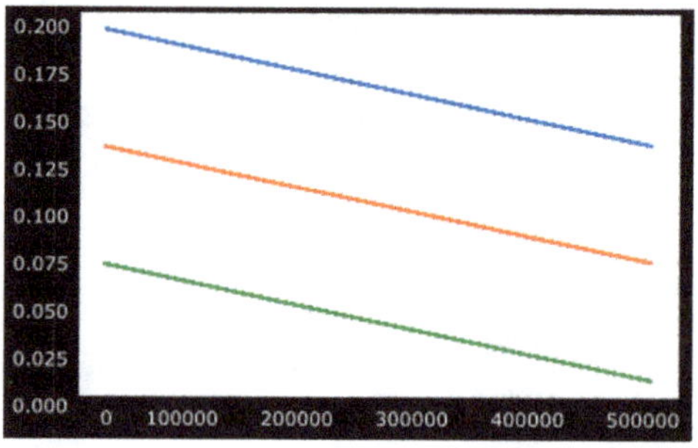

Fig. 8.3 Output of ramp together function

block shows how we might perform some parts of a procedure on a CPU and some on external hardware. In general, we would need to split this up intentionally to think about what is critical in terms of timing and what is not. Here we do not have a software loop synchronizing our sweeps on our various channels by multiple calls to an underlying function.

Example: QCoDeS QDAC2 Hardware Voltage Ramping

```
def hardware_sweep(fast_chan_nums, fast_starts, fast_ends, fast_steps, time_per_fast_scan):
    sweeps = []

    for c in range(len(fast_chan_nums)):
        ch = fast_chan_nums[c]
        st = fast_starts[c]
        sp = fast_ends[c]

        sweeps.append(sweep_factory(ch, st, sp, fast_steps, time_per_fast_scan))

    for sw in sweeps:
        sw.start()

    read = read_simple(int(sample_rate_per_channel * time_per_fast_scan))

    return read

def sweep_factory(chan, start, stop, fast_steps, time_per_fast_scan):
    dwell_s = time_per_fast_scan / fast_steps

    return getattr(qdac, f'ch{chan.zfill(2)}').dc_sweep(
        repetitions=1,
        start_V=start,
        stop_V=stop,
        points=fast_steps,
        dwell_s=dwell_s
    )
```

We would use this in the following way.

```
data = hardware_sweep(['1'], [0], [1], 10, 10)
plt.plot(data)
```

In these code examples, you can see how we make use of the QCoDeS provided classes *QDAC2* and *nidaqmx* to coordinate control and measurement. Though, don't get too excited. These are not qubit control or readout yet. For that, we need to do a bit more work, or use a few more tools (Fig. 8.4).

Another software tool targeted to ion trap quantum computing is ARTIQ [2]. ARTIQ involves its own high-level Python-based programming language. After a program is written in this language which describes an experiment, the code is compiled and executed on an FPGA with nanosecond timing. ARTIQ can also orchestrate synchronized operation between such FPGAs. It is a part of the Sinara project which also has developed hardware including Analog to Digital converters (ADCs), digital to analog converters (DACs), Direct Digital Synthesis (DDS), AWG modules and more. We could accomplish a voltage ramp on a single channel of a Zotino board from -10v to 0v with a frequency of 1.28kHz using ARTIQ with the following code.

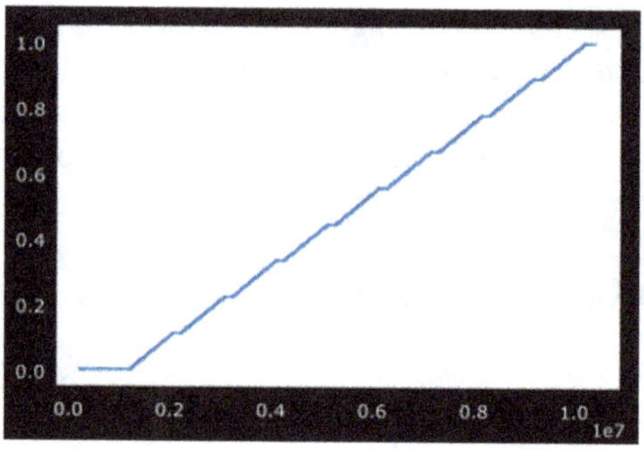

Fig. 8.4 Output of hardware ramp function

A Zotino is a 32-channel, 16-bit DAC. It has an update rate of 1MSPS. The code is broken into two parts. The first encapsulated in *def build(self):* specifies code that runs on the host device, in this case adding drivers for the core device and Zotino board as attributes. The second part is decorated with *@kernel* and runs on the FPGA. In this part, we specify the ramp as a list of voltages in "machine units" which will be understood by the hardware. We then prevent underflow with a fixed delay before writing each of our voltages out on a DAC channel in a loop. This write is done in two steps, the first with a call to *write_dac_mu* and the second loading the buffer to the DAC channel with *load()*. A 7 us delay is required to allow the voltage to be reached, and at least 800ns to prevent underflow.

Example: ARTIQ Ramping

```
from ARTIQ.experiment import *

class ramp_generator(EnvExperiment):

    def build(self):

        self.setattr_device("core")
        self.setattr_device("zotino0")

    @kernel
    def run(self):
        self.core.reset()
        n_steps = 50
        voltages_mu = [((1<<16)//n_steps) * i for i in range(n_steps)]

        self.core.break_realtime()

        self.zotino0.init()
        delay(200*us)

        while(1):
            for voltage in voltages_mu:

                self.zotino0.write_dac_mu(0, voltage)
                self.zotino0.load()

                delay(7*us)
```

This code clearly delineates between two code execution environments. Since it has this awareness of the control system architecture, it is again low level. A pattern is emerging which one might not notice if they lived in the world of QisKit, Cirq, Pennylane and other higher level packages which are decoupled from lower level considerations. The pattern is that there is another layer in the quantum software stack that has a high level of dependency on control system details. We propose to add this layer to the bottom of di Matteo et. al's abstraction hierarchy. After all, it is a part of the stack in which many of us work.

To capture the different ways in which users of different types such as quantum software engineers and experimentalists might interact within a quantum computing stack, consider the following possible stack in which end users interact with high-level algorithms and innovate in algorithm development, programmers write and innovate in software, while experimentalists tune, calibrate and innovate on hardware. A compiler lowers domain-specific and generic code to lower level representations. There are multiple "paths" for different user types to use the hardware and associated niche tooling in a way that reflects existing communities which operate differently (Fig. 8.5).

The other package that QSTL (a group of experimentalists) uses frequently is Keysight's Labber, which makes qubit control very easy through both a graphical user interface and Python API. Labber includes drivers for various instruments that might be found in a research lab. Some of these drivers implement single-qubit and multi-qubit pulse generation, for example. It is therefore as easy as mapping your ports properly and clicking a few buttons in the Labber GUI to generate a control pulse for your qubit(s).

Labber also allows for users to write their own drivers, for example to play custom waveforms on the output port of an AWG, or to write software on top of their provided drivers through the Python API. QSTL's Free Open Source Software (FOSS) on top of Labber includes Python classes which allow easy operation of 8 different instruments including sourcemeters, AWGs, digitizers, data acquisition systems, DACs and lock-in amplifiers. The Python package also includes a library of 7 pre-written experiments. For example, a set of hardware ramps between various voltages with ns precision can be implemented quickly and expressively using the following code snippet. The package uses JavaScript Object Notation (JSON) format files to parameterize experiments.

Example: Labber Hardware Ramping with LabberWrapper

```
from labberwrapper.experiments.mw_experiment import Piece, Piecewise, hardware_piecewise_microwave

# define the SET to be measured
SET1 = AWG_SET(1)

# load the experiment config
config = json.load(open('./labberwrapper/experiment_configs/mw_experiment.json', 'r'))
jschema_mw = json.load(open('./labberwrapper/json_schemas/mw_experiment.json', 'r'))

# validate the config
validate(instance=config, schema=jschema_mw)
```

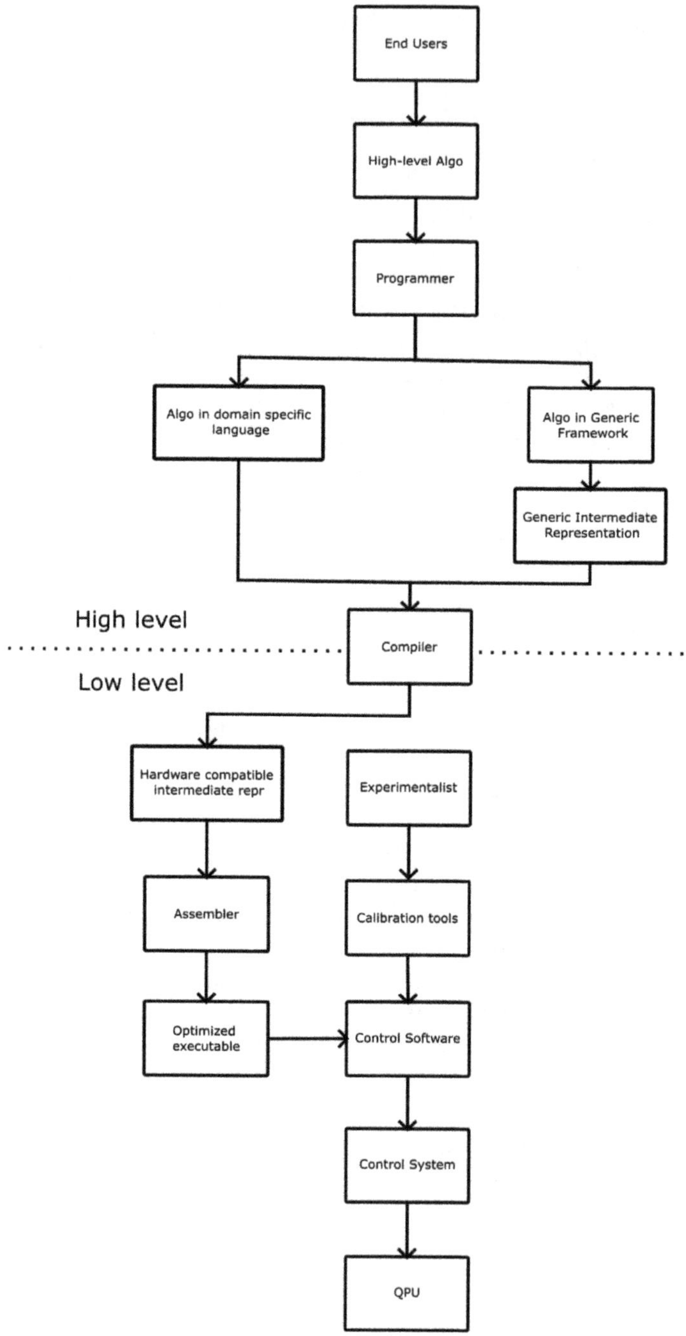

Fig. 8.5 Programming model

```
# run the experiment
hardware_piecewise_microwave(
    single_electron_transistor=SET1,
    piecewise=Piecewise(
        pieces=[
            Piece(volts=1, time_ns=10),
            Piece(volts=2, time_ns=10),
            Piece(volts=1, time_ns=10)
        ],
        ramp_time_ns=config['ramp_time'],
        resolution_ns=1
    ),
    samples=config['samples'],
    records=config['records'],
    averages=config['averages'],
    buffer_size=config['buffer_size']
)
```

The way in this package separates the hardware model from the more expressive layers above is by allowing experiment, device and instrument details to be specified in JSON. For example, we would specify the ports relating to a device such as a quantum dot as:

```
{
    "plunger_1": 1,
    "plunger_2": 2,
    "barrier_middle": 3,
    "ai_ch_num": 1
}
```

A specific experiment such as a voltage sweep would be parameterized by more key value pairs. The specific values for things like barrier voltages in quantum dot experiment must be experimentally determined based on measurements, for example, of a device's charge stability diagram. This is all involved in device tuning which has its own automation software and methods [3–5].

```
{
    "fast_ch": [1],
    "bias_v": 0.7,
    "plunger_v": 0.2,
    "acc_v": 0.12,
    "vb1_v": 0.14,
    "vb2_v": 0.15,
    "fast_vstart": 0,
    "fast_vend": −0.5,
    "fast_steps": 50,
    "fast_step_size": 0.05,
    "fast_ch_name": "Vx"
}
```

Such an experiment configuration would be checked against validation rules before being used to perform any actual experiment. This introduces a verification step to the experiment that might save an experimentalist from accidentally frying a device. This validation involves a schema which the configuration is checked against.

```
{
    "$id": "json_schemas/experiment_schemas/1D_&_2Dsweep.json",
    "$schema": "https://json-schema.org/draft/2020-12/schema",
    "title" : "1D & 2D sweep values",
    "required" : [
        "fast_ch",
        "bias_v",
        "plunger_v",
        "acc_v",
        ...
    ],
    "type": "object",
    "properties": {
        "slow_ch": {
            "type": "array"
        },
        "fast_ch": {
            "type": "array"
        },
        "bias_v": {
            "type": "number",
            "minimum": -2.5,
            "maximum": 2.5
        },
        "plunger_v": {
            "type": "number",
            "minimum": -2.5,
            "maximum": 2.5
        },
        ...
    }
}
```

Another open source repository at Labber-software/Drivers has 96 additional Labber drivers. While these abstractions are helpful and make our code more expressive, the abstractions provided by Keysight's QCS are even moreso. The package includes a library of Paulis which we can reference instead of crafting our own RF waveforms, if we like. Below is an example of how we might perform active reset using this gate library.

Example: Quantum Control System (QCS)

```
program = qcs.Program()

qubits = qcs.Qudits([1])
gates_to_conditionally_apply = [qcs.GATES.I, qcs.GATES.x]
times_to_repeat = 2

program.add_measurement(qubits, reset=reset, repeat_reset=times_to_repeat)
```

QCS is a bit more high level in the traditional sense that it introduces more abstraction. It provides an expressive syntax for us to specify our quantum programs in. QCS also integrates with Trueq from Quantum Benchmark to provide error diagnostics and correction. For example, we can use this tool to implement benchmarking protocols including randomized benchmarking. Randomized benchmarking estimates the average gate errors by implementing long sequences of randomly sampled gates and looking at how the fidelity of the final state decreases as the length of the random sequence increases. Originally, Haar-random[2] gates were used but lately the community has used uniformly randomly sampled Clifford gates.

Here, the first line after the import generates a collection of circuits on which we will run Streamlined Randomized Benchmarking (SRB) on qubit 1. Instead of [1] we could also have provided a list containing more single-qubit indexes in which case SRB would be performed on each, or containing lists of two qubits in which case two-qubit SRB would be performed. The second line finds the Pauli matrix compiled into the first circuit. The final line draws the circuit for us.

```
import trueq as tq

circuits = tq.make_srb([1], [4, 200], 30)
compiled_pauli = circuits[0].key.compiled_pauli
circuits[0].draw()
```

Furthermore, Trueq has high-level abstractions for algorithms and subroutines. We can create a 5-qubit Quantum Fourier Transform (QFT) circuit for example by the following simple code.

```
import numpy as np
import trueq as tq

circuit = tq.algorithms.qft(range(5))
circuit.measure_all()
```

The library constructs circuits in cycles which are each a set of gates that can happen at the same time, i.e. on different qubits. A cycle contains labeled gates. For example, we could create a one-cycle circuit:

```
    cycle = {(0, 1): tq.Gate.cnot, (2, 3, 4): tq.Gate.t}
    circuit = tq.Circuit([cycle])
```

Trueq also integrates with Cirq, PyQuil, QisKit, and QASM, some of the most used software frameworks for quantum computing worldwide. Therefore, one can write an algorithm in Cirq, PyQuil, QisKit, or QASM and then have error correction added by Trueq before the algorithm is finally executed by lower level QCS code. This would be an example of a full-stack quantum program execution.

Let's consider a peer and competitor to QCS which we've already mentioned: Qua. Qua is low-level in the sense that it can interact directly with Quantum Machines control systems, but it includes multiple components including simulators which can be used to execute Qua code instead of the physical control system, for example. An example of an amplitude sweep Rabi oscillation could be implemented in Qua by the below code.

[2] Alfred Haar was once an asistant to Hilbert [6].

Example: Qua Programming

```
from qm.qua import *

start, stop, step = −1, 1, 0.1
with program() as prog:
    a = declare(fixed)

    with for_(a, start, a < stop − 0.0001, a + step):
        play("x90"*amp(a), "qubit")

        align("qubit", "resonator")
        measure("readout", "resonator", None)
```

The call to $play()$ tells the simulator or the OPX to modulate the waveform. The second parameter to $play()$ tells what analog output port to use. This should be mapped beforehand. This can be done by mapping Qua elements to simulation channels, or physical channels. The general syntax is $play(pulse * amp(A), element)$ where A can be scalar or a matrix. $pulse$ and $element$ are JSON specifying a pulse (with a shape as an array of values, parameters) and a qubit or readout resonator (with a frequency, operations, port, etc). The call to $align$ makes sure that all the elements provided as parameters wait to finish their currently running statements before the program continues. $measure$ is comprised of three steps: playing the measurement pulse, waiting for a time before sampling the returning pulse, and finally processing the acquired data including demodulation.

The way in which Qua separates the hardware model from the higher level abstraction layers is by allowing hardware details to be specified in configuration objects which we would use to specify the $element$ provided to a $pulse$ call. For example, we could specify the configuration of a readout resonator as:

```
'resonator': {
    'RF_inputs': {
        'port': ('oct1', 1)
    },
    'RF_outputs': {
        'port': ('oct1', 1)
    },
    'intermediate_frequency': rr_IF,
    "time_of_flight": 150,
    "smearing": 0,
    'operations': {
        'readout': 'readout_pulse'
    },
}
```

Similarly for a qubit,

```
'qubit1': {
    'mixInputs': {
        'lo_frequency': qubit_LO,
        'mixer': 'mixer_qubit',
        'Q': ('con1', 1),
        'I': ('con1', 2)
    },
```

```
    'intermediate_frequency': qubit_IF,
    'operations': {
        'pi': 'pi_pulse',
        'pi_half': 'pi_half_pulse'
    }
}
```

Another tool at a similar level in the stack to Qua is QUASAR which have both been described as "at the hardware control level" [2], which also assumes some things about the hardware of the control system, in particular that there is a classical co-processor. QUASAR defines an instruction set architecture which contains both traditional operations as well as quantum operations on addressable qubits. A RISC-V core is primarily responsible for executing the program but quantum instructions are forwarded to a QUASAR co-processor. It is ideal for hybrid quantum-classical program execution.

As one can see, there are multiple APIs and libraries that accomplish similar things which are in some cases proprietary (like Labber) and it has been noted [2] that there is currently need for more interoperability among the existing tools.

One way in which these things can be unified is by common intermediate program representations which are then parsed and understood by the various lower level components. [7] identifies seven assembly-like imperative low-level quantum programming languages which are referred to [7, p. 230] as intermediate representations. These are QCL, LanQ, Scaffold, QASM, OpenQASM, cQASM and Jaqual. To this list we would add Blackbird. QIR is also a notable quantum intermediate representation from Microsoft that aims to provide a common interface between other languages.

These intermediate representations are however not entirely interoperable themselves. For example, Blackbird is intended to represent circuits for continuous-variable quantum computers while the others work in the gate model. An example of quantum teleportation in Blackbird therefore looks very different from examples in the other modality.

Example: Quantum Teleportation in Blackbird

```
name StateTeleportation
version 1.0

Squeezed(-{sq}) | 1
Squeezed({sq}) | 2
BSgate(pi/4, 0) | (1, 2)

BSgate(pi/4, 0) | (0, 1)
MeasureX | 0
MeasureP | 1

Xgate(sqrt(2)*q0) | 2
Zgate(sqrt(2)*q1) | 2
```

Here is an example of teleportation in OpenQASM.

Example: Quantum Teleportation in OpenQASM

```
OPENQASM 3;
include "stdgates.inc";

qubit[3] q;
bit c0;
bit c1;
bit c2;

reset q;
U(0.3, 0.2, 0.1) q[0];
h q[1];
cx q[1], q[2];

barrier q;

cx q[0], q[1];
h q[0];

c0 = measure q[0];
c1 = measure q[1];

if(c0==1) z q[2];
if(c1==1) x q[2];

c2 = measure q[2];
```

There has been an informal proposal for a way to make these types of algorithms more interoperable by allowing algorithms to be executed in either modality [8]. Since OpenQASM 3.0, these assembly languages can also represent classical contexts and hybrid quantum-classical algorithms.

Regardless of how nice these circuit representations are, it should be clear that they always need to be parsed and understood by a system that is lower level and understands the control system that is actually going to facilitate the execution of the program. To this end parser implementations have been developed such as [9]. This is only one example of all of the parsing, compiling and transpiling technologies that constitute the "glue" between the layers of our abstraction hierarchy.

As we look at these parts of the stack, however, we begin to drift from the focus of this chapter. Indeed, for books that cover high-level quantum programming extensively we recommend [10–12]. We will also cover this a little bit in our last chapter. [7] identifies eighteen high abstraction quantum imperative programming languages. Languages like these can be used to write a quantum program which is then lowered to OpenQASM or QIR for example. The predominant model that these use is the gate model, however many tools also exist on top of quantum annealer backends.

8.3 Case Study: Open Quantum Design

An example of a stack that shows clearly where all these software components come into the picture is that shared by Open Quantum Design, "the first" fully open source quantum computing stack [13], an effort from the Institute for Quantum

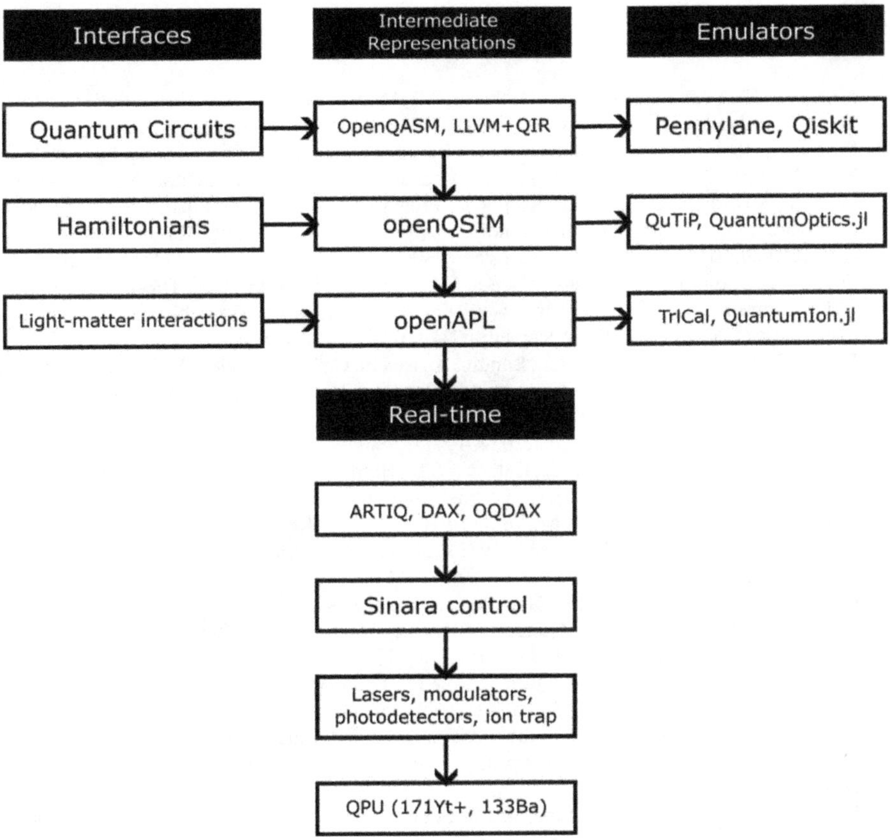

Fig. 8.6 Open quantum design stack

Computing and Perimeter Institute. It shows clearly where each element of the stack comes in including ARTIQ, Sinara, OpenQASM and more. The QPU may be based on 171Yt+ (Bloodstone processor) or 133Ba ions (Beryl processor). The computer can be programmed in the gate model, by specifying Hamiltonians or light-matter interactions. Several emulators/simulators can be used to execute the code after it is lowered to an intermediate representation. The representation and the simulator used depends on the model (circuit, Hamiltonian, light-matter). When it comes to the actual quantum control, ARTIQ and Sinara are used to control optical components such as the control lasers, photodetectors, etc. (Fig. 8.6).

References

1. Matteo, O.D., Núñez-Corrales, S., Reinhardt, S.P., Mattson, T., et al.: An Abstraction Hierarchy Toward Productive Quantum Programming (2024). https://arxiv.org/abs/2405.13918
2. Shammah, N., Roy, A.S., Almudever, C.G., Bourdeauducq, S., Butko, A., Cancelo, G., Clark, S.M., Heinsoo, J., Henriet, L., Huang, G., Jurczak, C., Kotilahti, J., Landra, A., LaRose, R., Mari, A., Nowrouzi, K., Ockeloen-Korppi, C., Prawiroatmodjo, G., Siddiqi, I., Zeng, W.J.: Open Hardware Solutions in Quantum Technology. APL Quantum **1**(1), 011501 (2024). https://doi.org/10.1063/5.0180987. http://arxiv.org/abs/2309.17233. ArXiv:2309.17233 [physics]
3. Darulová, J.: Automated Tuning of Gate-defined Quantum Dots. Doctoral Thesis, ETH Zurich (2020). https://doi.org/10.3929/ethz-b-000473107. https://www.research-collection.ethz.ch/handle/20.500.11850/473107. Accepted: 2021-03-17T08:57:52Z
4. Durrer, R., Kratochwil, B., Koski, J., Landig, A., Reichl, C., Wegscheider, W., Ihn, T., Greplova, E.: Automated tuning of double quantum dots into specific charge states using neural networks. Phys. Rev. Appl. **13**(5) (2020). https://doi.org/10.1103/physrevapplied.13.054019. http://dx.doi.org/10.1103/PhysRevApplied.13.054019
5. Ziegler, J., McJunkin, T., Joseph, E., Kalantre, S.S., Harpt, B., Savage, D., Lagally, M., Eriksson, M., Taylor, J.M., Zwolak, J.P.: Toward robust autotuning of noisy quantum dot devices. Phys. Rev. Appl. **17**, 024069 (2022) https://doi.org/10.1103/PhysRevApplied.17.024069
6. (Szeged), B.S.N.: Alfred Haar (1885–1933). In: Results in Mathematics, vol. 8. Birkhäuser Verlag, Basel (1985)
7. Cruz-Lemus, J.A., Serrano, M.A.: Quantum software tools overview. In: Serrano, M.A., Pèrez-Castillo, R., Piattini, M. (Eds.), Quantum Software Engineering, pp. 229–250. Springer International Publishing, Cham (2022). https://doi.org/10.1007/978-3-031-05324-5_12
8. Edwards, M.: A High-Level Quantum Templating Language (2024). https://marcusedwards-20301.medium.com/a-high-level-quantum-templating-language-8a98d710f0b5
9. Kim, S., Edwards, M.: Enabling the Verification and Formalization of Hybrid Quantum-Classical Computing with Openqasm 3.0 compatible qasm-ts 2.0 (2024). https://arxiv.org/abs/2412.12578
10. Exman, I., Pérez-Castillo, R., Piattini, M., Felderer, M. (eds.): Quantum Software: Aspects of Theory and System Design. Springer Nature Switzerland, Cham (2024). https://doi.org/10.1007/978-3-031-64136-7. https://link.springer.com/10.1007/978-3-031-64136-7
11. Serrano, M.A., Pérez-Castillo, R., Piattini, M.: Quantum Software Engineering. Springer, Berlin (2022). https://link.springer.com/book/10.1007/978-3-031-05324-5
12. Wong, H.Y.: Introduction to Quantum Computing: From a Layperson to a Programmer in 30 Steps. Springer International Publishing, Cham (2024). https://doi.org/10.1007/978-3-031-36985-8
13. Welcome - Open Quantum Design. https://docs.openquantumdesign.org/

Chapter 9
Hardware Architectures

Abstract We have covered everything that you need to build and operate a simple quantum computer, but what about scaling it up? In practice, it is generally believed that we will need thousands of qubits in order to run useful, error-corrected quantum algorithms. In this chapter, we cover several prevalent architectures for different types of quantum computers that each make promises about connectivity and scalability.

9.1 Learning Objectives

We will aim to learn about the architecture of the QPU specifically in this chapter, rather than the quantum computer as a whole. Our learning objectives are (Fig. 9.1):

- Understand the high-level considerations that go into the design of a QPU
- Get a sense of the landscape of QPUs that have been fabricated and progress in the field
- Understand the connectivity and scalability constraints of different approaches to making QPUs
- Learn through an example about how to run an algorithm on a particular type of spin qubit processor.

9.2 QPU Architecture versus Quantum Computer Architecture

In the last sections we talked about computer architecture in a sense that was similar to how we think about computer architecture in the classical world. In this section, we will abandon this premise and consider quantum computer architecture in the sense that quantum computer architects usually do: focusing on the QPUs themselves. We will review architectural considerations relating to each of the types of quantum computers we have learned about throughout the book. This is by no means an exhaustive review, but it aims to be well-rounded.

© The Author(s), under exclusive license to Springer Nature Switzerland AG 2026
M. S. Edwards, *An Introduction to Quantum Computing for Computer Engineers*,
https://doi.org/10.1007/978-3-032-03650-6_9

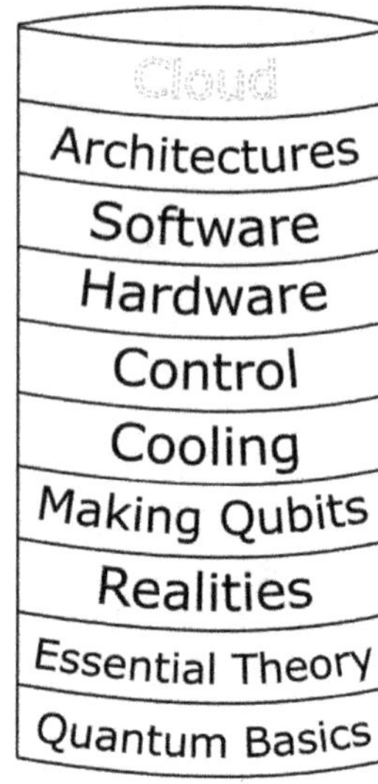

Fig. 9.1 Learning Objective for this chapter: architectures

One requirement for a useful quantum computer is for it to be based on a scalable qubit technology. This means that the qubits must be well-defined two-level systems, and that they must interact with each other. It is a challenge to achieve these things, with the interconnect bottleneck problem for example being an issue in spin qubits. The interconnect bottleneck is described by Rent's exponent, a measure of optimization in the wiring scheme. It should be possible to fit as many as 1000 spin qubits in a $5 \times 5\,\mu\text{m}^2$ grid [1]. However, the issue becomes how to route control lines to all of these tightly packed qubits on a 2D surface? It is also a challenge to couple this many qubits to one another. Typically interactions are nearest-neighbor. In spin qubit, ion trap and neutral atom-based quantum computers, one strategy is to shuttle qubits physically from one place to another to enable interaction between them, but this incurs significant overhead [2–4].

9.3 High-Level Considerations

High-level considerations for quantum computer architectures have been explored in the past [5]. In particular, QCs have been described as collections of physical qubits connected by teleportation-based interconnects. It has been said that communication with the classical controls is limited strictly to measurement results.

However, as one should know by this point of the book, this is an oversimplification. Indeed, in many types of quantum computing (including both superconducting and quantum dot) it is only feasible to achieve a qubit by careful tuning of each quantum dot or SQUID which is a device- and fabrication-dependent procedure involving many classical control voltages. Measurements of various kinds are required to reach operable regimes not limited to qubit measurement: measurement of a charge stability diagram, or DC transport across a SET, for example. Classical controls also provide the fast pulses required to execute gates, whether these be laser pulses or voltages.

The model in [5, p. 26] is however correct from a standpoint of information flow once the low-level considerations of the quantum computer implementation are ignored and we are only concerned with an actual computation. Overall, the model is also sufficient to describe the context for many layout considerations. How do we group our physical qubits into cells which are connected through long-range interconnects? At what point is it useful to separate the qubits into more cells, versus less? Some QC architectures attempt to address this. For example, in [1], Li et al. envision a cross-bar architecture of quantum dots with local qubit interaction via the Heisenberg exchange mechanism as well as long-range entanglement. In the next section we will go into this architecture and more in detail.

The ability to transmit quantum information between processors, sites within processors, etc., is a major consideration of quantum computer architecture. This can be addressed in several type of QC implementations (spin qubit, ion trap, neutral atom, etc.) by shuttling. In some imagined optical architectures it is possible to move quantum information between processors using flying qubits. In others, transducers allow for optical communication over fiber optic cable.

Another major consideration is fault-tolerance. We discussed error correction in an earlier chapter, but our architecture must facilitate error correction. For example, it is difficult if not impossible to create a toric grid of qubits due to the mainly 2D nature of the layers involved in any nanofabrication process. It is however feasible to implement other codes by making sure we have ancillas connected to nearby data qubits in our architecture so that we can perform the CNOTs necessary to implement our syndrome measurement.

High-level considerations for a QPU.

All of DiVincenzo's criteria are still relevant. However, of these, we are particularly interested in the scalability of the physical system when thinking about architecture.

It is very important that any proposed architecture can work at scale, otherwise we will never reach the thousands of qubits required for useful fault-tolerant quantum computing.

9.4 Survey of Architectures

9.4.1 Superconducting Architectures

According to [6], superconducting quantum computer architectures based on charge qubits, flux qubits and phase qubits have emerged as leading candidates for scalable quantum processor architecture.

We learned a bit in Chap. 4 about superconducting qubits like flux qubits and transmons. We also learned that they could be coupled by various types of couplings. A flux tunable coupling was described which could allow us to control the interaction between neighboring qubits, for example.

We however did not go into how these types of couplings could be used to realize a scalable architecture of superconducting qubits. Google's Sycamore and Willow processors, for example, realize large grids of coupled superconducting qubits, with the latest processor (Willow) boasting 105 qubits [7] (Fig. 9.2).

The most remarkable part of the demonstration of Willow is that as qubits are added to a surface code implemented on Willow, the noise decreases. The processor also represents five years of advancement since Sycamore, which realized a smaller grid of 53 qubits (Fig. 9.3).

In these depictions of the Willow and Sycamore architectures, the rectangles represent adjustable couplers and the stars represent qubits.

While these architectures from Google have been some of the most notable, however, IBM has outpaced these in terms of raw qubit count. In the IBM roadmap, a 1121 qubit processor was promised by 2023 [8]. The company passed the 100 qubit milestone in 2021 [9]. The 1121 qubit Condor was indeed unveiled during the

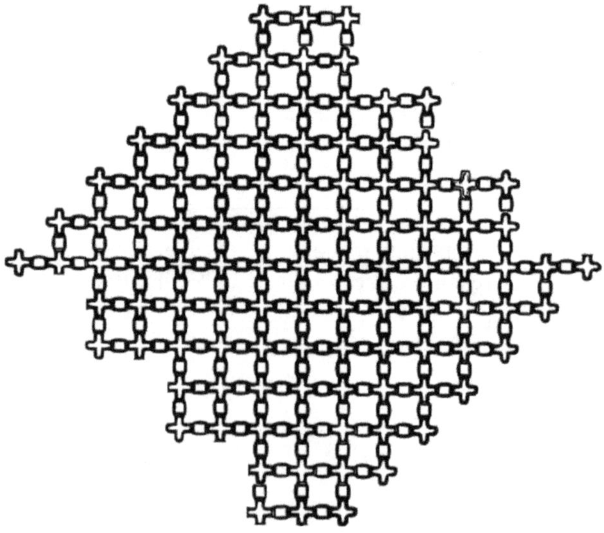

Fig. 9.2 Willow architecture

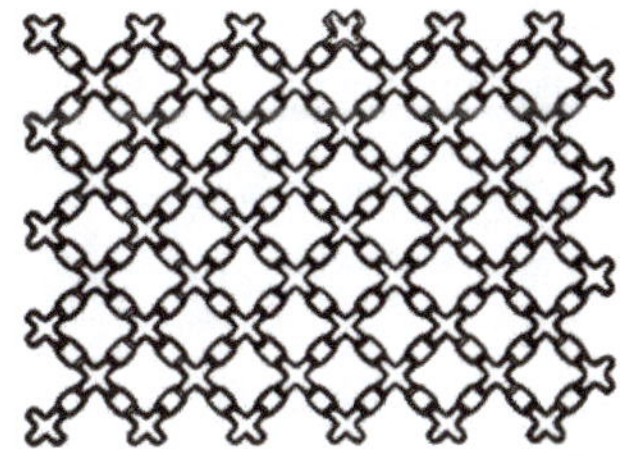

Fig. 9.3 Sycamore architecture

IBM Quantum Summit 2023. The roadmap means that IBM has been doubling their processors' qubit counts by twice or more every year since 2019.

This however comes with its own difficulties. The Condor has over a mile of cryogenic I/O wiring [10]. This is where innovations like D-Wave's Quantum Flux Parametron readout on their 5,000 qubit processor architecture become important as this reduced the number of I/O lines needed to readout all of those qubits. We have usually considered annealer-based QC separately from other superconducting implementations in this book, but at the end of the day it is also based on superconducting qubits. Just remember that these qubits cannot be compared apples-to-apples with Google's or IBM's since they perform such a different function.

Let's look at how an architecture like that of Sycamore or Willow can be used to realize quantum error correction (Fig. 9.4).

In this diagram, we visualize data qubits (open circles) and syndrome measurement qubits (filled circles). As discussed in Chap. 3, measuring a $ZZZZ$ stabilizer involves performing the following circuit with the CNOTs' controls being the data qubits surrounding a measurement qubit a on either side (Fig. 9.5).

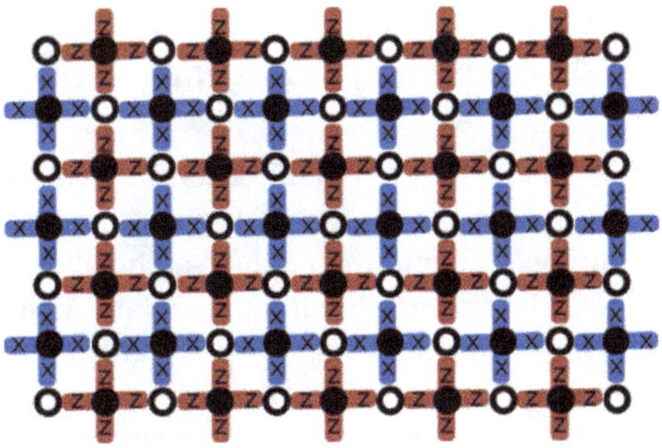

Fig. 9.4 2D surface code

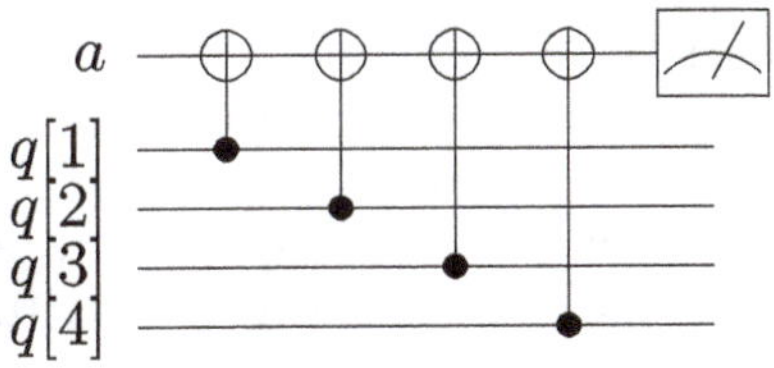

Fig. 9.5 ZZZZ measurement circuit

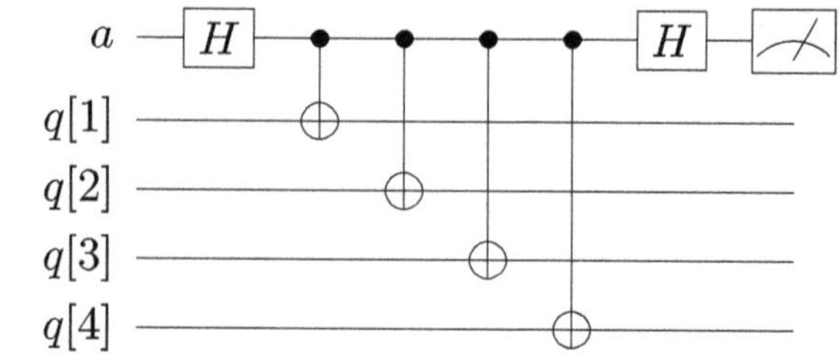

Fig. 9.6 XXXX measurement circuit

Measuring a XXXX stabilizer involves operations that are theoretically equivalent to the following circuit (Fig. 9.6).

These two-qubit gates are, of course, facilitated by couplings between qubits. The red couplings are used for $ZZZZ$ stabilizer measurements and the blue are used for XXXX stabilizer measurements.

It is easy to see how the grid-like superconducting quantum processor architectures lend themselves to 2D surface codes. It is however also easy to see that the couplings are nearest-neighbor. This coupling architecture means we do not have the coveted all-to-all connectivity. We also cannot rely on shuttling to help us when it comes to superconducting implementations. Instead, any dynamic routing of our qubits must be done by SWAP gates. There is an overhead associated with this kind of routing, which is generally a part of a quantum program's compilation. This kind of compilation is targeted to a specific backend architecture and limited by its connectivity map. Connectivity between superconducting chips remains an open problem but it is generally believed that it is important to be able to couple between processors housed in separate dilution refrigerators.

9.4.2 Photonic Architectures

Abughanem [11] lists and describes twenty three unique approaches to photonic quantum computing that have been realized by various entities. The approaches to photonic quantum computer architecture taken by these entities include the use of Field Programmable Photonic Gate Arrays (FPPGAs), fully connected photonic circuits, universal interferometers (capable for example of executing arbitrary 12×12 unitary operators), multi-mode photonics processors with architectures made of tiled unit cells, dynamically programmable loop-based interferometers and more.

For example, the Quix architecture involves a unit cell composed of thermo-optic tunable elements and waveguides which realize a tunable beam-splitter and phase shifter [12]. Here the thermo-optic tunable elements are depicted in blue and the waveguides in black (Fig. 9.7).

The close proximity of the waveguides in these two places realize 50:50 directional couplers. By tiling this unit cell 190 times, a 20-mode photonics processor is realized.

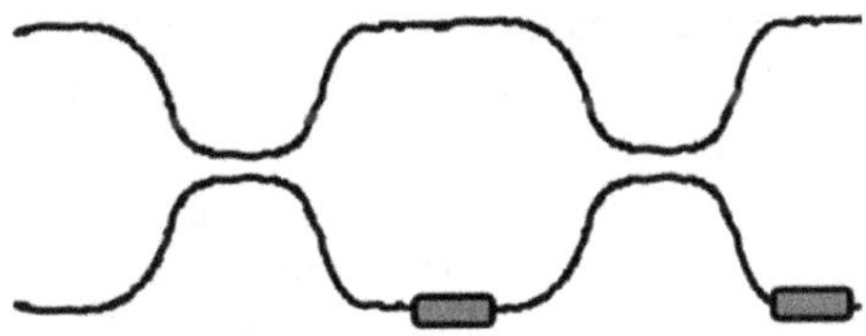

Fig. 9.7 Quix unit cell

Ascella uses a similar architecture of waveguides and thermal phase shifters and has achieved a 12-mode photonic processor thereby [13].

Boson samplers such as Xanadu's Borealis [14] are another class of architectures that are prevalent in the photonic quantum computing world. Boson sampling can be used to demonstrate quantum advantage, performing a sampling from a complex distribution faster than a classical computer could hope to. Boson sampling is however not useful as a quantum computer for general applications since a boson sampler is not a universal quantum computer. Another application-specific photonic circuit is TuringQ's subset sum problem solving three-dimensional waveguide network [15].

In addition to the Borealis demonstration by Xanadu, which was ground-breaking in that it was the first *programmable* Gaussian Boson Sampler, Xanadu published in Nature in January 2025 on another ground-breaking achievement, which they call the "world's first scalable, networked and modular quantum computer" [16, 17]. This modular photonic quantum computer design, known as "Aurora", is based on GKP states. Qubits encoded in GKP states are relatively robust to noise. The Aurora computer has 12 qubits and is a universal quantum computer, unlike a Gaussian Boson Sampler.

Photonic Inc. has publicly proposed an approach to long-range connectivity between their processors involving the use of optically active T-centers which interface with optical fiber at telecom wavelengths [18]. Routing on-chip can be achieved by SWAP operations, and between chips by their novel fiber link.

Error correction has recently been demonstrated in an integrated photonic chip [19] and a $86\% \pm 3.2\%$ state fidelity is achieved. The demonstration is based on an alternative approach to quantum computing called measurement-based quantum computing [20].

9.4.3 Annealer Architectures

D-Wave's annealers have particular architectures which govern which qubits may be coupled to which others [21]. The layout of a D-Wave QPU is important to consider when formulating objective functions for the computer to solve. For example, the Pegasus architecture includes 5640 qubits with 40,484 couplers.

D-Wave quantum computers use two different orientations of qubits. The orientations are vertical and horizontal. D-Wave QPUs also incorporate three different kinds of couplers. The types of couplers are external, internal and odd. External couplers couple qubits in the same row or column between unit cells while internal couplers couple horizontal and vertical qubits that "overlap" within a unit cell and odd couplers couple similarly aligned qubits.

The D-Wave Chimera topology, for example, has the following arrangement of qubits and couplers. Here the vertical qubits are depicted by long red rectangles, the horizontal ones by blue rectangles and the external couplers are in black. Assume internal couplers to exist wherever vertical and horizontal qubits appear to overlap. The figure shows two unit cells, each with eight qubits and sixteen internal couplers (Fig. 9.8).

Each horizontal qubit in a unit cell is coupled to all of the vertical qubits in the cell, and each vertical qubit is coupled to all the horizontal ones. Vertical qubits are not coupled to each other, and neither are horizontal qubits coupled to each other within a unit cell. Therefore, if you draw the connectivity graph with the horizontal qubits in one column and the vertical qubits in another column, the graph is fully connected between the columns and looks just like a Restricted Boltzmann Machine (RBM).

In the later Pegasus topology, the qubits are also shifted horizontally or vertically. This is where odd couplers come in, coupling the similarly aligned qubits. A unit cell contains twenty four qubits.

You can't do error correction on annealers in the same sense as you would do in a gate model quantum computing scenario, since the operations are not available and the function of the qubits is not the same. Instead, you might perform flux drift compensation or calibration refinement [22]. With each cooldown, around 3% of the qubits are also found not to behave properly [21]. These are simply made off-limits to users.

9.4.4 NMR Architectures

In NMR-based quantum computing, the architecture of the quantum processor per se is based on the molecule that hosts the spins which are used to encode qubits. For example, we might use C_{13}, H_1, F_{19}, N_{15} whose nuclei are spin-$\frac{1}{2}$ [23, p. 328]. Irregardless of the number of molecules in the molecular bath that is used, the qubits in different molecules from each other generally interact negligibly. So, while these molecules are all used in parallel computation, they each perform the entire com-

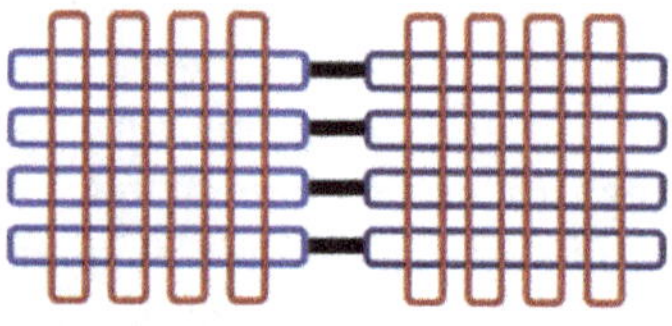

Fig. 9.8 Chimera architecture

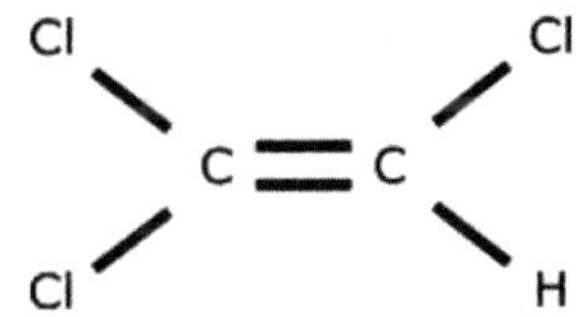

Fig. 9.9 Trichloroethylene molecule

putation and the "processor architecture" is entirely determined by one molecule's morphology (Fig. 9.9).

The interatomic bonds in the molecule used provide the physical basis for the J-couplings that enable two-qubit gates and connectivity between the qubits. In trichloroethylene, for example, the hydrogen and the two carbon atoms' nuclei provide the spins used to encode the qubits. We therefore have a connectivity map of the following sort (Fig. 9.10).

The size and morphology of the molecules used therefore limits what can practically be done with NMR quantum computers. However, error correction was demonstrated with NMR very early on in 1998 [24], using trichloroethylene.

The error correction algorithm demonstrated was a phase flip error correction algorithm. A phase flip is a sign flip that would for example take $|+\rangle$ to $|-\rangle$ by a change of sign.

Error correction algorithms can be decomposed into four parts. First, the system is initialized to a known state (usually $|\mathbf{0}\rangle$). Then redundancy is introduced via a codeword so that error may be diagnosable and recoverable. Next errors are accumulated, usually by random processes. After errors accumulate on the original data qubit and possibly the redundancy qubits, the data in decoded. Finally a correction is applied contingent on the error syndrome.

A simple example of a bit flip error-correcting circuit that can correct bit flip errors on a single data qubit and is also robust to any one bit flip error on the codeword qubits is the following (Fig. 9.11).

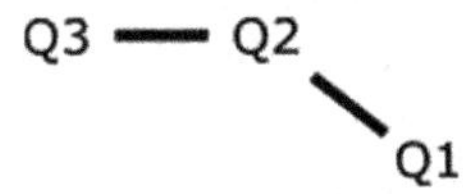

Fig. 9.10 Trichloroethylene qubit connectivity

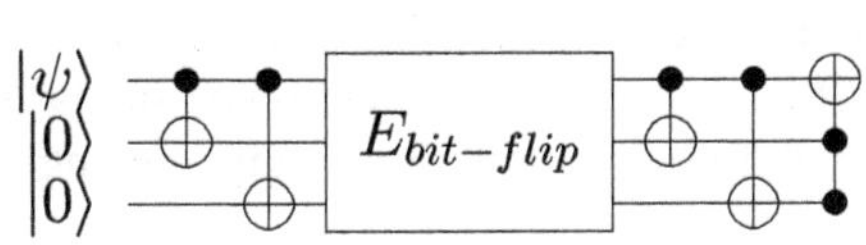

Fig. 9.11 Small bit flip correction circuit

It is easy to see from the corresponding truth table that regardless of any single-bit flip error the initial and final states of the first qubit are identical in all cases.

codeword	error	final state
000	000 (none)	000
111	111 (none)	100
000	100	011
111	011	111
000	010	010
111	101	110
000	001	001
111	110	101

It is easy to see that the circuit is not robust to multiple errors by posing a simple counterexample.

codeword	error	final state
000	011	100

This gives us an idea of the limitations of such an error-correcting code.

A similar error-correcting code that is more relevant to the work of Cory et al. is a phase (sign) flip-correcting circuit. To account for sign flips we simply convert from the Hadamard basis which translates our sign flips to bit flips and use the circuit for bit flip errors with this one change (Fig. 9.12).

In both of these circuits the first two CNOT gates encode the codeword. The E represents an error channel. The last two CNOTs decode an error syndrome which then is used to conditionally apply a correction to the data qubit via a Toffoli gate.

9.4.5 Spin Qubit Architectures

Spin qubit architectures that have been proposed include a scalable architecture based on a one-dimensional array of semiconductor spin qubits [25], more recently the two-dimensional grid-based cross-bar architecture [1] as well as others [26]. In a linear array of quantum dots, error correction has also been demonstrated [27]. Spin qubit architectures are not limited to semiconductor spin qubits but extend also to donors [28], for example. Something particularly interesting is the approach presented by Li et al. to incorporate long-range entanglement between unit cells, something which

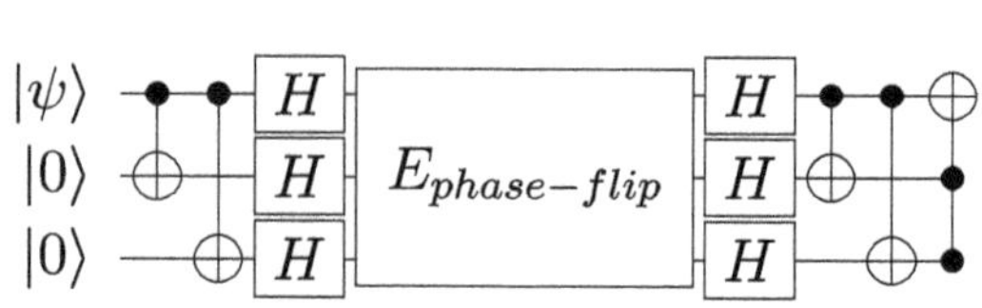

Fig. 9.12 Small phase flip-correcting circuit

we did not see in the superconducting or NMR architectures we reviewed. Another recent example of a spin qubit architecture was realized in Germanium in 2024 [29].

Zajac et. al's one-dimensional array of semiconductor spin qubits was realized in 2016 by the following metalization pattern (Fig. 9.13).

In this figure, the red gates in the top row represent plungers under which electrons are accumulated and the green metalizations are barriers. This arrangement is very similar to what we designed ourselves (if you were following along!) in Chap. 4. The metalization pattern in the second (bottom) row contains charge sensors. This is an example of a linear array realized by three unit cells of the group's scalable architecture. In theory, other (linear) tilings of their unit cell could yield shorter or longer arrays of quantum dots and charge sensors. The unit cell itself consists of three quantum dots and one charge sensor and spans across both rows as shown by the outlined unit cell in the figure. Generally, a single charge sensor can resolve signals relating to multiple qubits if all their dots are distanced *differently* from the sensor. Otherwise, two or more sensors can be used to triangulate the source of a signal from a dot that has a particular distance to each one. The authors comment that scaling limitations are not likely to arise from fabrication considerations but rather from limitations in cryogenic control electronics [25]. Good thing we covered control electronics in Chap. 7.

The cross-bar architecture has the following grid-like structure [1]. In 2018, Li et al. defined the qubits in their cross-bar architecture using the spin states in silicon-based quantum dots (Fig. 9.14).

In this architecture, there are more dots than there are qubits. The array is sparsely initialized so that there are qubits in every other dot (the purple dots in our figure). This is to reduce cross-talk and to allow for long-range entanglement to be achieved by shuttling. Shuttling involves allowing an electron to tunnel from one dot to an adjacent dot, thereby physically relocating the qubit. This operation requires two dots, one from which the electron tunnels and one to which it tunnels. That is why we leave empty dots for electrons to tunnel to—the problem of solving for combinations of shuttles between dots to achieve a particular coupling between qubits is still a problem that incurs overhead and has merited research [2–4].

In the cross-bar architecture, there are three kinds of shared control lines. These are vertical lines (denoted as rows R0 through R2), horizontal lines (denoted as columns C0 through C2) and diagonal lines (denoted as qubit lines Q_{-3} through Q_3). The vertical and horizontal lines are used as barriers which can be tuned to allow for the

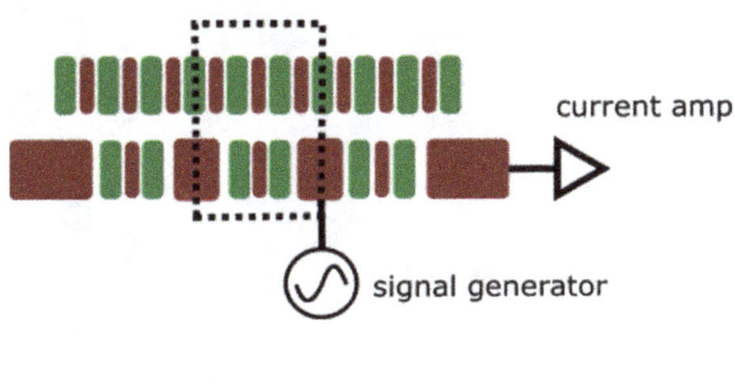

Fig. 9.13 One-dimensional dot array

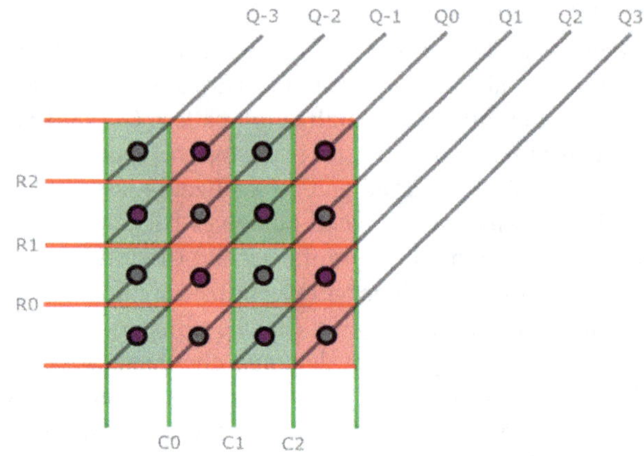

Fig. 9.14 Cross-bar architecture

desired shuttling operation. The diagonal lines also assist with shuttling by way of DC voltages. For example, if we wanted for the qubit between row lines R0 and R1 and between column lines C0 and C1 to shuttle one dot to the left, we would set the voltages $Q_{-1} > Q_{-1} > Q_0$ and lower the voltage on C0. In this way, any two qubits can be moved into proximity of one another so that the Heisenberg exchange is non-negligible (Fig. 9.15). However, it is not always possible to parallelize two shuttling operations since this might require a DC voltage on a diagonal line to be both greater than a neighboring diagonal line's voltage for the purposes of one shuttling operation, and less than it for the purposes of the other.

Recent work has also accounted for readout-aware routing in an architecture that incorporates readout qubits [30]. This was motivated by the fact that previous work assumed an ideal measurement process that did not require an ancilla qubit. However, we know that in order to implement Pauli-Spin-Blockade-based measurement we indeed do need an ancilla.

This architecture therefore introduces periodically spaced readout dots (black) which serve only as ancillas for readout, among the other dots (purple). The idea is

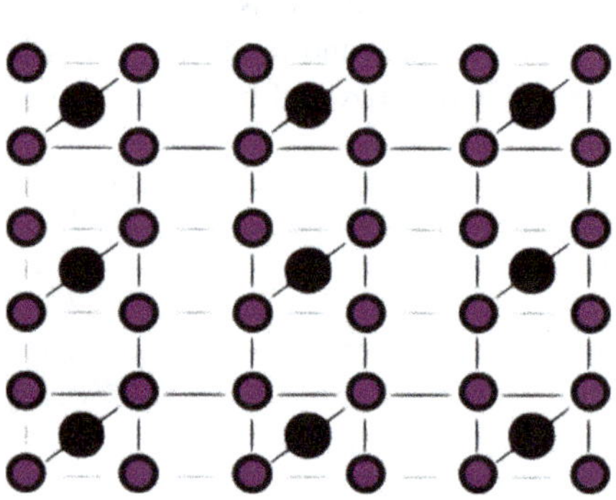
Fig. 9.15 Spin qubit architecture incorporating readout qubits

that any qubit can be shuttled within proximity of a readout dot in order to enable the readout of any qubit.

In a similar vein, a sparsely initialized architecture was realized in Germanium in 2024 [29] with ten non-readout quantum dots. The authors note that increased qubit connectivity lowers both physical qubit overhead and logical qubit error rates. A shuttling strategy similar to that of the other architectures already discussed is taken to couple distant qubits by shuttling them into close proximity.

The experimental demonstration of this architecture involved four readout dots (green) and ten others (purple) separated by twelve barriers (blue) (Fig. 9.16).

As mentioned previously, error correction has been demonstrated in a linear array of quantum dot-based spin qubits [27]. The error correction used in this demonstration was a nine-qubit Shor code. Concatenating a bit flip code with a phase flip code yields the nine-qubit Shor code.

It is worth mentioning that while superconducting architectures have been scaled far beyond these demonstrations of spin qubit architectures, there remains a distinct possibility that semiconductor spin qubits will overcome superconducting qubits in this area. A reason for this is that quantum dots are much smaller than Josephson junctions.

9.4.6 *Ion Trap Architectures*

There are two main technologies that underlie ion trap architectures. The first is the Paul trap, which we have already discussed. The second is the Penning trap, which is a bit different. A Penning trap uses a magnetic field to confine the ion which has cyclotron motion around the axis of the B field. Penning traps, which use static fields, have lower heating rates and therefore higher coherence, but the challenge with them is how to address the ions given their cyclotron motion. Penning trap innovations are mostly happening only in academic and national lab settings at the moment, with examples like NIST and ETH Zurich looking into microfabricated Penning traps [31].

Regarding connectivity, when it comes to Paul and Penning traps, the way to entangled qubits is using motional modes. The qubit is largely unperturbed by shuttling. Coherence is therefore very good in terms of single ion coherence. Heating, however, still causes problems, affecting the ion chain's motion. With n connected

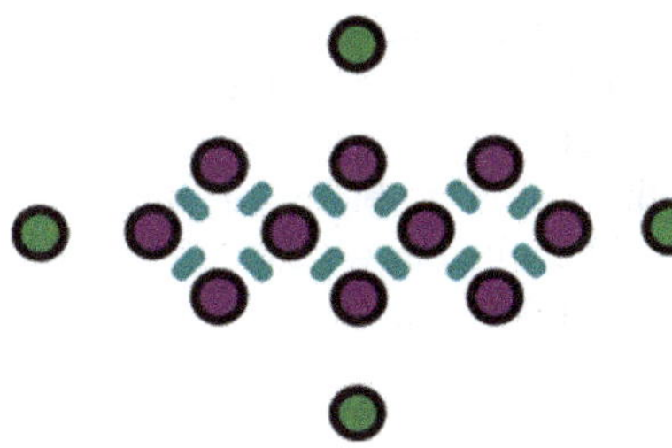

Fig. 9.16 Ten quantum dot architecture realization

oscillators, we have n normal modes of motion. We can detune to address each mode with our lasers. Motional mode coherence is only on the order of tens of ms. Reasonable gate times are tens of microseconds to hundreds of microseconds.

Most ion trap architectures these days are based on chip trap technology. Chip traps can be microfabricated, as they do not involve the charged rods that we discussed earlier regarding the original Paul trap designs. Instead, chip traps make fabrication easier. Researchers figured out that we can keep the topology of the Paul and Penning traps [32], but etch the electrodes on semiconductors and then coat in a conductor (gold most often). So, Si is etched and coated in gold, so we can lay DC electrodes and RF lines in convenient architectures.

In a typical trap, the ions are suspended about 70 um above trap surface. Integrated optics however, channeled through fibers and not free space, allow light to be integrated into the trap. There is therefore no need for careful laser alignment calibration. This is another technology that underlies emerging ion trap architectures. This is not the only technology, as Oxford Ionics uses microwaves to talk with the ions, for example.

We can't grow ion chains infinitely because the axial motion is low frequency and susceptible to 1/f noise. The axial heating that comes from surface dynamics cannot be overcome and limits how long we can grow chains. We might have a reasonable expectation to get to 50-ion chains.

A limited number of ions can fit in a trap. In order to facilitate interactions between any two qubits, there are proposals such as the T-trap [33] which is a shuttling-based trap architecture that allows any two ions to be shuttled into close proximity adjacency. In the figure, electrodes are blue and trapping zones, where ions can be shuttled and held, are purple. Voltage sequences on the electrodes may be used to shuttle ions between adjacent trapping zones. For example, an ion that is in between two others which we wish to couple may be moved out of the way by shuttling it into "storage" in the horizontally oriented trapping zones while the other two are left in the vertically oriented zones. The ions in remaining in the vertically oriented zones may then be shuttled into adjacent zones (Fig. 9.17).

Another architecture is the "racetrack" architecture [34]. Here the orange circles represent gated qubits and purple circles represent qubits that are sitting in storage during gates. The dark gray half-loops are RF electrodes. The top blue zones are trapping zones used for sorting ions and not for applying gates. The green trapping zones are "conveyor belt" regions used for ion storage. Depicted is the right half of the elliptical "racetrack" which is mirrored in the left half (Fig. 9.18).

IonQ wants to build long chains, with about 50 ions, with all-to-all connectivity and photonic interconnects between chains to enable entanglement.

Grid architectures are also proposed that have small zones. We shuttle ions between zones to then entangle them. This routing via shuttling becomes a classical optimization problem for our compiler. There are scaling limitations from classical side, whether we use a Satisfiability-Modulo-Theories (SMT) solver, heuristics-based solver, etc.

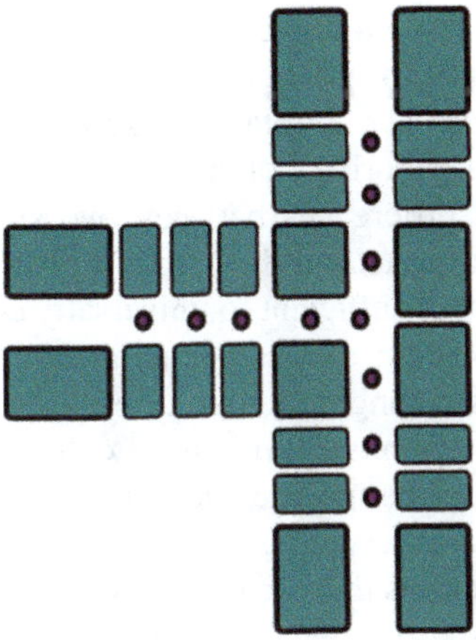

Fig. 9.17 T-trap architecture

Fig. 9.18 Racetrack architecture

Distance 31 bit flip and phase flip codes were demonstrated using the racetrack architecture [34] and a Steane color code has also been demonstrated on a Honeywell ion trap quantum computer [35] with a similar architecture.

9.5 Tutorial: SpinQ

We alluded to the fact that quantum program compilation is architecture-dependent. Well, there are software tools for compiling quantum programs for particular architectures. One such tool is SpinQ, which targets scalable spin qubit architectures [4]. SpinQ targets architectures that support shuttling, and this is what makes it unique. In general, neutral atom and ion trap shuttling involves different constraints than spin qubit shuttling. Multi-agent path-finding can be used to solve these sorting problems, but this is not scalable due to the computational overhead. SpinQ is designed specifically to target programs to spin qubit processors.

In a cross-bar architecture [1], Z rotations are achieved by a well-timed qubit shuttling to and from a neighboring column. The g-factor, precession frequency and spin quantization axis differences between two sites allows us to rotate our state by any angle by shuttling to and then from a given site after a specific time. Remember, two-axis control is sufficient to universally control a qubit. However, X and Y rotations can be achieved in a cross-bar architecture also by control signals on the control lines. All qubits belonging to odd or even columns are rotated together. Therefore, we need to move qubits in and out of columns appropriately to address them. For example, if there is one specific qubit in an even column we wish to rotate, we first rotate it with all the other qubits in even columns. Then, we move it to an odd column. We rotate the qubits that are still in even columns back to their original states by applying the transpose of the operation. Finally, we move the target qubit back to where it began.

How to run an algorithm on a spin qubit processor?

SpinQ takes a quantum algorithm together with an architectural configuration and produces a compiled circuit. This is done in steps.

1. Gate decomposition
2. Initial placement
3. Routing and scheduling.

The routing for two-qubit gates involves finding the shortest Manhattan distance between the two qubits involved and performing shuttle-based SWAPs. The configuration where qubits are on every second site is maintained as much as possible.

In the first pass of the scheduler, X, Y and Z gates are parallelized. In a second pass, any conflicts in the parallelization are resolved by making gates sequential. This has a time complexity of O(n) in the number of gates.

While SpinQ is specific to the cross-bar architecture, beSnake is a follow-on technology that is more flexible in terms of the targeted architecture [2]. beSnake uses a Breadth-First Search (BFS) exploration when deciding how to shuttle qubits.

Now let's take a look at how we might perform the necessary shuttling to do two-qubit gates given a few example configurations (Fig. 9.19).

Here the colored dots are sites with qubits and the gray dots are empty sites. Say we wanted to perform a two-qubit gate on the two blue qubits, and another on the two orange qubits. Then it would be possible to parallelize moving the blue qubit in the first row down, and the blue qubit in the bottom row up. We could do this and then shuttle the orange qubits together either by moving one down or the other up one tile.

Consider a case where there is more of a densely populated grid. In a case like this, if we were using beSnake, the compiler would use a lookahead to explore each

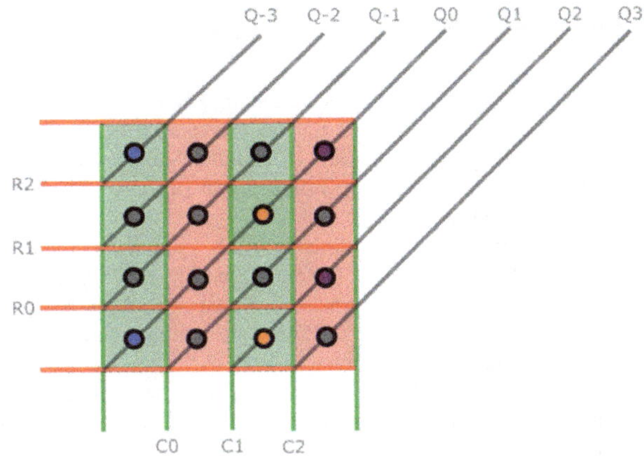

Fig. 9.19 Example cross-bar configuration

possible combination of shuttle operations that could bring us to our goal. If one qubit is in the way of another, we also explore all the ways it can be shuttled out of the way. Say we wanted to perform a two-qubit gate with the two blue qubits. Then you can see that there is a purple qubit, just above the blue qubit in the lowest row, that is in the way (Fig. 9.20).

When considering where to move the troublesome qubit that is in the way, the compiler would see that it is not possible to shuttle it to the left as this space is also occupied. It might choose therefore to shuttle to the right instead, in which case the path is now clear for the blue qubits to be shuttled into proximity of one another.

In cases like this, if we are using beSnake, we could also perform a SWAP gate on the blue qubit in the last row and the qubit that is in its way instead of one of our shuttling operations. This kind of logical SWAP-based routing is not supported in SpinQ.

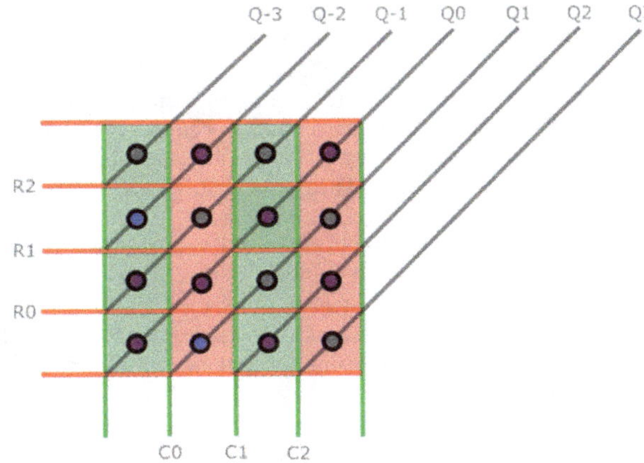

Fig. 9.20 Dense cross-bar configuration

References

1. Li, R., Petit, L., Franke, D.P., Dehollain, J.P., Helsen, J., Steudtner, M., Thomas, N.K., Yoscovits, Z.R., Singh, K.J., Wehner, S., Vandersypen, L.M.K., Clarke, J.S., Veldhorst, M.: A Crossbar Network for Silicon Quantum Dot Qubits. Science Advances **4**(7), eaar3960 (2018). https://doi.org/10.1126/sciadv.aar3960. http://arxiv.org/abs/1711.03807. ArXiv:1711.03807 [cond-mat, physics:quant-ph]
2. Paraskevopoulos, N., Almudever, C.G., Feld, S.: beSnake: A Routing Algorithm for Scalable Spin-Qubit Architectures (2024). https://doi.org/10.48550/arXiv.2403.16090 [quant-ph]
3. Paraskevopoulos, N., Hamel, D., Sarkar, A., Almudever, C.G., Feld, S.: ArtA: Automating Design Space Exploration of Spin Qubit Architectures (2024). https://doi.org/10.48550/arXiv.2407.18151 [quant-ph] version: 1
4. Paraskevopoulos, N., Sebastiano, F., Almudever, C.G., Feld, S.: SpinQ: Compilation Strategies for Scalable Spin-Qubit Architectures (2024). https://doi.org/10.48550/arXiv.2301.13241 [quant-ph]
5. Metodi, T.S., Chong, F.T.: Quantum Computing for Computer Architects. Synthesis Lectures on Computer Architecture. Springer International Publishing, Cham (2006). https://doi.org/10.1007/978-3-031-01718-6. https://link.springer.com/10.1007/978-3-031-01718-6
6. Huang, H.L., Wu, D., Fan, D., Zhu, X.: Superconducting quantum computing: a review. Sci. Chin. Inf. Sci. **63**(8), 180501 (2020). ArXiv:2006.10433 [quant-ph]
7. Acharya, R., Aghababaie-Beni, L., Aleiner, I., Andersen, T.I., Ansmann, M., Arute, F., Arya, K., Asfaw, A., Astrakhantsev, N., Atalaya, J., Babbush, R., Bacon, D., Ballard, B., Bardin, J.C., Bausch, J., Bengtsson, A., Bilmes, A., Blackwell, S., Boixo, S., Bortoli, G., Bourassa, A., Bovaird, J., Brill, L., Broughton, M., Browne, D.A., Buchea, B., Buckley, B.B., Buell, D.A., Burger, T., Burkett, B., Bushnell, N., Cabrera, A., Campero, J., Chang, H.S., Chen, Y., Chen, Z., Chiaro, B., Chik, D., Chou, C., Claes, J., Cleland, A.Y., Cogan, J., Collins, R., Conner, P., Courtney, W., Crook, A.L., Curtin, B., Das, S., Davies, A., De Lorenzo, L., Debroy, D.M., Demura, S., Devoret, M., Di Paolo, A., Donohoe, P., Drozdov, I., Dunsworth, A., Earle, C., Edlich, T., Eickbusch, A., Elbag, A.M., Elzouka, M., Erickson, C., Faoro, L., Farhi, E., Ferreira, V.S., Burgos, L.F., Forati, E., Fowler, A.G., Foxen, B., Ganjam, S., Garcia, G., Gasca, R., Genois, ., Giang, W., Gidney, C., Gilboa, D., Gosula, R., Dau, A.G., Graumann, D., Greene, A., Gross, J.A., Habegger, S., Hall, J., Hamilton, M.C., Hansen, M., Harrigan, M.P., Harrington, S.D., Heras, F.J.H., Heslin, S., Heu, P., Higgott, O., Hill, G., Hilton, J., Holland, G., Hong, S., Huang, H.Y., Huff, A., Huggins, W.J., Ioffe, L.B., Isakov, S.V., Iveland, J., Jeffrey, E., Jiang, Z., Jones, C., Jordan, S., Joshi, C., Juhas, P., Kafri, D., Kang, H., Karamlou, A.H., Kechedzhi, K., Kelly, J., Khaire, T., Khattar, T., Khezri, M., Kim, S., Klimov, P.V., Klots, A.R., Kobrin, B., Kohli, P., Korotkov, A.N., Kostritsa, F., Kothari, R., Kozlovskii, B., Kreikebaum, J.M., Kurilovich, V.D., Lacroix, N., Landhuis, D., Lange-Dei, T., Langley, B.W., Laptev, P., Lau, K.M., Guevel, L.L., Ledford, J., Lee, K., Lensky, Y.D., Leon, S., Lester, B.J., Li, W.Y., Li, Y., Lill, A.T., Liu, W., Livingston, W.P., Locharla, A., Lucero, E., Lundahl, D., Lunt, A., Madhuk, S., Malone, F.D., Maloney, A., Mandrá, S., Martin, L.S., Martin, S., Martin, O., Maxfield, C., McClean, J.R., McEwen, M., Meeks, S., Megrant, A., Mi, X., Miao, K.C., Mieszala, A., Molavi, R., Molina, S., Montazeri, S., Morvan, A., Movassagh, R., Mruczkiewicz, W., Naaman, O., Neeley, M., Neill, C., Nersisyan, A., Neven, H., Newman, M., Ng, J.H., Nguyen, A., Nguyen, M., Ni, C.H., O'Brien, T.E., Oliver, W.D., Opremcak, A., Ottosson, K., Petukhov, A., Pizzuto, A., Platt, J., Potter, R., Pritchard, O., Pryadko, L.P., Quintana, C., Ramachandran, G., Reagor, M.J., Rhodes, D.M., Roberts, G., Rosenberg, E., Rosenfeld, E., Roushan, P., Rubin, N.C., Saei, N., Sank, D., Sankaragomathi, K., Satzinger, K.J., Schurkus, H.F., Schuster, C., Senior, A.W., Shearn, M.J., Shorter, A., Shutty, N., Shvarts, V., Singh, S., Sivak, V., Skruzny, J., Small, S., Smelyanskiy, V., Smith, W.C., Somma, R.D., Springer, S., Sterling, G., Strain, D., Suchard, J., Szasz, A., Sztein, A., Thor, D., Torres, A., Torunbalci, M.M., Vaishnav, A., Vargas, J., Vdovichev, S., Vidal, G., Villalonga, B., Heidweiller, C.V., Waltman, S., Wang, S.X., Ware, B., Weber, K., White, T., Wong, K., Woo, B.W.K., Xing, C., Yao, Z.J., Yeh, P., Ying, B.,

Yoo, J., Yosri, N., Young, G., Zalcman, A., Zhang, Y., Zhu, N., Zobrist, N.: Quantum Error Correction Below the Surface Code Threshold (2024). https://doi.org/10.48550/arXiv.2408.13687 [quant-ph]

8. Gambetta, J.: IBM's roadmap for scaling quantum technology. In: IBM Quantum Computing Blog (2020). https://www.ibm.com/quantum/blog/ibm-quantum-roadmap
9. Jerry Chow Oliver Dial, J.G.: IBM Quantum Breaks the 100-Qubit Processor Barrier. IBM Quantum Computing Blog (2021). https://www.ibm.com/quantum/blog/127-qubit-quantum-processor-eagle
10. Gambetta, J.: IBM quantum system two: the era of quantum utility is here. In: IBM Quantum Computing Blog (2023). https://www.ibm.com/quantum/blog/quantum-roadmap-2033
11. AbuGhanem, M.: Photonic Quantum Computers (2024). https://doi.org/10.48550/arXiv.2409.08229 [quant-ph]
12. Taballione, C., Anguita, M.C., de Goede, M., Venderbosch, P., Kassenberg, B., Snijders, H., Kannan, N., Vleeshouwers, W.L., Smith, D., Epping, J.P., van der Meer, R., Pinkse, P.W.H., van den Vlekkert, H., Renema, J.J.: 20-mode universal quantum photonic processor. Quantum **7**, 1071 (2023). https://doi.org/10.22331/q-2023-08-01-1071
13. Maring, N., Fyrillas, A., Pont, M., Ivanov, E., Stepanov, P., Margaria, N., Hease, W., Pishchagin, A., Lemaître, A., Sagnes, I., Au, T.H., Boissier, S., Bertasi, E., Baert, A., Valdivia, M., Billard, M., Acar, O., Brieussel, A., Mezher, R., Wein, S.C., Salavrakos, A., Sinnott, P., Fioretto, D.A., Emeriau, P.E., Belabas, N., Mansfield, S., Senellart, P., Senellart, J., Somaschi, N.: A versatile single-photon-based quantum computing platform. Nat. Photon. **18**(6), 603–609 (2024)
14. Madsen, L.S., Laudenbach, F., Askarani, M.F., Rortais, F., Vincent, T., Bulmer, J.F.F., Miatto, F.M., Neuhaus, L., Helt, L.G., Collins, M.J., Lita, A.E., Gerrits, T., Nam, S.W., Vaidya, V.D., Menotti, M., Dhand, I., Vernon, Z., Quesada, N., Lavoie, J.: Quantum computational advantage with a programmable photonic processor. **606**(7912), 75–81 (2022). https://doi.org/10.1038/s41586-022-04725-x
15. Xu, X.Y., Huang, X.L., Li, Z.M., Gao, J., Jiao, Z.Q., Wang, Y., Ren, R.J., Zhang, H.P., Jin, X.M.: A scalable photonic computer solving the subset sum problem. Sci. Adv. **6**(5), eaay5853 (2020). https://doi.org/10.1126/sciadv.aay5853
16. Xanadu.: Xanadu Introduces Aurora: World's First Scalable, Networked and Modular Quantum Computer (2025). https://www.xanadu.ai/press/xanadu-introduces-aurora-worlds-first-scalable-networked-and-modular-quantum-computer
17. Aghaee Rad, H., Ainsworth, T., Alexander, R.N., Altieri, B., Askarani, M.F., Baby, R., Banchi, L., Baragiola, B.Q., Bourassa, J.E., Chadwick, R.S., Charania, I., Chen, H., Collins, M.J., Contu, P., D'Arcy, N., Dauphinais, G., De Prins, R., Deschenes, D., Di Luch, I., Duque, S., Edke, P., Fayer, S.E., Ferracin, S., Ferretti, H., Gefaell, J., Glancy, S., González-Arciniegas, C., Grainge, T., Han, Z., Hastrup, J., Helt, L.G., Hillmann, T., Hundal, J., Izumi, S., Jaeken, T., Jonas, M., Kocsis, S., Krasnokutska, I., Larsen, M.V., Laskowski, P., Laudenbach, F., Lavoie, J., Li, M., Lomonte, E., Lopetegui, C.E., Luey, B., Lund, A.P., Ma, C., Madsen, L.S., Mahler, D.H., Mantilla Calderón, L., Menotti, M., Miatto, F.M., Morrison, B., Nadkarni, P.J., Nakamura, T., Neuhaus, L., Niu, Z., Noro, R., Papirov, K., Pesah, A., Phillips, D.S., Plick, W.N., Rogalsky, T., Rortais, F., Sabines-Chesterking, J., Safavi-Bayat, S., Sazhaev, E., Seymour, M., Rezaei Shad, K., Silverman, M., Srinivasan, S.A., Stephan, M., Tang, Q.Y., Tasker, J.F., Teo, Y.S., Then, R.B., Tremblay, J.E., Tzitrin, I., Vaidya, V.D., Vasmer, M., Vernon, Z., Villalobos, L.F.S.S.M., Walshe, B.W., Weil, R., Xin, X., Yan, X., Yao, Y., Zamani Abnili, M., Zhang, Y.: Scaling and networking a modular photonic quantum computer. Nature **638**(8052), 912–919 (2025). https://doi.org/10.1038/s41586-024-08406-9. https://www.nature.com/articles/s41586-024-08406-9
18. Inc, P., Afzal, F., Akhlaghi, M., Beale, S.J., Bedroya, O., Bell, K., Bergeron, L., Bonsma-Fisher, K., Bychkova, P., Chaisson, Z.M.E., Chartrand, C., Clear, C., Darcie, A., DeAbreu, A., DeLisle, C., Duncan, L.A., Smith, C.D., Dunn, J., Ebrahimi, A., Evetts, N., Pinheiro, D.F., Fuentes, P., Georgiou, T., Guha, B., Haenel, R., Higginbottom, D., Jackson, D.M., Jahed, N., Khorshidahmad, A., Shandilya, P.K., Kurkjian, A.T.K., Lauk, N., Lee-Hone, N.R., Lin, E., Litynskyy, R., Lock, D., Ma, L., MacGilp, I., MacQuarrie, E.R., Mar, A., Khah, A.M., Matiash, A., Meyer-Scott, E., Michaels, C.P., Motira, J., Noori, N.K., Ospadov, E., Patel, E.,

Patscheider, A., Paulson, D., Petruk, A., Ravindranath, A.L., Reznychenko, B., Ruether, M., Ruscica, J., Saxena, K., Schaller, Z., Seidlitz, A., Senger, J., Lee, Y.S., Sevoyan, O., Simmons, S., Soykal, O., Stott, L., Tran, Q., Tserkis, S., Ulhaq, A., Vine, W., Weeks, R., Wolfowicz, G., Yoneda, I.: Distributed Quantum Computing in Silicon (2024). https://doi.org/10.48550/arXiv.2406.01704 [quant-ph]

19. Zhang, H., Wan, L., Paesani, S., Laing, A., Shi, Y., Cai, H., Luo, X., Lo, G.Q., Kwek, L.C., Liu, A.Q.: Encoding Error Correction in an Integrated Photonic Chip. PRX Quant. **4**(3), 030340 (2023)
20. Raussendorf, R., Briegel, H.J.: Quantum Computing Via Measurements Only (2000). https://arxiv.org/abs/quant-ph/0010033
21. D-Wave QPU Architecture: Topologies-D-Wave System Documentation. https://docs.dwavesys.com/docs/latest/c_gs_4.html
22. Error-Correction Features-D-Wave System Documentation. https://docs.dwavesys.com/docs/latest/c_qpu_error_correction.html
23. Nielsen, M.A., Chuang, I.L.: Quantum Computation and Quantum Information 10th Anniversary. Cambridge University Press, Cambridge (2010)
24. Cory, D.G., Price, M.D., Maas, W., Knill, E., Laflamme, R., Zurek, W.H., Havel, T.F., Somaroo, S.S.: Experimental quantum error correction. Phys. Rev. Lett. **81**(10), 2152–2155 (1998). https://doi.org/10.1103/PhysRevLett.81.2152. https://doi.org/10.1103/PhysRevLett.81.2152
25. Zajac, D., Hazard, T., Mi, X., Nielsen, E., Petta, J.: Scalable gate architecture for a one-dimensional array of semiconductor spin qubits. Phys. Rev. Appl. **6**(5), 054013 (2016). https://doi.org/10.1103/PhysRevApplied.6.054013
26. Veldhorst, M., Eenink, H.G.J., Yang, C.H., Dzurak, A.S.: Silicon CMOS architecture for a spin-based quantum computer. Nat. Commun. **8**(1), 1766 (2017)
27. Jones, C., Fogarty, M.A., Morello, A., Gyure, M.F., Dzurak, A.S., Ladd, T.D.: Logical qubit in a linear array of semiconductor quantum dots. Phys. Rev. X **8**(2), 021058 (2018)
28. Vandersypen, L.M.K., Bluhm, H., Clarke, J.S., Dzurak, A.S., Ishihara, R., Morello, A., Reilly, D.J., Schreiber, L.R., Veldhorst, M.: Interfacing spin qubits in quantum dots and donors–hot, dense, and coherent. npj Quant. Inform. **3**(1), 1–10 (2017). https://doi.org/10.1038/s41534-017-0038-y
29. Wang, C.A., John, V., Tidjani, H., Yu, C.X., Ivlev, A., Déprez, C., van Riggelen-Doelman, F., Woods, B.D., Hendrickx, N.W., Lawrie, W.I.L., Stehouwer, L.E.A., Oosterhout, S., Sammak, A., Friesen, M., Scappucci, G., de Snoo, S.L., Rimbach-Russ, M., Borsoi, F., Veldhorst, M.: Operating semiconductor quantum processors with hopping spins (2024). http://arxiv.org/abs/2402.18382 [cond-mat]
30. Thombre, R., Edwards, M., Salfi, J., Di Matteo, O.: Towards Readout-Aware Layout Synthesis for Spin Qubit Systems with Double Quantum Dot Readouts . In: 2024 IEEE International Conference on Quantum Computing and Engineering (QCE), pp. 589–590. IEEE Computer Society, Los Alamitos (2024). https://doi.org/10.1109/QCE60285.2024.10419
31. NIST.: Compact Penning Traps (2015). https://www.nist.gov/pml/quantum-measurement/atomic-spectroscopy/compact-penning-traps. Last Modified: 2016-08-25T13:49-04:00
32. Jain, S., Sägesser, T., Hrmo, P., Torkzaban, C., Stadler, M., Oswald, R., Axline, C., Bautista-Salvador, A., Ospelkaus, C., Kienzler, D., Home, J.: Penning micro-trap for quantum computing. Nature **627**(8004), 510–514 (2024)
33. Hensinger, W.K., Olmschenk, S., Stick, D., Hucul, D., Yeo, M., Acton, M., Deslauriers, L., Rabchuk, J., Monroe, C.: T-junction ion trap array for two-dimensional ion shuttling, storage and manipulation. Appl. Phys. Lett. **88**(3), 034101 (2006). ArXiv:quant-ph/0508097
34. Moses, S., Baldwin, C., Allman, M., Ancona, R., Ascarrunz, L., Barnes, C., Bartolotta, J., Bjork, B., Blanchard, P., Bohn, M., Bohnet, J., Brown, N., Burdick, N., Burton, W., Campbell, S., Campora, J., Carron, C., Chambers, J., Chan, J., Chen, Y., Chernoguzov, A., Chertkov, E., Colina, J., Curtis, J., Daniel, R., DeCross, M., Deen, D., Delaney, C., Dreiling, J., Ertsgaard, C., Esposito, J., Estey, B., Fabrikant, M., Figgatt, C., Foltz, C., Foss-Feig, M., Francois, D., Gaebler, J., Gatterman, T., Gilbreth, C., Giles, J., Glynn, E., Hall, A., Hankin, A., Hansen, A., Hayes, D., Higashi, B., Hoffman, I., Horning, B., Hout, J., Jacobs, R., Johansen, J., Jones,

L., Karcz, J., Klein, T., Lauria, P., Lee, P., Liefer, D., Lu, S., Lucchetti, D., Lytle, C., Malm, A., Matheny, M., Mathewson, B., Mayer, K., Miller, D., Mills, M., Neyenhuis, B., Nugent, L., Olson, S., Parks, J., Price, G., Price, Z., Pugh, M., Ransford, A., Reed, A., Roman, C., Rowe, M., Ryan-Anderson, C., Sanders, S., Sedlacek, J., Shevchuk, P., Siegfried, P., Skripka, T., Spaun, B., Sprenkle, R., Stutz, R., Swallows, M., Tobey, R., Tran, A., Tran, T., Vogt, E., Volin, C., Walker, J., Zolot, A., Pino, J.: A race-track trapped-ion quantum processor. Phys. Rev. X **13**(4) (2023). https://doi.org/10.1103/physrevx.13.041052

35. Ryan-Anderson, C., Bohnet, J., Lee, K., Gresh, D., Hankin, A., Gaebler, J., Francois, D., Chernoguzov, A., Lucchetti, D., Brown, N., Gatterman, T., Halit, S., Gilmore, K., Gerber, J., Neyenhuis, B., Hayes, D., Stutz, R.: Realization of Real-Time Fault-Tolerant Quantum Error Correction. Phys. Rev. X **11**(4), 041058 (2021)

Chapter 10
Cloud Quantum Computing

Abstract Once we have a fully functional quantum computer in our lab, it is important that we make it accessible and easy to use through the cloud. We give a brief overview of existing quantum cloud solutions offered by Xanadu, IBM, D-Wave, Amazon, etc., and how to use them, with examples. Toward the end of the chapter, we implement our own quantum cloud using common web development tools and techniques.

10.1 Learning Objectives

Your first interaction with quantum computing was likely through the cloud, whether it was through IBM's Quantum Experience platform or you were programming Xanadu's web accessible X8 computer. We will learn about the services that might provide access to quantum computers or simulators in the cloud through several learning objectives (Fig. 10.1):

- Understand the landscape of quantum cloud offerings
- See examples of quantum programs written for each platform
- Gain practical skills by developing our own minimal cloud quantum simulator.

10.2 Web Development

Web development is a discipline in its own, with its own principles and references. In this chapter we will cover the basics needed to develop your own cloud quantum computing service using modern web development tools and techniques. This will not be an exhaustive introduction to software engineering, however. For a more complete reference which goes into depth on programming models related to quantum computing, please see Quantum Software Engineering by Serrano et al. [1]. Before delving into these quantum-specific software engineering concepts, it would also be a good idea to familiarize yourself with software engineering basics from sources

© The Author(s), under exclusive license to Springer Nature Switzerland AG 2026

M. S. Edwards, *An Introduction to Quantum Computing for Computer Engineers*,
https://doi.org/10.1007/978-3-032-03650-6_10

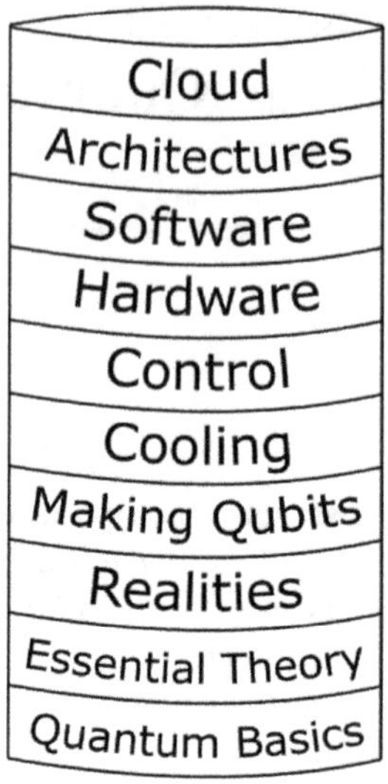

Fig. 10.1 Learning objective for this chapter: cloud

like the classic titles Clean Code and Clean Architecture from Robert Martin[1] [2, 3]. Many of the principles we will employ in the development of our cloud quantum computing service are derived from these foundational references. When it comes to code, though, the best way to learn is by doing, and if you follow along you should be off to a great start!

This topic has a place in this book because it has been recognized almost across the board as a necessary part of a quantum computing ecosystem by every major player in the quantum computing space.

The landscape of cloud quantum computing.

Each company including D-Wave, Xanadu, Google, IBM, Amazon, Microsoft, Rigetti and more have developed their own cloud quantum computing systems. Without these systems it would be incredibly inconvenient for users to interact with their quantum computers.

The control software that we discussed in Chap. 8 may be enough for an experimentalist in a lab to work with, but when we want to support a larger community of users we need a way to ingest requests, handle this load and return meaningful results to many users at once. The software covered in this chapter dovetails with the software covered in Chap. 8 by passing user-provided algorithms and experimental parameters to the control software sitting on the classical control computer beside

[1] Martin's books and methodologies are so widely known and adopted that software engineers everywhere refer to him endearingly as "Uncle Bob".

the quantum computer. This could look like providing QASM or QIR, for example, to the low-level software sitting a layer closer to the metal.

These cloud systems are more than frontends for quantum computers, however. They are full-stack engineering projects in themselves which require the development and integration of components including compilers, parsers, application programming interfaces (APIs), controllers, models, databases and more. This is easily enough to worry about for a dedicated team of experts at any company. It is also most often where companies create community engagement through open source initiatives and free software. There are communities of voluntary contributors around IBM's Qiskit, Xanadu's Pennylane, Google's Cirq, D-Wave's Ocean, Amazon's BraKet, Rigetti's PyQuil and Microsoft's Q# which all have free open source components. This is not an exhaustive list.

As a savvy reader may have already noticed, I have eluded to the Model-View-Controller (MVC) model of software applications. We will indeed adopt this model in particular areas of our project. We will also look at how to build a future-proof scalable backend with a microservices architecture, which is a common approach to building high-available backends in industry. Famously, Facebook began as a PHP monolith (the opposite of a microservice) and was then fragmented in a later effort into a scalable service mesh. Microservices do just that: break a monolithic project, maybe contained in one source repository, into a number of smaller "services" that may be spun up at any time and in any number in response to the specific quantity and type of requests coming into the system from users.

On the backend we will work with Python, which is largely dominant in the quantum computing space. Though IonQ also writes Rust and Rigetti writes Lisp and you might want to familiarize yourself with these languages if you are looking to work at either of these companies, Python is the best bet to broaden your horizons. For parallelization we will make use of a Celery event queue. We will build a Flask-based application since Flask is lightweight compared to alternatives like Django which are perhaps better suited to projects where all of the functionality is web-centric. These packages basically allow us to write a server application that serves webpages and content to our frontend. In our case, this content might be a simple web form or it might be the result of a measurement from an actual quantum processing unit. We will introduce and make use of Docker, Kubernetes and Google Cloud Platform on the DevOps side of things, which can be enough for a dedicated Developer Operations or "DevOps" engineer to worry about at any given company. Our tutorial will provide code and examples for users of Mac OS and Ubuntu, but should not be hard to generalize to Unix and Linux-based operating systems.

10.3 Some Cloud Offerings

10.3.1 D-Wave Ocean

To give an idea of what is offered by these companies, let's start with D-Wave's Ocean ecosystem. In a previous chapter we have already covered how D-Wave's computing model requires problems to be expressed as quadratic unconstrained Boolean optimizations (QUBOs) before they can be minor-embedded and solved via adiabatic potential energy minimization, or annealing, which is argued to make use of quantum tunneling. In order to facilitate this workflow, a number of intermediate technologies are provided. These include packages containing QUBO models, a binary quadratic model constructor, a REST interface, a quantum circuit builder, preprocessing tools, visualizers and more. Seven of the fifteen open source packages available to developers in the Ocean ecosystem relate to the construction and use of nonlinear and quadratic models. This is very specific to the platform. The Ocean SDK is a top-level repository for the Ocean tools which helps with the installation of the entire toolset. It uses git submodules to organize the various packages in the ecosystem, keep them up-to-date and easy to install together. This is a common pattern which we will also see how to use in this chapter. Like most of the packages we will mention, D-Wave's are written in Python.

For example, we could make use of pyqubo to solve for Ising Hamiltonian parameters that would encode a particular problem.

Example: PyQUBO

```
from pyqubo import Constraint, Spin

# Declare qubits
A0, A1, B1, B0 = Spin("A0"), Spin("A1"), Spin("B1"), Spin("B0")

# Optional strength of the constraint, or importance of the corresponding problem
M = 1.0

# Tell the compiler that (A0+A1*B1+B0-1)**2 is a constraint which should be zero when
# the solution is not broken. This is equivalent to requiring that A0+A1*B1+B0 is one
constraint = Constraint((A0+A1*B1+B0-1)**2, 'error_condition')

# Define the Hamiltonian in terms of its objective function and constraint
exp = A0+A1*B1+B0 + M * constraint

model = exp.compile()
qubo = model.to_qubo()
```

These parameters can then be submitted to D-Wave via the Ocean API.

10.3.2 Xanadu AI

Xanadu's ecosystem revolves around packages like StrawberryFields and Pennylane. Yes, the company playfully names their packages after classic Beatles songs! StrawberryFields is a Python package that provides users with a programming interface to the company's photonic quantum computing hardware. As opposed to the packages we will review which deal strictly with the gate model of quantum computing, StrawberryFields allows users to compose photonic circuits involving elements like beamsplitters and photon detectors. Pennylane, on the other hand, is the company's package for gate model quantum computing with an emphasis on quantum machine learning. For a good introduction to quantum machine learning, see the informative reference [4] and the seminal quantum approximate optimization algorithm paper [5]. Pennylane is a more direct competitor to IBM's Qiskit. Xanadu has many more (78 in total) packages including an implementation of their own quantum assembly language called Blackbird and Gaussian boson sampling focused package named The Walrus, for example. They have an active open source community and getting involved in the development of their ecosystem is made easy by good documentation, git issues labeled based on their difficulty/beginner contributor friendliness and responsive reviewers.

Let's take a quick look at what it would be like to use Pennylane to implement an algorithm from a well-known quantum computing paper on efficient state tomography [6]. This implementation comes from a larger codebase that won the 2024 installation of Xanadu's annual in-person hackathon event, which is considered very competitive [7]. The main idea behind the 2010 paper by Cramer et al. that introduced MSVT is that state tomography can be done efficiently using local measurements and unitary control if the state being characterized is a Matrix Product State (MPS) [8]. So, let's look at how to implement state tomography on one qubit.

Quantum State Tomography (QST) [9] typically scales exponentially in the system size. In order to fully characterize a quantum state, QST uses a complete set of observables' measurements. After these measurements are performed, the state must be reconstructed from the individual measurement outcomes collected over many shots. This task of reconstruction has undesirable scaling characteristics.

Some common approaches to state tomography are to use the Pauli basis or the Bell basis, for example. Measurements are performed in orthonormal bases, but prior to each such measurement, the state is also rotated a specific amount. This allows us to collect more information about the state. In fact, the more different rotations we make before our measurements, the more accurate our tomographic state reconstruction can be. Observables are defined, for example, using phases of $\frac{2\pi r}{p}$ which are repeat for p values of r. For each phase, the state is measured with respect to each basis operator. The basis must be complete, meaning that its operators span the entire state space from which we are sampling. In the Pauli basis, we would measure X, Y and Z by collapsing the system to an eigenstate of each Hamiltonian, as seen in Chapter 2. Most quantum computers that have been made available via the cloud are limited, however, to z-basis measurements. We therefore must perform rotations to map the eigenstates of our various basis operators to the computational basis states $|0\rangle$ and $|1\rangle$. If we are considering Pauli basis QST we will be interested in using the Hadamard gate to map the X eigenstates $|+\rangle$ and $|-\rangle$ to the computational basis states, and $HS^{\dagger}$ to map the Y eigenstates to $|0\rangle$, $|1\rangle$.

Pennylane does not have a native implementation of the $S^{\dagger}$ gate, which gives us an opportunity to take a look at how a gate is defined in Pennylane. We can define our own $S^{\dagger}$ with the below code snippet.

```
class SDG(Operation):
    "$\textcolor{mauve}{""}$
    S dagger implementation for pennylane.
    $\textcolor{mauve}{""}$"
    num_params = 0
    num_wires = 1
    par_domain = "R"

    @property
    def matrix(self):
        return np.array(
            [
                [1, 0],
                [0, −complex(0, 1)]
            ]
        )

    def decomposition(self):
        return [qml.U1(3 ∗ np.pi / 2, wires=self.wires)]
```

We can see that a gate defined in Pennylane has a *decomposition*, a *matrix* property, a property specifying how many parameters it takes (if the gate is parameterized), the number of wires (or qubits) the gate applies to and a parameter domain (which will help Pennylane to determine whether the *Operator* is differentiable, etc.). In the matrix property we specify an exact matrix representation of the *Operator*. Notice how in addition to Sympy, Pennylane is built on Numpy which it uses for its arrays and more. In the *decomposition*, we decompose the gate into native gates such as the U1 gate, which allows us to specify the ϕ in a gate of the form $\begin{bmatrix} 1 & 0 \\ 0 & \exp(i\phi) \end{bmatrix}$.

For this example implementation we limit ourselves to 1-qubit MPS sites and 1-qubit state tomography. The 1-qubit QST implementation is provided below.

Example: Tomography with Pennylane

```
def tomography(state_prep, qubit, phases):
    "$\textcolor{mauve}{""}$
    Performs measurements in each of the three orthogonal axes of the Bloch sphere to determine the qubit's state.

    :param state_prep: The state prep to inject into each of our tomography circuits.
    :param qubit: The target qubit.
    :param phases: The number of relative phases to measure with respect to each axis.
    that it has been split up over time.

    :return: The results from each circuit.
    $\textcolor{mauve}{""}$"
```

```
results = []
bloch_vectors = []

exp_vector = range(0, phases)
index = 0

while index < len(exp_vector):
    phase = 2 * np.pi * index / (len(exp_vector) - 1)
    index += 1

    @qml.qnode(dev)
    def x_circuit(phase, qubit):
        state_prep()
        qml.U1(phase, wires=qubit)
        qml.Hadamard(wires=qubit)
        qml.measure(qubit)
        return qml.probs(qubit)

    @qml.qnode(dev)
    def y_circuit(phase, qubit):
        state_prep()
        qml.U1(phase, wires=qubit)
        SDG(wires=qubit)
        qml.Hadamard(wires=qubit)
        qml.measure(qubit)
        return qml.probs(qubit)

    @qml.qnode(dev)
    def z_circuit(phase, qubit):
        state_prep()
        qml.U1(phase, wires=qubit)
        qml.measure(qubit) # pennylane lets us measure other things!
        return qml.probs(qubit)


    results.append(x_circuit(phase, qubit))
    results.append(y_circuit(phase, qubit))
    results.append(z_circuit(phase, qubit))

bloch_vectors = analyze_tomographic_results(
    qubit,
    exp_vector,
    results,
    [qml.PauliX(wires=0), qml.PauliY(wires=0), qml.PauliY(wires=0)]
)

return results, bloch_vectors
```

Notice the patterns we use in this code snippet which are specific to Pennylane. In particular, we use the decorator *qml.qnode* to decorate our functions which become self-encapsulated quantum programs to be run on the quantum backend. In this decorator, we specify the backend we intend the program, or circuit, to run on. We might instantiate a backend with the following line of code if we wanted to use the default qubit simulator.

```
dev = qml.device('default.qubit', wires=2, shots=1000)
```

Pennylane boasts of some backend agnosticism, however, and we may have just as well wanted to run our code on a Cirq simulator, or a real hardware backend.

```
dev = qml.device('cirq.simulator', wires=2)
```

If we had specified a real hardware backend, a request would be made by the Python package sitting on your computer to the RESTful web API sitting on a web server. Before these requests get there it is likely load-balanced. It might sit on a queue for a time if there are many such requests coming in, and it might even be throttled if there are too many. We could for example have used an IonQ QPU backend.

```
dev = qml.device('ionq.qpu', backend="aria-1", wires=2)
```

Pennylane supports eight built-in as well as two community provided simulation-related devices which can be specified as backends. It also supports execution on remote hardware backends and additional simulators through plugins, including IBM Quantum Experience, Amazon Braket, Cirq, Qrack, Microsoft QDK, Qulacs, IonQ, AQT, Rigetti, Quantum Inspire, ProjectQ, StrawberryFields, Orquestra and Honeywell backends. All a developer needs to do from their laptop is provide a descriptor and configuration for the appropriate backend, and their Pennylane code will be executed remotely.

Pennylane allows us to specify what sort of results we want to roll up from our measurement outcomes too, including the most traditional choice of an outcome probability histogram using *qml.probs(qubit)*. We could however have chosen *qml.expval* or *qml.sample* which would have returned the expectation value of a specific measurement and a set of raw samples from a supplied observable respectively.

After performing the tomography circuits, we need to reconstruct the state. One way to do this is to reconstruct a set of Bloch vectors.

This example has given us insight into what it is like to work with a gate model quantum programming framework. It has not highlighted the strength of the package in quantum machine learning specifically. So, let's revisit the Ising Hamiltonian by taking a look at one of the challenges posed in Xanadu's 2023 hackathon, restated here.

Example: Variational Quantum Eigensolver with Pennylane

The following code will simulate the effect of a magnetic field on a closed spin chain. The spin chain is comprised of qubits at each of N sites which are initialized in either the $|0\rangle$ or $|1\rangle$ states along the transverse z-axis-oriented magnetic field. The magnetic field strength is given by h.

$$H = -\sum_{i=1}^{N} Z_i \otimes Z_{i+1} - h \sum_{i}^{N} X_i$$

It is possible to calculate the effects of a strong magnetic field on the ground energy of such a spin chain using the Variational Quantum Eigensolver (VQE) [10] algorithm. We do so for a chain of length four, where the $N+1$ label is associated with the first qubit.

```
dev = qml.device("default.qubit", wires=4)

def create_Hamiltonian(h):
    "$\textcolor{mauve}{""}$
    Function in charge of generating the Hamiltonian of the statement.

    Args:
        h (float): magnetic field strength

    Returns:
        (qml.Hamiltonian): Hamiltonian of the statement associated to h
    $\textcolor{mauve}{""}$"
    coeffs = []
    observables = []

    for i in range(4):
        if i != 3:
            j = i + 1
        else:
            j = 0
        coeffs.append(−1)
        observables.append(qml.PauliZ(wires=[i]) @ qml.PauliZ(wires=[j]))

    for k in range(4):
        coeffs.append(−h)
        observables.append(qml.PauliX(wires=[k]))

    return qml.Hamiltonian(coeffs, observables)
```

See how we can instantiate a Hamiltonian with a list of weighted observables in Pennylane. This is very handy for when we want to measure Hamiltonians that aren't simple Paulis.

```
@qml.qnode(dev)
def model(params, H):
    "$\textcolor{mauve}{""}$
    To implement VQE you need an ansatz for the candidate ground state!
    Define here the VQE ansatz in terms of some parameters (params) that
    create the candidate ground state. These parameters will
    be optimized later.

    Args:
        params (numpy.array): parameters to be used in the variational circuit
        H (qml.Hamiltonian): Hamiltonian used to calculate the expected value

    Returns:
        (float): Expected value with respect to the Hamiltonian H
    $\textcolor{mauve}{""}$"
    for w in range(4):
        qml.RX(params[w ∗ 2], wires=w)
        qml.RZ(params[w ∗ 2 + 1], wires=w)

    return qml.expval(H)
```

Here is where we begin to implement VQE. VQE requires an ansatz—or initial guess—for the ground state energy. We provide such a guess here in terms of a number of parameters. We will see how these parameters can be used in an optimization loop.

```
def train(h):
    "@\textcolor{mauve}{""}@
    In this function you must design a subroutine that returns the
    parameters that best approximate the ground state.

    Args:
        h (float): magnetic field strength

    Returns:
        (numpy.array): parameters that best approximate the ground state.
    @\textcolor{mauve}{""}@"


    def cost(params):
        "@\textcolor{mauve}{""}@ Define a cost function that only depends on params, given alpha and beta fixed
            @\textcolor{mauve}{""}@"

        return model(params, create_Hamiltonian(h))

    # Initialize parameters, choose an optimization method and number of steps
    init_params = np.array([0.2, 0.2, 0.2, 0.2, 0.2, 0.2, 0.2, 0.2], requires_grad=True)
    opt = qml.AdamOptimizer(stepsize=0.1)
    steps = 100

    # set the initial parameter values
    params = init_params

    for i in range(steps):
        # update the circuit parameters

        params = opt.step(cost, params)

    return params
```

Now we get to the QML. This code highlights the setup for quantum machine learning algorithms with Pennylane, including defining a cost function which our ML attempts to minimize (the energy), and varying parameters until that cost function is minimized. For a given magnetic field strength, then, this code can approximate our ground state variationally. The Adam optimizer is a useful tool that performs gradient descent to minimize cost functions, which is one of the most ubiquitous ML algorithms. Xanadu provides various optimizers, most but not all of which are gradient-descent-based.

Gradient descent can be performed in neural networks of various types, but what is always necessary is that the neurons have a specific nonlinear activation function which summarizes their many inputs from a previous layer into a single digital output. The previous layer could be the input layer, or another layer of neurons, to which the current layer may be fully or sparsely connected. The goal of the optimization is to approximate an overall transfer function to minimize the cost function designated. That means as the machine "learns", its true goal is to achieve the following, where C is the cost function and w are the weights in the neural network.

$$\frac{\partial C}{\partial w}(w) = 0$$

This optimization is achieved using a method called back propagation. We define some helpful function to summarize the information within the neural network, where (L) denotes a layer in the network and σ is the activation. Here $b^{(L)}$ is a bias on the layer L and $w^{(L)}a^{(L-1)}$ are the weighted values from the previous layer.

$$z^{(L)} = w^{(L)}a^{(L-1)} + b^{(L)}$$
$$a^{(L)} = \sigma(z^{(L)})$$

Then we can define the derivative of the cost in relation to the weights for a layer explicitly using the chain rule.

$$\frac{\partial C}{\partial w^{(L)}} = \frac{\partial z^{(L)}}{\partial w^{(L)}} \frac{\partial a^{(L)}}{\partial z^{(L)}} \frac{\partial c^{(L)}}{\partial a^{(L)}}$$

We can then define the derivative of the full cost function for a system trained on multiple input sets, where C_k is the observed cost of processing a particular input set k.

$$\frac{\partial C}{\partial w^{(L)}} = \frac{1}{n} \sum_{k=0}^{n-1} \frac{\partial C_k}{\partial w^{(L)}}$$

This is the gradient of the cost which is "descended" after each dataset is used to train the system. The direction of the gradient simply informs the direction of an adjustment of the parameters, weights and biases of the system.

These are classical machine learning concepts, but what is quantum machine learning? Quantum machine learning can be broken down into a number of sub-categories. The first is quantum-inspired machine learning, wherein classical machine learning algorithms are designed with inspiration from quantum physics. Secondly, we have hybrid quantum classical machine learning where a part of our optimization loop is implemented classically and a part is done in a quantum algorithm. An example of this would be QAOA, in which we specify a cost Hamiltonian H_C whose ground state encodes the solution to our optimization problem. A classical optimizer uses feedback from the quantum system to optimize the parameters to a set of quantum gates until the expectation value of H_C is minimized.

10.3.3 Google Quantum AI

Google announced their package for quantum computing called Cirq in 2018 [11], just a year before Google Quantum AI demonstrated quantum supremacy for the first time [12]. Cirq includes tools to work with quantum circuits, specify and manage hardware constraints, simulate quantum algorithms using density matrices or wave functions, and run experiments on real superconducting-based quantum processors.

With weekly open source meetings, Cirq also fosters a growing open source community. Unlike some competitors' open source platforms, Cirq is not Google-specific. Instead it is comprised of a backend agnostic layer that works seamlessly with a number of backend providers that allow for communication with Google, IonQ, Pasqal and Rigetti backends. What is provided on top of these backends is an extensive frontend layer with noise models, hardware-awareness and circuit transformation (i.e. transpilation) capabilities.

To get a feel for Cirq, consider the implementation below of Grover's algorithm [13, p.88], which is useful for searching for binary strings. In a classical setting, when searching for a string, we might check each string using a function that returns whether it is the string we are looking for $f(x) = (x == y)$. This would require $O(N)$ operations. However, with a quantum computer, we can achieve the same in $O(\sqrt{N})$. The function $f(x)$ is replaced by a quantum oracle (a set of gates) which tags the state we are looking for by flipping its sign. This can be a full-rank diagonal matrix with one negative entry. We perform amplitude amplification by inversion about the mean, which is positive since most states do not have negative signs. The positive states' amplitudes all become smaller while the negative state's amplitude becomes larger and positive when they flip across this mean. Inversion about the mean is implemented by the diffuser $2\,|\psi\rangle\,\langle\psi| - I$. We repeatedly perform reflections until the state approaches the solution. It is the number of iterations that we count as our operations, i.e. if we start with an equal superposition state with amplitudes $\alpha = \frac{1}{\sqrt{N}}$, we will need to perform $k = \frac{\pi}{4\theta} - \frac{1}{2} = \frac{\pi\sqrt{N}}{4}$ iterations with $\frac{\pi}{2} - \theta$ the angle between the target state and the initial state.

Example: Grover's Algorithm with Cirq

```
num_qubits = 2

qubits = cirq.LineQubit.range(num_Qubits)
ancilla = cirq.NamedQubit("Ancilla")

class Oracle(cirq.Gate):
    def __init__(self):
        super(Oracle, self)

    def _num_qubits_(self):
        return 2

    def _unitary_(self):
        return np.array([
            [1.0, 0.0, 0.0, 0.0],
            [0.0, 1.0, 0.0, 0.0],
            [0.0, 0.0, 1.0, 0.0],
            [0.0, 0.0, 0.0, -1.0]
        ]) / np.sqrt(2)

    def _circuit_diagram_info_(self, args):
        return "O"
```

```
oracle = Oracle()

def iteration():
    circuit = cirq.Circuit()

    # prepare initial superposition
    circuit.append(cirq.H.on_each(*qubits))

    # apply oracle
    circuit.append(oracle.on(*qubits))

    # apply diffuser
    circuit.append(cirq.H.on_each(*qubits))
    circuit.append(cirq.Z.on_each(*qubits))
    circuit.append(cirq.CZ.on_each(*qubits))
    circuit.append(cirq.H.on_each(*qubits))

    # measure input register
    circuit.append(cirq.measure(*qubits, key="result"))

    return circuit
```

Note that if you look at Google's docs you will find a different way to implement Grover's algorithm using an oracle that conditionally flips an ancilla instead of flipping the target state's sign. This approach is equally valid. Also note that we can create an n-qubit diffuser with the following algorithm.

```
def diffuser(qubits):
    circuit = cirq.Circuit()

    # |psi> -> |00..0>
    circuit.append(cirq.H.on_each(*qubits))

    # |00..0> -> |11..1>
    circuit.append(cirq.X.on_each(*qubits))

    # multi-controlled-Z gate
    mcz = cirq.ControlledGate(sub_gate=cirq.Z, num_controls=num_qubits-1)
    circuit.append(mcz.on(*qubits))

    # |11..1> -> |00..0>
    circuit.append(cirq.X.on_each(*qubits))

    # |00..0> -> |psi>
    circuit.append(cirq.H.on_each(*qubits))

    return circuit
```

We could run this circuit in the Microsoft Azure Quantum cloud for example by following these steps. First, we connect to the Azure quantum service. We choose a backend "target" and provide its identifier to the *service.run* method.

```
from azure.quantum.cirq import AzureQuantumService
provider = AzureQuantumService(workspace)

result = service.run(
    program=circuit,
    repetitions=500,
    target="honeywell.hqs-lt-s1",
    timeout_seconds=200
)
```

We can also use an asynchronous model to execute our circuits in the background.

```
# non-blocking call
job = service.create_job(
    program=circuit,
    repetitions=500,
    target="honeywell.hqs-lt-s1",
)

# status can be checked anytime
print(job.status())

# blocking call
res = job.results()
```

Azure gives us access to Honeywell hardware, but if we want to run on AQT, IonQ, Pasqal or Rigetti backends we can simply use Cirq.

```
import cirq
from cirq.aqt.aqt_sampler import AQTSampler
from cirq.aqt.aqt_device import get_aqt_device

# replace these with real values
access_token = 'jdhfl4y787f4hhufh4r8hhu93osa'
workspace = 'MY_WORKSPACE'
resource = 'MY_RESOURCE'

aqt_sampler = AQTSampler(workspace=workspace, resource=resource, access_token=access_token)

device, qubits = get_aqt_device(2)
aqt_sweep = aqt_sampler.run(circuit, repetitions=200)
```

10.3.4 IBM Qiskit

The last package we will cover is Qiskit, which is the most widely used and starred package covered here (with 5k stars to Cirq's 4.2k, Pennylane's 2.3k, Ocean's 395) at the time of this book's writing. Qiskit includes the quintessential quantum assembly language for gate model quantum computers, tools for working with circuits and operators, transpiling circuits, optimizing for hardware, lowering to pulse sequences via Qiskit Pulse, executing on hardware backends, post-processing results from executions and more. Like Cirq, Qiskit includes a noise modeling component called Aer, simulators including statevector, density matrix, stabilizer, matrix product state, unitary and superoperator simulators.

Building an algorithm with Qiskit is similar to doing so in Pennylane and Cirq, and backend execution is similarly built right into Qiskit for IBM backends. Consider the following example.

The Bernstein-Vazirani [13, p. 50] algorithm is similar to Grover's algorithm in that we are searching for something. However, in this case, we are trying to find a hidden integer value. What we have is an oracle that returns $y \cdot x = \sum_i y_i x_i \bmod 2$ for input x and hidden integer y. Using a classical function to calculate $f(x) = y \cdot x$ we would have to calculate $f(1), f(2), f(2^i), \ldots, f(2^{n-1})$ to reveal each bit of y. With the quantum Bernstein-Vazirani algorithm we can find the value with one query.

The first step of the algorithm is to put the input in an equal superposition with Hadamards.

$$H^{\otimes} |0\rangle = \frac{1}{\sqrt{2^n}} \sum_x |x\rangle$$

We can encode the output of $y \cdot x$ using the phase with an oracle made of Pauli Z matrices.

$$\begin{aligned}(-1)^{y \cdot x} &= (-1)^{y_0 x_0}(-1)^{y_1 x_1} \ldots (-1)^{y_{n-1} x_{n-1}} \\ &= ((1 - y_0)I + y_0 Z)((1 - y_1)I + y_1 Z) \ldots ((1 - y_{n-1})I + y_{n-1} Z)\end{aligned}$$

We apply the oracle to our superposition to obtain

$$\frac{1}{\sqrt{2^n}} \sum_x (-1)^{y \cdot x} |x\rangle$$

Finally, we apply Hadamards again.

$$H^{\otimes n} \frac{1}{\sqrt{2^n}} \sum_x (-1)^{y \cdot x} |x\rangle = |y\rangle$$

Example: Bernstein-Vazirani with Qiskit

```
#initialization
import numpy as np

# importing Qiskit
from qiskit import QuantumCircuit, ClassicalRegister, QuantumRegister

num_qubits = 5
y = 6
bin_y = bin(y)[2:].zfill(num_qubits)

qr = QuantumRegister(n_qubits)
cr = ClassicalRegister(n_qubits, name='result')

bv_circuit = QuantumCircuit(qr, cr)
```

```
bv_circuit.h(qr)

# The oracle
for i in range(n_qubits):
    if (bin_y[i] == '1'):
        bv_circuit.z(qr[i])

bv_circuit.h(qr)

bv_circuit.measure(qr, cr)
```

We can then execute the experiment on an IBM quantum computer with the following snippet.

```
from qiskit.primitives import StatevectorSampler
from qiskit.transpiler.preset_passmanagers import generate_preset_pass_manager

from qiskit_ibm_runtime import SamplerV2 as Sampler
from qiskit_ibm_runtime import QiskitRuntimeService

service = QiskitRuntimeService(channel="ibm_quantum")

# get the least busy operational quantum hardware backend
backend = service.least_busy(operational=True, simulator=False)
target = backend.target

# transpile the circuit for the backend
pass_manager = generate_preset_pass_manager(target=target, optimization_level=3)
isa_circuit = pass_manager.run(bv_circuit)

# run the circuit
sampler = Sampler(mode=backend)
job = sampler.run([isa_circuit])
job_result = job.result()

# print the measurement outcomes
print(job_result[0].data.res.get_counts())
```

We could alternatively have specified a backend by name, such as "ibm_brisbane".

10.4 Building Our Own Quantum Cloud

All of these features and software components sit on top of the control systems covered in previous chapters. A "full-stack" engineer usually refers to someone who works on frontend, backend and middleware, but as you can see there is still a lot more to building a quantum computer. Therefore, to build a useful quantum computer you really need to be much more than a full-stack software engineer. This book should however give you an idea of what it is like to work as a full-stack engineer at a quantum computing company, even if in this case "full-stack" is a bit of a misnomer. Maybe "web-stack" is a better descriptor for the kind of project we will undertake in the following pages.

Now that we have seen some brief examples of how existing packages work, let's learn what it might take to implement our own minimal cloud quantum computing service. We will begin with how one might layout their project and what to take into consideration at this beginning stage.

10.4.1 Git and Submodules

One of several popular approaches to laying out a project with a significant amount of code that lends itself well to working with microservices is to use git submodules to separate the code that belongs to each service. First let's cover some git basics.

The git model of software collaboration involves what are called repositories and branches. A repository is like a folder that contains source code. You can host a repo on github, gitlab or various other sites, but in order to work effectively with the code you need to have a local clone of the repo too. Everyone working on a project clones (downloads) the repository onto their computer after which point their work begins to diverge from their colleagues'. The point of git is to track and merge these changes properly so that there are no lasting conflicts in any two sets of changes.

```
git clone git@github.com:comp-phys-marc/demo.git
```

There are different methodologies for how to organize your branches, like git flow which basically says you should organize your work into features and associate each branch with a feature, a hotfix, or the like. In all of the approaches, the basics are the same. A branch keeps track of all your changes to the code after a certain point called the root of the branch.

```
git checkout -b feature/add-phase-gate
```

You can always revert to the root, or to any commit you make along the way as you work on your branch. These commits are like save points. They should each contain a coherent set of changes that can be described in a short, concise summary. All the commits on a branch should contribute to the goal of that branch, whether it is to implement a new feature or fix a software bug. Before you can commit something, you can selectively add files to your commit by staging them with the *add* command. You can also choose to just add every change you've made under the repository root directory using the $-A$ flag or add everything under the working directory using ..

```
git add .
git commit -m "I added the phase gate implementation"
```

As you are working on your branch, the main branch will likely be updated by others who finish their respective work. As this happens, you will want to do what is called merging from the main branch to make sure you keep up to date.

```
git merge origin/main
```

You will also want to push your changes to the remote repo so that you have a backup of your work and the github or gitlab version of the code matches your local source code.

```
git push origin feature/add-phase-gate
```

When you are finished working on a branch, it is often the case that developers will use the git squash command to combine commits and cleanup their commit history before it gets merged into main. This is however not required.

```
git reset --hard HEAD~10 # go back 10 commits
git merge --squash HEAD@{1} # merge in everything up to the latest commit made before the reset
git commit # combine all 10 commits into one
```

What is required is that a pull request is made. A pull request asks that maintainer(s) of the main branch, usually someone or some people who review(s) your code, check(s) that everything you wrote looks good and then merges your changes into the main branch. This can't be done if your changes conflict with some changes that have been introduced to main in the meantime. So, you need to resolve these changes beforehand. Pull requests aren't implemented in git directly, but are a feature of hosts like github and gitlab, so the best way to create a PR is to use the host's UI or to use a tool like github cli.

```
gh pr create --title "S gate added" --body "Add matrix implementation and native gate decomposition of S
    gate" --base main
```

Where submodules come in is when you have a project that can be divided into multiple sub projects. Instead of maintaining one massive repository, you can maintain several smaller ones, which are all referenced from a top-level repository whose function is really just to allow developers to quickly update all the source repos together, and clone/update all the source on their computers quickly. D-Wave Ocean and IBM Qiskit both use this method to allow users to work with their myriad repositories more easily, and so will we.

To create a repo we just run the following in the project root directory.

```
git init
git add .
git commit -m "init commit"
git remote add origin git@github.com:comp-phys-marc/submodule.git
git push origin main
```

Then to add this repository to a top-level repo as a submodule, we run the following in the top-level repo.

```
git submodule add git@github.com:comp-phys-marc/submodule.git <location-in-top-level>
```

Going forward we can maintain our submodules with the following commands. A developer new to the project can clone the project with all the submodules by cloning the top-level repo by adding this flag:

```
git clone git@github.com:comp-phys-marc/top-level.git --recurse-submodules
```

We can update the submodules in the future with:

```
git submodule update --init --recursive
```

We will create submodules for an API service, simulation service and user service to start.

10.4.2 DevOps

We have a way to manage our source code. This is great. We have repositories for our user service, simulation service, and API service. The user service will handle all user management tasks such as creating accounts, logging in, etc. The simulation service will host, run and scale our stabilizer simulator for us. The API service will process requests from the frontend and facilitate all communication with the other services.

Using containers to horizontally scale your services is an alternative model for parallel computing to process and thread parallelism which spins up extra servers in response to demand. These virtual machines are hosted on one or two physical servers in a datacenter, but each behave like a fully functioning linux server. Port forwarding is used to allow the containers to access the internet through the physical server's connection.

To orchestrate communication between these, we make use of a message bus technology such as RabbitMQ, Celery and Kafka. A remote procedure execution protocol like RPC is used by these.

Using Google Cloud Platform (GCP) we can deploy our services using a few bash scripts. The deploy script for each service will be similar to the one for the simulation service.

```
export PROJECT_ID="$(gcloud config get-value project -q)"
cd ../simulationservice
docker build -t gcr.io/${PROJECT_ID}/simulationservice:v1 .
docker push gcr.io/${PROJECT_ID}/simulationservice:v1
gcloud compute instances update-container stabilizersimulation --container-image
    gcr.io/${PROJECT_ID}/simulationservice:v1
```

This script does a few things. The first thing is to use the gcloud cli to get the ID of the configured project. We must have configured the project earlier. In this script, the commands are identified by lines begun with >. If you had previously created a gcloud account and project, you might see the following.

```
> gcloud auth list
  * my account

> gcloud config set account 'my account'

> gcloud projects list
  - stabilizer simulator

> gcloud config set project 'stabilizer simulator'
```

From inside the simulation service directory, the deploy script then uses the docker build and push commands. The lifecycle of a Docker container generally involves the following steps [14].

A DockerFile specifies how a Docker image should be built. It should specify all of the dependencies and environmental configurations required for the application inside to run. For example, our simulation service might have the following DockerFile, which specifies the version of Python we need, the list of Python packages (in requirements.txt) we require, and installs everything from our project root into the Docker container's app directory (or volume). Finally, the DockerFile specifies the command used to run the service, which in our case is a Celery entrypoint that starts the simulation service as a worker that can receive and process messages on an event/message bus.

```
FROM python:3.11
ADD requirements.txt /app/requirements.txt
ADD ./simulationservice/ /app/
WORKDIR /app/
RUN pip3 install -r requirements.txt
ENTRYPOINT celery -A app worker --loglevel=info --pool=solo --without-heartbeat -n
    simulationservice@%h -Q simulation
```

Once the DockerFile is defined, we can build our Docker image, which captures all of this configuration in a format that can be shared. Like GitHub is used for sharing our code, we can also share the Docker images in Docker Hub or upload the images to our own image repositories. The -t flag allows us to assign a name and version, otherwise known as a tag, to the build.

```
docker build -t simulationservice:latest ./DockerFile
```

If we want to download an image from an existing repository we can also use the pull command to clone it to our local computer.

```
docker pull simulationservice:latest
```

A container can be run on a local machine for testing just like it can be run in the cloud (on a server somewhere). To do this yourself, use the run command. The -d "detach" flag runs the container in the background so that the terminal is free for you to use at the same time.

```
docker run -d --name mysim simulationservice:latest
```

There are several commands that pertain to running Docker containers such as stopping the container, pausing it and unpausing it. The container can also be deleted by removing it.

```
docker stop mysim
docker restart mysim
docker pause mysim
docker unpause mysim
docker rm mysim
```

The more interesting and useful commands for our use case relate to the updating of our remote instances. If we have played with a container locally for a while and have implemented updates which we deem valuable to the configuration or environment, we can create a new tag and commit these to it.

```
docker commit mysim simulationservice:v2
```

Then we may push our updates to remote repositories using the push command we've already seen.

```
docker push gcr.io/${PROJECT_ID}/simulationservice:v2
```

Once our v2 simulation service is in our Google Cloud repository, we can tell GCP to deploy the new image.

```
gcloud compute instances update-container stabilizersimulation --container-image
      gcr.io/${PROJECT_ID}/simulationservice:v2
```

We can automate the deployment of multiple services by running each deploy script from a "master" deploy script.

```
export PROJECT_ID="$(gcloud config get-value project -q)"

./deploy_api.sh
./deploy_simulation.sh
./deploy_user.sh
```

Similarly, a tool called Docker Compose can help us orchestrate multiple services together. We might have a Docker Compose config at the root of our top-level repository, specifying the DockerFiles, volumes and contexts for each of our builds. This specifies our multi-container application.

```
services:
  userservice:
    build:
      context: ./userservice/
      dockerfile: dockerfile
    volumes:
      - ./userservice/app
  simulationservice:
```

```
    build:
      context: ./simulationservice/
      dockerfile: dockerfile
    volumes:
      - ./simulationservice/app
  apiservice:
    build:
      context: ./apiservice/
      dockerfile: dockerfile
    volumes:
      - ./apiservice/app
```

We could then build all our images with the following short command.

```
docker-compose build
```

You would be able to work with these built images in the same way as if they had been individually built with Docker at this stage. There are some further benefits to Compose, however, and we will highlight that scaling our services becomes extremely easy, i.e. to 5 instances.

```
docker-compose scale simulationservice=5
```

Similar to Docker Compose is a tool named Kubernetes, which we will only touch on briefly. However, Google provides a service called Google Kubernetes Engine (GKE) which allows us to use the Kubernetes cli tool kubectl to do some of the same things in the remote environment, such as creating a cluster of several services.

```
gcloud container clusters create stabilizercluster --num-nodes=2
```

We can deploy an image to such a cluster.

```
kubectl run stabilizersimulator-deployment --image=gcr.io/${PROJECT_ID}/simulationservice:v2 --port
    8080
```

Then it is easy to manage the port forwarding so that the container is accessible from the outside world. We call this exposing a deployment.

```
kubectl expose deployment stabilizersimulator-deployment --port 80 --target-port 8080
```

Finally, we can scale the cluster as we wish so that requests may be consumed in parallel by any number of services. Of course, keep in mind that Google charges you for compute time and you cannot scale up without incurring costs.

```
kubectl scale deployment stabilizersimulator-deploymen --replicas=3
```

10.4.3 Simulator

In this section we will look at the design of a simple type of quantum simulator called a stabilizer simulator. The packages for quantum computing which we have already become familiar with are more equipped than what we will build here, but this will give us an introduction to the software engineering side of quantum simulators. We will take an Object Oriented Programming approach (OOP) since it will serve the reader in most industry settings.

Circuits consisting of the gates CNOT, H, S and measurements (all of which we have encountered before) are considered so-called "stabilizer circuits". These are stated by the Gottesman-Knill theorem to be easily simulatable by classical computers. H and S together are the generators of the Clifford group which, as we learned before, includes the Pauli operators.

We can implement our own stabilizer simulator relatively easily and it will be able to perform whatever quantum algorithms we are interested in which are classically simulable. We can take inspiration from the linearity of quantum mechanics in our class design, and we can keep track of basis vectors separately from their complex coefficients in order to keep the calculation of probabilities simple and straightforward, for example. To save time and space we will always work in the computational basis. While this is sufficient to represent any state, it is of course an arbitrary choice and could be relaxed in a more complex implementation of a simulator.

```
class Ket:
"@\textcolor{mauve}{""}@
A class that represents the data associated with a single computational basis ket in a quantum state.
@\textcolor{mauve}{""}@"

def __init__(self, coeff=None, val=None):
    "@\textcolor{mauve}{""}@
    Initializes a ket with a value and coefficient.

    :param coeff: The ket's complex coefficient.
    :param val: The qubit bitstring value.
    @\textcolor{mauve}{""}@"
    self.set_coefficient(coeff)
    self.set_val(val)
```

Here we use simple setters to keep track of the statevector and the complex coefficient of a single Ket, as described. Let's implement the setters so that they can take user-friendly inputs, using *string* bitstrings and the *complex* type in Python instead of tensors. The methods themselves will handle the scaffolding of the tensor product state.

```
    def set_val(self, val):
    "@\textcolor{mauve}{""}@
    Sets the qubit string value for the ket.

    :param val:
    :return:
    @\textcolor{mauve}{""}@"
    self._val = None
    self.num_qubits = len(val)
```

```
        if isinstance(val, str):
            for bit in val:
                if bit not in ['0', '1']:
                    self._val = None
                    raise ValueError("state value {0} is not entirely 1's and 0's".format(val))
                if bit == '0':
                    if self._val is None:
                        self._val = ZERO
                    else:
                        self._val = np.kron(self._val, ZERO)
                if bit == '1':
                    if self._val is None:
                        self._val = ONE
                    else:
                        self._val = np.kron(self._val, ONE)

    def set_coefficient(self, coeff):
        "@\textcolor{mauve}{""}@
        Sets the coefficient of the ket's term in the overall quantum state.

        :param coeff: The coefficient of the ket.
        :raises: ValueError
        @\textcolor{mauve}{""}@"
        if isinstance(coeff, complex):
            self._coefficient = coeff
        else:
            raise ValueError("setting coefficient of incorrect type was attempted")
```

While Python is not a strongly typed language, it is often a good idea to use *isinstance* to check or assert the type of inputs as we do here. Recent Python versions also support type hints which are encouraged, but optional reminders of what should be passed to a function or method. We will not use them here for the sake of brevity.

np.kron here performs the tensor product, or Kronecker product, needed to build our state. Recall that $|0000\rangle$ is really shorthand for $|0\rangle \otimes |0\rangle \otimes |0\rangle \otimes |0\rangle$ which is shorthand for $\begin{bmatrix}1\\0\end{bmatrix} \otimes \begin{bmatrix}1\\0\end{bmatrix} \otimes \begin{bmatrix}1\\0\end{bmatrix} \otimes \begin{bmatrix}1\\0\end{bmatrix}$ which is equal to a 16-length column vector with only one 1 in the first row and zeroes for the rest of the entries.

We can implement an API reminiscent of Qiskit, Pennylane and Cirq by adding methods to the *Ket* class for each of the gates. Two examples are provided.

```
    def y(self, qubit):
        "$\textcolor{mauve}{""}$
        Performs a Pauli Y gate on the target qubit.

        :param qubit: The target qubit.
        :return: The ket after the operation.
        $\textcolor{mauve}{""}$"
        targeted_gate = correct_dimensionality(
            np.array([[0, -1],
                      [1, 0]]),
            qubit,
            self.num_qubits
        )
        self._coefficient = self._coefficient * complex(0, 1)
        self._val = targeted_gate @ self._val
        return self

    def z(self, qubit):
        "$\textcolor{mauve}{""}$
```

```
    Performs a Pauli Z gate on the target qubit.

    :param qubit: The target qubit.
    :return: The ket after the operation.
    $\textcolor{mauve}{""}$"
    targeted_gate = correct_dimensionality(
        np.array([[1, 0],
                  [0, −1]]),
        qubit,
        self.num_qubits
    )
    self._val = targeted_gate @ self._val
    return self
```

Notice how we handle the change to the coefficient and the statevector. This pattern will work for the Paulis and Hadamard gates. We will however need to take a different tack with each of CNOT and S.

```
def cx(self, source, target):
    "$\textcolor{mauve}{""}$
    Performs a Controlled X gate on the target qubit with the
    source qubit as controller.

    :param source: The source qubit.
    :param target: The target qubit.
    :return: The ket after the operation.
    $\textcolor{mauve}{""}$"

    cx_matrix = [[0. for _ in range(len(self._val))] for _ in range(len(self._val))]

    for i, row in enumerate(cx_matrix):
        label = f'{i:0{self.num_qubits}b}'
        if label[source] == '1':
            label = label[0:target] + ('0' if label[target] == '1' else '1') + label[target+1:]
        one_position = int(label, 2)
        row[one_position] = 1.

    cx_gate = np.array(cx_matrix)
    self._val = cx_gate @ self._val
    return self
```

We can solve for the matrix of a CNOT operating on an n-qubit system by simply thinking about the operation that the gate should have on each basis state. Our CNOT operation maps each basis state to one basis state (with the target flipped if the source is 1), unlike the Hadamard which maps basis states to superpositions of basis states. Therefore, our matrix has one 1 per row. The position of the 1 in the first row tells us which state to map to the first basis state, the position of the 1 in the second row gives us which state is mapped to the second basis state, and so on. The bitstring of the state (i.e. 101 in $|101\rangle$) is conveniently equal to the position (i.e. 5) of the 1 in our row, only in base 2.

It is easy to tell whether our target qubit is in state $|1\rangle$ or state $|0\rangle$ since we only consider one basis Ket at a time.

```
def vector_to_bitstring(vector):
    "@\textcolor{mauve}{""}@
    Converts a vector representation of a state to a bitstring.

    :param vector: The state vector.
```

```
:return:
@\textcolor{mauve}{""}@"
num_qubits = np.log2(len(vector))
bitstring = f'{0:0{int(num_qubits)}b}'
for i, row in enumerate(vector):
    if row[0] == 1:
        bitstring = f'{i:0{int(num_qubits)}b}'
return bitstring
```

This makes it extremely easy to implement S and $S^{\dagger}$.

```
def s(self, qubit):
    "@\textcolor{mauve}{""}@
    Performs an S phase shift gate on the target qubit.

    :param qubit: The target qubit.
    :return: The ket after the operation.
    @\textcolor{mauve}{""}@"
    if vector_to_bitstring(self._val)[qubit] == '1':
        self._coefficient = self._coefficient * complex(0, 1)
    return self

def sdg(self, qubit):
    "@\textcolor{mauve}{""}@
    Performs an S dagger phase shift gate on the target qubit.

    :param qubit: The target qubit.
    :return: The ket after the operation.
    @\textcolor{mauve}{""}@"
    if vector_to_bitstring(self._val)[qubit] == '1':
        self._coefficient = self._coefficient * -complex(0, 1)
    return self
```

The Hadamard is the only single-qubit gate with very different behavior.

```
def h(self, qubit):
    "$\textcolor{mauve}{""}$
    Performs a Hadamard gate on the target qubit.

    :param qubit: The target qubit.
    :return: The two resulting kets.
    $\textcolor{mauve}{""}$"
    targeted_gate = correct_dimensionality(
        np.array([[1, 1],
                  [1, -1]]),
        qubit,
        self.num_qubits
    )
    self._coefficient = self._coefficient / np.sqrt(2)
    self._val = targeted_gate @ self._val
    pos_vals, neg_vals = to_sum_of_basis_kets(self._val)

    res = [Ket(self._coefficient, pv) for pv in pos_vals] + [Ket(-self._coefficient, nv) for nv in neg_vals]
    return res
```

All that remains to our simple simulator is a class to hold our *Kets* together in an overall quantum *State*. This class is straightforward and mostly passes through operations to the *Kets* themselves, such as $x()$.

```
class State:
    "@\textcolor{mauve}{""}@
    A class that represents a full quantum state and maintains a set of underlying kets.
```

```
@\textcolor{mauve}{""}@"

def __init__(self, ket_list=[], num_qubits=1, symbol='psi'):
    "@\textcolor{mauve}{""}@
    Initializes a quantum state with a given set of kets and number of qubits.

    :param ket_list: The kets.
    :param num_qubits: The total number of qubits.
    :param symbol: The identifier for this quantum state.
    @\textcolor{mauve}{""}@"

    self.kets = []
    self.num_qubits = num_qubits
    self.symbol = symbol

    for ket in ket_list:
        self.add_ket(ket)

def x(self, qubit):
    "@\textcolor{mauve}{""}@
    Performs a Pauli X gate on the target qubit.

    :param qubit: The target qubit.
    :return: The full quantum state after the operation.
    @\textcolor{mauve}{""}@"
    for ket in self.kets:
        ket.x(qubit)
    return self
```

The only oddity in our stabilizer simulator is the Hadamard which can map a basis Ket to more than one basis Ket.

```
def h(self, qubit):
    "@\textcolor{mauve}{""}@
    Performs a Hadamard gate on the target qubit.

    :param qubit: The target qubit.
    :return: The full quantum state after the operation.
    @\textcolor{mauve}{""}@"
    new_kets = []
    for ket in self.kets:
        hadamard_result = ket.h(qubit)
        new_kets.extend(hadamard_result)
    self.kets = new_kets
    return self
```

To clean up our state, we might like to regularly call a method that normalizes the state, resulting in a $State$ with one Ket object per unique basis vector.

```
def normalize(self):
    "@\textcolor{mauve}{""}@
    Normalizes the current quantum state.

    :return: Returns the full normalized quantum state.
    @\textcolor{mauve}{""}@"
    total_probability = 0
    unique_kets = []
    for ket in self.kets:
        already_found = False
        for unique_ket in unique_kets:
            if ket.get_val() == unique_ket.get_val():
                already_found = True
                unique_ket.set_coefficent(unique_ket.get_coefficient() + ket.get_coefficient())
```

```
        if not already_found:
            unique_kets.append(ket)
    for unique_ket in unique_kets:
        total_probability += unique_ket.get_probability()
    norm_factor = 1 / sqrt(total_probability)
    if total_probability != 1:
        for unique_ket in unique_kets:
            unique_ket.set_coefficent(unique_ket.get_coefficient() * norm_factor)
    self.kets = unique_kets

    return self
```

10.4.4 Application Programming Interface

Frameworks we can use to implement an API layer in Python include Pyramid, Flask and Django. Out of these, Flask is the most lightweight. So, we will make use of Flask to implement a quick API layer that can accept an algorithm from a web request and pass it through to a simulator. In particular, we accept POST requests. Usually, we would implement CRUD (Create, Read, Update and Delete) actions using POST, GET, PUT and DELETE requests respectively.

```
import json
import signal
import sys
from flask import Flask, request, abort
from flask_cors import CORS
from kombu import Queue
from auth import AuthGuard, REFRESH_TOKEN_KEY, TOKEN_KEY, ExpiredSignatureError
from gevent import monkey
from gevent.pywsgi import WSGIServer
from settings import Settings
from emulatorcommon.message_bus import MessageBus

monkey.patch_all()

app = Flask(__name__)
CORS(app)

settings = Settings()

bus = MessageBus(settings)
rabbit = bus.connection

rabbit.conf.task_routes = {
    'user.tasks.*': {'queue': 'user'},
    'simulation.tasks.*': {'queue': 'simulation'},
}
rabbit.conf.task_default_queue = 'default'
rabbit.conf.task_queues = (
    Queue('user', routing_key='user.tasks.#'),
    Queue('simulation', routing_key='simulation.tasks.#'),
)


@app.route('/simulate', methods=['POST'])
def simulate():

    data = json.loads(request.data.decode("utf-8"))
    user_id = None
```

```
        if 'code' not in data or 'name' not in data:
            abort(400)

        try:
            user_id = AuthGuard.authenticate(data)

        except Exception as ex:
            abort(401)

        response = retry_if_necessary('simulation.tasks.execute',
                                      args=[user_id, data['code'], data['name'], data['experiment_id'],
                                            data['execution_type']],
                                      queue='simulation')

        if response is not None:
            return json.dumps(response), response['status']
        else:
            abort(403)

    def retry_if_necessary(task, args, queue, retries=1):
        attempts = 0
        response = rabbit.send_task(task, args=args, queue=queue).wait()

        if response['status'] == 500:
            print('Error occurred: {}'.format(response['message']), sys.stderr)
            print('Attempting retry {} of {}'.format(attempts, retries), sys.stderr)
            while attempts < retries:
                response = rabbit.send_task(task, args=args, queue=queue).wait()
                attempts += 1

        return response
```

We follow best practices here when we implement limited retries, and return standard meaningful error response codes. We return 401 when we run into authentication issues, 400 for a bad request, etc. These codes and their meanings can be referred to on Mozilla's developer web docs here. We leave the implementation of the auth guard out of this book as authentication is not a topic that we wish to dive into in detail.

In the case of our implementation of a scalable stabilizer simulator, we make use of a package called Celery which abstracts for us a protocol called RPC which defines how web microservices communicate with a message bus. This allows messages to be published on "topics" to which services may subscribe. The services then consume the messages based on their availability.

But how does our service consume a message containing the QASM code we wish to simulate? The first step is to parse the QASM code. This can be done using one of several approaches. [15] compares several OpenQASM 3.0 parsers including IBM's Python and ANTLR-based parser, their Rust-based parser and QASM-TS which is a TypeScript-based parser. In summary, the Rust-based parser is the most efficient but not very feature complete. It only supports a small subset of the OpenQASM 3.0 grammar, failing to parse all but 2/11 example QASM files. The Python and TypeScript parsers both support the full spec, but the TypeScript implementation is roughly an order of magnitude faster than the Python.

On the side of the simulation service therefore, we consume a message from the bus and pass the QASM to a parser.

```
import inspect from IPython.utils.capture
import capture_output from settings
import MessageBus from qedlib.parser.parser
import run_qasm

@conn.task(name="simulation.tasks.execute")
def execute_qasm(user_id, qasm, name, experiment_id, execution_type):

    try:

        with capture_output() as captured:

            result = run_qasm(qasm, execution_type)

        readout = captured.stdout

        return {
            "message": f""{name} simulation successful."',
            "result": json.dumps(readout),
            "status": 200
        }

    except Exception as e:
        error_message = str(e)
        return {
            "message": f'An error occurred: {error_message}',
            "status": 500
        }
```

The *run_qasm* function essentially translates each line of the QASM to a function that we implemented before, using the below logic for example.

```
operator = line.split(SPACE)[0]
method = None

for register_name in registers:
    for gate in self._gates[register_name]:
        if gate.lower() == operator.lower():
            method = self._gates[register_name][gate]
```

This allows us to simulate the quantum algorithm in the cloud. But what if we want to execute something on a real quantum computer? We can also accept different execution types and we could, say, pass the QASM to Qiskit to run on an available quantum computer using the approach we saw in the previous section. To get a Qiskit circuit from a string of QASM, we can simply use:

```
circuit = QuantumCircuit.from_qasm_str(qasm)
```

References

1. Serrano, M.A., Pérez-Castillo, R., Piattini, M. (eds.): Quantum Software Engineering. Springer International Publishing, Cham (2022). https://doi.org/10.1007/978-3-031-05324-5
2. Martin, R.C. (ed.): Clean Code: A Handbook of Agile Software Craftsmanship. Prentice Hall, Upper Saddle River, NJ (2009)
3. Martin, R.C., Martin, R.C.: Clean architecture: a craftsman's guide to software structure and design. Robert C. Martin series. Prentice Hall, London, England (2018). OCLC: on1004983973

4. Biamonte, J., Wittek, P., Pancotti, N., Rebentrost, P., Wiebe, N., Lloyd, S.: Quantum machine learning. Nature **549**(7671), 195–202 (2017)
5. Farhi, E., Goldstone, J., Gutmann, S.: A quantum approximate optimization algorithm (2014). https://doi.org/10.48550/arXiv.1411.4028. ArXiv:1411.4028 [quant-ph]
6. Cramer, M., Plenio, M.B., Flammia, S.T., Gross, D., Bartlett, S.D., Somma, R., Landon-Cardinal, O., Liu, Y.K., Poulin, D.: Efficient quantum state tomography. Nat. Commun. **1**(1), 149 (2010). ArXiv:1101.4366 [quant-ph]
7. QHack 2024 Highlights | PennyLane Blog. https://pennylane.ai/blog/2024/03/qhack-2024-highlights/
8. Perez-Garcia, D., Verstraete, F., Wolf, M., Cirac, J.: Matrix product state representations. Quantum Inf. Comput. **7**(5&6), 401–430 (2007). https://doi.org/10.26421/QIC7.5-6-1. http://www.rintonpress.com/journals/doi/QIC7.5-6-1.html
9. Hashim, A., Nguyen, L.B., Goss, N., Marinelli, B., Naik, R.K., Chistolini, T., Hines, J., Marceaux, J.P., Kim, Y., Gokhale, P., Tomesh, T., Chen, S., Jiang, L., Ferracin, S., Rudinger, K., Proctor, T., Young, K.C., Blume-Kohout, R., Siddiqi, I.: A practical introduction to benchmarking and characterization of quantum computers (2024). https://arxiv.org/abs/2408.12064
10. Peruzzo, A., McClean, J., Shadbolt, P., Yung, M.H., Zhou, X.Q., Love, P.J., Aspuru-Guzik, A., O'Brien, J.L.: A variational eigenvalue solver on a photonic quantum processor. Nature Communications **5**(1) (2014). https://doi.org/10.1038/ncomms5213. http://dx.doi.org/10.1038/ncomms5213
11. Announcing Cirq: An Open Source Framework for NISQ Algorithms. http://research.google/blog/announcing-cirq-an-open-source-framework-for-nisq-algorithms/
12. Arute, F., Arya, K., Babbush, R., Bacon, D., Bardin, J.C., Barends, R., Biswas, R., Boixo, S., Brandao, F.G.S.L., Buell, D.A., Burkett, B., Chen, Y., Chen, Z., Chiaro, B., Collins, R., Courtney, W., Dunsworth, A., Farhi, E., Foxen, B., Fowler, A., Gidney, C., Giustina, M., Graff, R., Guerin, K., Habegger, S., Harrigan, M.P., Hartmann, M.J., Ho, A., Hoffmann, M., Huang, T., Humble, T.S., Isakov, S.V., Jeffrey, E., Jiang, Z., Kafri, D., Kechedzhi, K., Kelly, J., Klimov, P.V., Knysh, S., Korotkov, A., Kostritsa, F., Landhuis, D., Lindmark, M., Lucero, E., Lyakh, D., Mandrà, S., McClean, J.R., McEwen, M., Megrant, A., Mi, X., Michielsen, K., Mohseni, M., Mutus, J., Naaman, O., Neeley, M., Neill, C., Niu, M.Y., Ostby, E., Petukhov, A., Platt, J.C., Quintana, C., Rieffel, E.G., Roushan, P., Rubin, N.C., Sank, D., Satzinger, K.J., Smelyanskiy, V., Sung, K.J., Trevithick, M.D., Vainsencher, A., Villalonga, B., White, T., Yao, Z.J., Yeh, P., Zalcman, A., Neven, H., Martinis, J.M.: Quantum supremacy using a programmable superconducting processor. **574**(7779), 505–510 (2019). https://doi.org/10.1038/s41586-019-1666-5
13. Mermin, N.D.: Quantum Computer Science: An Introduction. Cambridge University Press (2007). Google-Books-ID: q2S9APxFdUQC
14. (1) Understanding the docker lifecycle: a simplified guide | LinkedIn. https://www.linkedin.com/pulse/understanding-docker-lifecycle-simplified-guide-anandu-omanakuttan-nazec/
15. Kim, S., Edwards, M.: Enabling the verification and formalization of hybrid quantum-classical computing with openqasm 3.0 compatible qasm-ts 2.0 (2024). https://arxiv.org/abs/2412.12578

Appendix A
Group Theory Class

```
import itertools
import numpy as np


# basic operations provided
def add(one, two):
    return one + two

def mul(one, two):
    return one * two

def additive_inverse(member):
    return - member

def multiplicative_inverse(identity, member):
    return identity / member


class Group:

    def __init__(self, operation, members, identity, inverse) -> None:
        if operation is not None:
            self.operation = operation
        else:
            raise ValueError("A composition operation must be provided with a Group.")
        if identity is not None:
            self.identity = identity
        else:
            raise ValueError("An identity element must be provided with a Group.")
        if isinstance(members, list):
            self.members = []
            self.inverses = {}
            for pair in members:
                if inverse:
                    pair = [pair, inverse(pair)]
                if not (isinstance(pair, list) and len(pair) == 2) \
                    or not (self.operation(*pair) == self.identity).all() \
                        or not (self.operation(*pair.reverse()) == self.identity).all():
                    raise ValueError("Group members must be provided in pairs of inverses or an inverse function
                            must be provided.")
                if not (self.operation(pair[0], self.identity) == pair[0]).all() \
                    or not (self.operation(pair[1], self.identity) == pair[1]).all():
                    raise ValueError("Group members composed with the identity must not change.")
                self.members += pair
```

© The Editor(s) (if applicable) and The Author(s), under exclusive license to Springer Nature Switzerland AG 2026

M. S. Edwards, *An Introduction to Quantum Computing for Computer Engineers*, https://doi.org/10.1007/978-3-032-03650-6

```
                self.inverses[str(pair[0])] = pair[1]
                self.inverses[str(pair[1])] = pair[0]

        self.commutative = None
        self.table = {}

    def generate_table(self):
        self.table = {}

        if self.commutative:
            for [first, second] in itertools.combinations(self.members, 2):
                res = self.operation(first, second)
                self.table[res] = [first, second]
        else:
            for [first, second] in itertools.permutations(self.members, 2):
                res = self.operation(first, second)
                self.table[res] = [first, second]


class Ring(Group):

    def __init__(self, additive_operation, multiplicative_operation, members, additive_identity,
            multiplicative_identity, additive_inverse_function) -> None:
        if additive_operation is not None:
            self.additive_operation = additive_operation
        else:
            raise ValueError("A Ring requires an additive operation.")
        if multiplicative_operation is not None:
            self.multiplicative_operation = multiplicative_operation
        else:
            raise ValueError("A Ring requires a multiplicative operation.")
        if additive_identity is not None:
            self.additive_identity = additive_identity
        else:
            raise ValueError("An additive identity element must be provided with a Ring.")
        if multiplicative_identity is not None:
            self.multiplicative_identity = multiplicative_identity
        else:
            raise ValueError("A multiplicative identity element must be provided with a Ring.")

        if isinstance(members, list):
            self.members = []
            self.inverses = {}

            # additive requirements
            for pair in members:
                if additive_inverse_function:
                    pair = [pair, additive_inverse_function(pair)]
                if not (isinstance(pair, list) and len(pair) == 2) \
                    or not (self.additive_operation(*pair) == self.additive_identity).all() \
                        or not (self.additive_operation(pair[1], pair[0]) == self.additive_identity).all():
                    raise ValueError("Ring members must be provided in pairs of inverses or there must be an inverse
                        function.")
                if not (self.additive_operation(pair[0], self.additive_identity) == pair[0]).all() \
                    or not (self.additive_operation(pair[1], self.additive_identity) == pair[1]).all():
                    raise ValueError("Ring members composed with the additive identity must not change.")
                self.members += pair
                self.inverses[str(pair[0])] = pair[1]
                self.inverses[str(pair[1])] = pair[0]

            # multiplicative requirements
            for member in members:
                if not (self.multiplicative_operation(member, multiplicative_identity) == member).all():
                    raise ValueError("Ring members multiplied with the multiplicative identity must not change.")

            for [first, second, third] in itertools.combinations(self.members, 3):
                if not np.allclose(self.multiplicative_operation(self.multiplicative_operation(first, second), third), \
```

```
                self.multiplicative_operation(first, self.multiplicative_operation(second, third))):
                raise ValueError("Ring multiplicative operation must be associative.")

            # multiplication left distributive with respect to additiion
            if not (self.multiplicative_operation(first, self.additive_operation(second, third)) \
                == self.additive_operation(self.multiplicative_operation(first, second),
                    self.multiplicative_operation(first, third))).all():
                raise ValueError("Ring multiplicative operation must be left distributive with respect to addition.")

            # multiplication right distributive with respect to additiion
            if not (self.multiplicative_operation(self.additive_operation(second, third), first) \
                == self.additive_operation(self.multiplicative_operation(second, first),
                    self.multiplicative_operation(third, first))).all():
                raise ValueError("Ring multiplicative operation must be right distributive with respect to
                    addition.")


class Commutator:

    def __init__(self, group):
        self.group = group
        if not isinstance(group, Group):
            raise TypeError("Commutators must be defined with respect to Groups.")

    def commute(self, first, second):

        if not isinstance(self.group, Ring):
            if self.group.commutative == True:
                return self.group.identity

            return self.group.operation(self.group.inverses[first], self.group.operation(self.group.inverses[second],
                self.group.operation(first, second)))
        else:
            return self.group.additive_operation(self.group.multiplicative_operation(first, second),
                self.group.multiplicative_operation(self.group.inverses[str(second)], first))


class PseudoGroup:
    """
    Relaxes the constraints of a Group. Useful for solving mod n inverses so that a proper Group can be constructed
        next.
    """

    def __init__(self, operation, members, identity) -> None:
        if operation is not None:
            self.operation = operation
        else:
            raise ValueError("A composition operation must be provided with a Group.")
        if identity is not None:
            self.identity = identity
        else:
            raise ValueError("An identity element must be provided with a Group.")
        if isinstance(members, list):
            self.members = members

        self.commutative = None
        self.table = {}

    def generate_table(self):
        self.table = {}

        if self.commutative:
            for [first, second] in itertools.combinations(self.members, 2):
                res = self.operation(first, second)
                self.table[res] = [first, second]
        else:
            for [first, second] in itertools.permutations(self.members, 2):
```

```
                res = self.operation(first, second)
                self.table[res] = [first, second]


class Abelian(Group):

    def __init__(self, operation, members, identity) -> None:
        super().__init__(operation, members, identity)
        self.commutative = True
```

Appendix B
Quantum Fourier Transform

As mentioned in Chap. 2, the Quantum Fourier Transform unitary is the following.

$$
\begin{aligned}
U_{QFT} &= \frac{1}{\sqrt{2^n}} \sum_{y=0}^{2^n-1} \exp\left(\frac{2\pi i x y}{2^n}\right) |y\rangle \\
&= 2^{-n/2} \otimes_{l=1}^{n} \left(|0\rangle + \exp(2\pi i 0.j_{n-l+1} \dots j) \, |1\rangle\right)
\end{aligned}
$$

This second line is new. Here, we use fractional binary notation whereby $0.j_l j_{l+1} \dots j_m = j_l 2^{-1} + j_{l+1} 2^{-2} + \dots + j_m 2^{-(m-l+1)}$. We can decompose this U_{QFT} into unitaries in the following way. S is related to the Paulis by $Z = SS$, for example. Decomposing gates into products of Paulis and vice versa is in general not straightforward but is covered in [3].

$$
R_k = \begin{bmatrix} 1 & 0 \\ 0 & \exp(2\pi i/2^k) \end{bmatrix} \quad S = \begin{bmatrix} 1 & 0 \\ 0 & i \end{bmatrix} \quad T = \begin{bmatrix} 1 & 0 \\ 0 & \exp(\frac{i\pi}{4}) \end{bmatrix}
$$

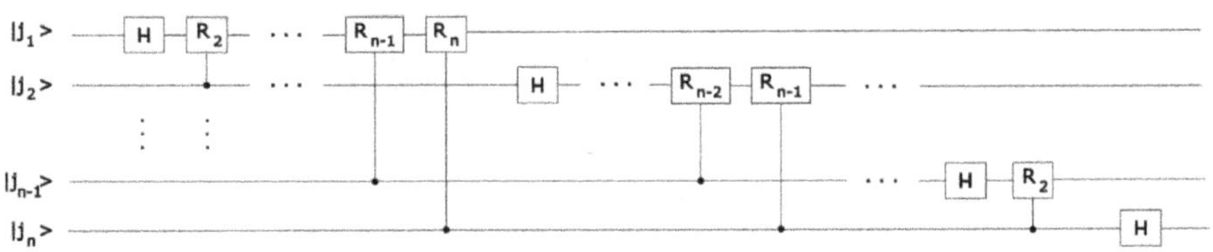

© The Editor(s) (if applicable) and The Author(s), under exclusive license to Springer Nature Switzerland AG 2026

M. S. Edwards, *An Introduction to Quantum Computing for Computer Engineers*,
https://doi.org/10.1007/978-3-032-03650-6

Find below our implementation of a 3-qubit QFT using Pennylane. Notice how we invert the order of the qubits at the end with a SWAP to get the results in the expected ordering.

```
@qml.qnode(dev)
def qft_3():
    qml.Hadamard (wires=[0])
    qml.ControlledQubitUnitary(np.array([[1, 0], [0, 0−1j]]), control_wires=[1], wires =[0]) # Controlled−S gate
    qml.ControlledQubitUnitary(np.array([[1, 0], [0, complex(np.cos(np.pi / 4), − np.sin(np.pi / 4))]]),
            control_wires=[2], wires=[0]) # Controlled−T gate
    qml.Hadamard(wires=[1])
    qml.ControlledQubitUnitary(np.array([[1, 0], [0, 0 − 1j]]), control_wires=[2] , wires=[1]) # Controlled−S gate
    qml.Hadamard(wires=[2])
    qml.SWAP(wires=[0, 2])

    return qml.probs(wires=[0, 1, 2])
```

Appendix C
Perturbation Theory

We will need to understand how to work with more tricky states such as degenerate states with eigenvalues that are associated with multiple eigenstates, for example. To attack problems with states that are harder to work with, we can use the approach called perturbation theory [2, p. 249].

The goal of perturbation theory is to provide approximations of solutions to problems that are very difficult to solve by starting with something exactly solvable and "perturbing" that simpler system.

Suppose we have solved the time-dependent Schrödinger equation $H^0\psi_n^0 = E_n^0\psi_n^0$. Then there is a complete set of orthonormal eigenfunctions ψ_n^0 such that $\langle\psi_n^0|\psi_m^0\rangle = \delta_{nm}$. After introducing a perturbative potential, we label the new eigen functions as $H\psi_n = E_n\psi_n$. Here $H = H_0 + \lambda H'$ where $\lambda << 1$ and H' is the perturbative potential we have introduced to the system.

$$\begin{aligned}\psi_n &= \psi_n^0 + \lambda\psi_n^1 + \lambda^2\psi_n^2 + \ldots\\ E_n &= E_n^0 + \lambda E_n^1 + \lambda^2 E_n^2 + \ldots\end{aligned}$$

E_n^1 is the first-order correction to the energy, E_n^2 the second-order correction, etc. ψ_n^1 is the first-order correction to the nth eigenvalue, ψ_n^2 the second-order correction, etc.

$$\begin{aligned}(H_0 + \lambda H')\psi_n &= (H_0 + \lambda H')(\psi_n^0 + \lambda\psi_n^1 + \lambda^2\psi_n^2 + \ldots)\\ &= (E_n^0 + \lambda E_n^1 + \lambda^2 E_n^2 + \ldots)(\psi_n^0 + \lambda\psi_n^1 + \lambda^2\psi_n^2 + \ldots)\end{aligned}$$

© The Editor(s) (if applicable) and The Author(s), under exclusive license to Springer Nature Switzerland AG 2026

M. S. Edwards, *An Introduction to Quantum Computing for Computer Engineers*,
https://doi.org/10.1007/978-3-032-03650-6

$$H^0\psi_n^0 + \lambda H^0\psi_n^1. + \lambda^2 H^0\psi_n^2$$
$$+ \lambda H'\psi_n^0 + \lambda^2 H'\psi_n^1 + \lambda^3 H'\psi_n^2$$
$$= E_n^0\psi_n^0 + \lambda E_n^0\psi_n^1 + \lambda^2 E_n^0\psi_n^2 + \lambda E_n^1\psi_n^0 + \lambda^2 E_n^1\psi_n^1 + \lambda^3 E_n^1\psi_n^2 + \lambda^2 E_n^2\psi_n^0$$
$$+ \lambda^3 E_n^0\psi_n^1 + \lambda^4 E_n^2\psi_n^2$$

$$H^0\psi_n^0 + \lambda(H^0\psi_n^1 + H'\psi_n^0) + \lambda^2(H^0\psi_n^2 + H'\psi_n^1) + \dots$$
$$= E_n^0\psi_n^0 + \lambda(E_n^0\psi_n^1 + E_n^1\psi_n^0) + \lambda^2(E_n^0\psi_n^2$$
$$+ E_n^1\psi_n^1 + E_n^2\psi_n^2) + \lambda^3(E_n^1\psi_n^2 + E_n^2\psi_n^1) + \dots$$

The first-order theory gives us the first-order correction for the energy of the perturbation. We can find it by taking the product of the first-order lambda equation with ψ_n^0 and integrating over volume.

$$\psi_n^0 H^0\psi_n^1 + \psi_n^0 H'\psi_n^0 = \psi_n^0 E_n^0\psi_n^1 + \psi_n^0 E_n^1\psi_n^0$$

$$\int \psi_n^0 H^0\psi_n^1 \, dv = \langle\psi_n^0 H^0|\psi_n^1\rangle$$
$$\int \psi_n^0 H^1\psi_n^1 \, dv = \langle\psi_n^0 H^1|\psi_n^0\rangle$$

etc.

$$\langle\psi_n^0 H^0|\psi_n^1\rangle + \langle\psi_n^0 H'|\psi_n^0\rangle = \langle\psi_n^0|E_n^0|\psi_n^1\rangle + \langle\psi_n^0|E_n^1|\psi_n^0\rangle$$
$$\langle\psi_n^0 H^0|\psi_n^1\rangle = \langle H^0\psi_n^0|\psi_n^1\rangle = \langle E_n^0\psi_n^0|\psi_n^1\rangle = E_n^0\langle\psi_n^0|\psi_n^1\rangle$$
$$E_n^0\langle\psi_n^0|\psi_n^1\rangle + \langle\psi_n^0 H'|\psi_n^0\rangle = E_n^0\langle\psi_n^0|\psi_n^1\rangle = E_n^1\langle\psi_n^0|\psi_n^0\rangle$$
$$E_n^1 = \langle\psi_n^0 H'|\psi_n^0\rangle$$

E_n^1 is the first-order correction. We can similarly solve for the second-order theory.

The Schrieffer-Wolff transformation is an operator version of second-order perturbation theory that diagonalizes a system Hamiltonian to first order in the interaction. It transforms H_0 + H' into a basis where it is diagonal to first order in the perturbation H'. H' is purely off-diagonal in the eigenbasis of H_0. This is achieved when S is chosen such that $H' + [S, H_0] = 0$. Let H' be an off-diagonal perturbation. Let $S^\dagger = -S$ and $H_{SW} = \exp(S)H_{env}\exp(-S)$. Let $\lambda = 1$ be a formal parameter. By the Baker-Campbell-Haussdorf lemma,

$$H_{\mathrm{SW}} = H_{\mathrm{env}} + \lambda[S, H_{\mathrm{env}}] + \frac{\lambda^2}{2}[S, [S, H_{\mathrm{env}}]]$$
$$= H_0 + \lambda H' + \lambda[S, H_0 + \lambda H'] + \frac{\lambda^2}{2}[S, [S, H_0 + \lambda H']]$$

Up to second order,

$$= H_0 + \lambda(H' + [S, H_0]) + \lambda^2([S, H'] + \frac{1}{2}[S, [S, H_0]])$$

When S is chosen appropriately, H_{SW} implifies. Let $H' + [S, H_0] = 0$. Then $[S, H_0] = -H'$.

$$H_{\text{SW}} = H_0 + \lambda^2([S, H'] - \frac{1}{2}[S, H'])$$
$$H_{\text{SW}} = H_0 \frac{1}{2}[S, H'] + O(H'^3)$$

We then solve for S in $[S, H_0] = -H'$ to obtain the diagonalized form of H_{SW}.

Let's look at one application of perturbtion theory that is mentioned in Chap. 3. For a more realistic look at electrons and holes in periodic solids we can start with the following Hamiltonian for an electron in a crystal with position r.

$$\left[\frac{p^2}{2m} + U(r)\right]\psi_{nk}(r) = \varepsilon_{nk}\psi_{nk}(r)$$

We define $\psi_{nk}(r) = \exp(ik \cdot r)\phi_{nk}(r)$ where $\phi_{nk}(r)$ is periodic with the crystal lattice. n is called the band index and k is the wavevector. Then we attempt to solve for ε_{nk}.

$$\begin{aligned}
\frac{p^2}{2m} &= -\frac{\hbar^2}{2m}\nabla^2 \\
\nabla \exp(ik \cdot r)\phi_{nk}(r) &= \exp(ik \cdot r)\nabla\phi_{nk}(r) + \phi_{nk}(r)(ik)\exp(ik \cdot r) \\
\nabla^2 \exp(ik \cdot r)\phi_{nk}(r) &= (ik)\exp(ik \cdot r)\nabla\phi_{nk}(r) + \exp(ik \cdot r)\nabla^2\phi_{nk}(r) \\
&+(ik)\exp(ik \cdot r)\nabla\phi_{nk}(r) \\
&+(-k^2)\phi_{nk}(r)\exp(ik \cdot r) \\
&-\frac{\hbar}{2m}\nabla^2\exp(ik \cdot r)\phi_{nk}(r) = -\frac{\hbar^2}{2m}(2ik)\exp(ik \cdot r)\nabla\phi_{nk}(r) \\
&+\frac{\hbar^2k^2}{2m}\phi_{nk}(r)\exp(ik \cdot r) \\
&-\frac{\hbar^2}{2m}\nabla^2\phi_{nk}(r)\exp(ik \cdot r) \\
&\frac{\hbar^2k^2}{2m}\phi_{nk}(r) - \frac{\hbar^2}{2m}(2ik)\nabla\phi_{nk}(r) - \frac{\hbar^2}{2m}\nabla^2\phi_{nk}(r) \\
&= \varepsilon_{nk}\phi_{nk}(r) - U(r)\phi_{nk}(r) \\
&\left[\frac{\hbar^2k^2}{2m} + \frac{\hbar}{m}k \cdot p - \frac{\hbar^2}{2m}\nabla^2 + U(r)\right]\phi_{nk}(r) = \varepsilon_{nk}\phi_{nk}(r)
\end{aligned}$$

This gives us the basic Hamiltonian for so-called $k \cdot p$ perturbation theory [1]. We can treat the $k \cdot p$ term as a perturbation H' to another term in the Hamiltonian which becomes H_0.

$$H' = \frac{\hbar}{m} k \cdot p$$

Keeping in mind $\langle \phi_m | \phi_n \rangle = \delta_{mn}$ and $\sum_m |\phi_m\rangle \langle \phi_m| = 1$ for a complete set we can solve the matrix form of the Hamiltonian using second-order degenerate perturbation theory. Let $\lambda = 1$ be a formal parameter.

$$H = H_0 + \lambda H'$$

$$H |\phi_n\rangle = E_n |\phi_n\rangle$$

$$E_n = \sum_m \lambda^m E_n^{(m)}$$

$$|\phi_n\rangle = \sum_m \lambda^m \left|\phi_n^{(m)}\right\rangle$$

Grouping terms by powers of λ yields the following.

$$H_0 \left|\phi_n^{(0)}\right\rangle = E_n^{(0)} \left|\phi_n^{(0)}\right\rangle$$

$$H_0 \left|\phi_n^{(1)}\right\rangle + H' \left|\phi_n^{(0)}\right\rangle = E_n^{(0)} \left|\phi_n^{(1)}\right\rangle + E_n^{(1)} \left|\phi_n^{(0)}\right\rangle$$

$$H_0 \left|\phi_n^{(2)}\right\rangle + H' \left|\phi_n^{(1)}\right\rangle = E_n^{(0)} \left|\phi_n^{(2)}\right\rangle + E_n^{(1)} \left|\phi_n^{(1)}\right\rangle + E_n^{(2)} \left|\phi_n^{(0)}\right\rangle$$

$E_n^{(0)}$ is d-degenerate with eigenkets $\left|\phi_{n,1}\right\rangle, \left|\phi_{n,2}\right\rangle, \ldots, \left|\phi_{n,d}\right\rangle$.

$$\left|\phi_n^{(0)}\right\rangle = c_1 \left|\phi_{n,1}\right\rangle + c_2 \left|\phi_{n,2}\right\rangle, \cdots + \cdots + c_d \left|\phi_{n,d}\right\rangle \; ; c_i \in \mathrm{C}$$

We act on both sides of the λ^1 equation by $\left\langle\phi_{n,i}\right| \; ; i = 1, 2, \ldots, d$ for each value of i. This gives us a $d \times d$ matrix eigenvalue equation. The d eigenvalues are the first-order energy corrections. The corresponding eigenvectors give the coefficients of the d 0th order unperturbed eigenkets.

$$\left\langle\phi_{n,i}\right| H_0 \left|\phi_n^{(1)}\right\rangle + \left\langle\phi_{n,i}\right| H' \left|\phi_n^{(0)}\right\rangle = \left\langle\phi_{n,i}\right| E_n^{(0)} \left|\phi_n^{(1)}\right\rangle + \left\langle\phi_{n,i}\right| E_n^{(1)} \left|\phi_n^{(0)}\right\rangle$$

$$\sum_{j=1}^{d} c_j \langle\phi_{n,i}| H' |\phi_{n,j}\rangle = E_n^{(1)} c_i$$

We now have d unperturbed eigenkets $\left|\phi_{n,i}^{(0)}\right\rangle$. We apply $\langle\phi_m| \neq \left\langle\phi_{n,i}^{(0)}\right|$ to either side of the λ^1 equation.

$$H_0 \left|\phi_{n,i}^{(1)}\right\rangle + H' \left|\phi_{n,i}^{(0)}\right\rangle = E_n^{(0)} \left|\phi_{n,i}^{(1)}\right\rangle + E_n^{(1)} \left|\phi_{n,i}^{(0)}\right\rangle$$

$$(E_m - E_n^{(0)})\langle\phi_m|\phi_{n,i}^{(1)}\rangle + \langle\phi_m| H' \left|\phi_{n,i}^{(0)}\right\rangle = 0$$

$$\left|\phi_{n,i}^{(1)}\right\rangle = \sum_{m;E_m \neq E_n^{(0)}} \frac{\langle\phi_m| H' \left|\phi_{n,i}^{(0)}\right\rangle}{E_n^{(0)} - E_m}$$

We apply $\left\langle\phi_{n,i}^{(0)}\right|$ to either side of the λ^2 equation.

$$(H_0 - E_n^{(0)}) \left|\phi_{n,i}^{(2)}\right\rangle + (H' - E_{n,i}^{(1)}) \left|\phi_{n,i}^{(1)}\right\rangle = E_{n,i}^{(2)} \left|\phi_{n,i}^{(0)}\right\rangle$$

$$E_{n,i}^{(2)} = \left\langle\phi_{n,i}^{(0)}\right| (H' - E_{n,i}^{(1)}) \left|\phi_{n,i}^{(1)}\right\rangle$$

$$E_{n,i}^{(2)} = \sum_m \left\langle\phi_{n,i}^{(0)}\right| H' |\phi_m\rangle \langle\phi_m|\phi_{n,i}^{(1)}\rangle - E_n^{(1)} \langle\phi_{n,i}^{(0)}|\phi_{n,i}^{(1)}\rangle$$

$$E_{n,i}^{(2)} = \sum_{m;E_m \neq E_n^{(0)}} \frac{|\langle\phi_m| H' \left|\phi_{n,i}^{(0)}\right\rangle|^2}{E_n^{(0)} - E_m}$$

In general, $k \cdot p$ models can involve infinite dimensional matrices. The 8×8 Kane and 4×4 Luttinger-Kohn models seen earlier in this book are simplified $k \cdot p$ models.

Appendix D
Shor's Algorithm Implementation

```
"""
Compiling Shor's Algorithm.

@Author: Marcus Edwards
@Date: 2025-03-03
"""

import sys
import random
import numpy as np
import pennylane as qml
import copy
import matplotlib.pyplot as plt
from pennylane import numpy as np
from functools import partial
from itertools import product


plt.style.use('pennylane.drawer.plot')

INPUT_QUBITS = 10
OUTPUT_QUBITS = 5
ANCILLA_QUBITS = 5
CARRY_QUBITS = 2


# Part 1. Helper functions


def get_gcd(l, N):
    """
    Find the GCD of two numbers.

    Args:
        l (int): First number.
        N (int): Second number.

    Returns:
```

© The Editor(s) (if applicable) and The Author(s), under exclusive license to Springer Nature Switzerland AG 2026
M. S. Edwards, *An Introduction to Quantum Computing for Computer Engineers*,
https://doi.org/10.1007/978-3-032-03650-6

```
        int: the GCD.
    """
    r = l % N

    while r:
        l = N
        N = r
        r = l % N

    return N


def is_coprime(l, N):
    """

    Determine if two numbers are coprime.

    Args:
        l (int): First number to check if is coprime with the other.
        N (int): Second number to check if is coprime with the other.

    Returns:
        bool: True if they are coprime numbers, False otherwise.
    """

    return get_gcd(l, N) == 1


def is_odd(r):
    """

    Determine if a number is odd.

    Args:
        r (int): Integer to check if is an odd number.

    Returns:
        bool: True if it is odd, False otherwise.
    """

    return r % 2 == 1


def is_not_one(x, N):
    """

    Determine if x is not +- 1 modulo N.

    Args:
        N (int): Modulus of the equivalence.
        x (int): Integer to check if it is different from +-1 modulo N.

    Returns:
        bool: True if it is different, False otherwise.
    """

    return x % N != N - 1 and x % N != 1


def shor(N, l=None, e=10**-17):
    """

    Return the factorization of a given integer.

    Args:
       N (int): integer we want to factorize.
       l (int): Optional choice of integer coprime to N.
```

```
        e (float): Optional error which the circuit is compiled to.

    Returns:
        array[int]: [p,q], the prime factors of N.
    """

    if l is None:
        l = random.randint(2, N - 2)
    if not is_coprime(l, N):
        p = get_gcd(l, N)
        q = N / p
    else:
        Us = get_matrix_l_mod_N(l, N)
        U = make_exponentiation_matrix_controlled(Us)
        r = get_period(U, e)
        if is_odd(r):
            [p, q] = shor(N)
        else:
            x = (l ** (r / 2)) % N
            if not is_not_one(x, N):
                [p, q] = shor(N)
            else:
                p = get_gcd(x - 1, N)
                q = get_gcd(x + 1, N)
    return [p, q]


# Part 2. Compiling CNOT circuits.


def make_exponentiation_matrix_controlled(Us):
    """

    Injects the control parameter into the modular exponentation circuits.
    :param Us: The modular exponentation circuits.
    :return: A function that calls all the controlled exponentation circuits.
    """
    def modular_exponentiation_matrix_controlled():
        index = 0
        for U in Us:
            U(c=INPUT_QUBITS - index - 1)
            index += 1

    return modular_exponentiation_matrix_controlled


def get_matrix_l_mod_N(l, N):
    """

    Synthesizes the U_{N,l} operator's matrix in terms of CNOTs.

    :param l: the integer coprime to N which we are finding the period of.
    :param N: the modulus being factored.
    :return: The matrix U_{N,l}.
    """
    i = 0
    unitaries = []
    while i < 2*OUTPUT_QUBITS:
        unitary = get_controlled_modular_multiplication_unitary(l, N, i)
        unitaries.append(unitary)
        i += 1
    return unitaries


def build_adder_circuit(a, c):
    """
```

```
    Builds a circuit to do modular multiplication via shifted additions.
    :param a: The pre-calculated l**(2**i).
    :param c: The controlling qubit.
    :return:
    """
    n = OUTPUT_QUBITS
    shift = 0 # not the i in the docstring... this one indexes a's bits

    # load b into ancilla register
    for j in range(OUTPUT_QUBITS):
        compileMultiControlledX((c, INPUT_QUBITS + j, INPUT_QUBITS + OUTPUT_QUBITS + j))

    assert build_adder_circuit_helper(format(int(a), '#07b').split('b')[1][::-1], n, shift, c)


def build_adder_circuit_helper(a, n, shift, c):
    """

    Recursive function that builds the modular multiplication circuit with shifted additions.
    :param a: The pre-calculated l**(2**j) as a big endian binary string.
    :param n: The size of the registers.
    :param i: The current index (from least to most significant digits).
    :param c: The controlling qubit.
    :return: whether finished.
    """
    if shift == n:
        return True
    else:
        if bool(a[shift]) is True:
            # add shifted b to the output register
            carry_qubit = INPUT_QUBITS + OUTPUT_QUBITS + ANCILLA_QUBITS + 0

            # note these variables are named based on the roles of the inputs to the MAJ and UMA
            output_qubit = INPUT_QUBITS + OUTPUT_QUBITS + 0 + shift
            ancilla_qubit = INPUT_QUBITS + 0

            MAJ((carry_qubit, output_qubit, ancilla_qubit), c)

            for _ in range(1 + shift, OUTPUT_QUBITS):
                carry_qubit = ancilla_qubit
                carry_qubit = ancilla_qubit
                output_qubit = output_qubit + 1
                ancilla_qubit = ancilla_qubit + 1

                MAJ((carry_qubit, output_qubit, ancilla_qubit), c)

            carry_output_qubit = INPUT_QUBITS + OUTPUT_QUBITS + ANCILLA_QUBITS + 1
            compileMultiControlledX((c, ancilla_qubit, carry_output_qubit))

            UMA((carry_qubit, output_qubit, ancilla_qubit), c)

            for _ in range(OUTPUT_QUBITS - 2, shift, -1):
                ancilla_qubit = carry_qubit
                output_qubit = output_qubit - 1
                carry_qubit = carry_qubit - 1
                UMA((carry_qubit, output_qubit, ancilla_qubit), c)

            if carry_qubit != INPUT_QUBITS + OUTPUT_QUBITS + ANCILLA_QUBITS + 0:

                ancilla_qubit = carry_qubit
                output_qubit = output_qubit - 1
                carry_qubit = INPUT_QUBITS + OUTPUT_QUBITS + ANCILLA_QUBITS + 0

                UMA((carry_qubit, output_qubit, ancilla_qubit), c)

        return build_adder_circuit_helper(a, n, shift + 1, c)
```

```
def is_non_singular(x):
    """

    Computes whether a matrix is singular and invertible.
    :param x: The matrix.
    :return: Whether it is singular.
    """
    non_singular = np.linalg.cond(x) < 1 / sys.float_info.epsilon
    return non_singular


def get_controlled_modular_multiplication_unitary(l, N, i):
    """

    Returns an in-place controlled modular multiplication unitary for power i.
    :param l: The integer coprime to N.
    :param N: The secret N.
    :param i: The power of l.
    :return: a partially initialized function to generate the modular multiplication circuit.
    """

    solns_exist = np.array([False for _ in range(OUTPUT_QUBITS)])
    U = []

    try:
        truth_table = { }

        for x in range(2**OUTPUT_QUBITS):
            truth_table[format(x, '#07b').split('b')[1][::-1]] = (
                format((x*(l**(2**i)) % N), '#07b').split('b')[1][::-1])

        for op_index, op_row in enumerate(range(OUTPUT_QUBITS)):
            b = []
            a = []
            for input in truth_table.keys():
                b.append(int(truth_table[input][op_row]))
                a.append([int(input[input_entry]) for input_entry in range(OUTPUT_QUBITS)])

            # row = np.linalg.lstsq(np.array(a), np.array(b))[0]

            soln_exists = False

            # solve linear system with binary variables
            for m, n, o, p, q in product([1, 0], repeat=OUTPUT_QUBITS):
                row = [m, n, o, p, q]
                row_works = True
                for index, input in enumerate(a):
                    if np.array(row) @ np.array(input).transpose() != b[index] or row in U or np.all(np.array(row)
                            == 0):
                        row_works = False
                        break
                if row_works:
                    soln_exists = True
                    break

            solns_exist[op_index] = soln_exists

            U.append(row)

        if not np.all(solns_exist) or not is_non_singular(U):
            raise Exception("No in-place solution exists.")

        return partial(CNOT_synth, A=np.array(U), n=len(row), m=2)

    except Exception as e:
```

```
    # if no in−place solution exists...
    if not np.all(solns_exist) or not is_non_singular(U):
        # define out−of−place multipliers using ancillas...

        # forward operation
        truth_table = { }

        for x in range(2 ** OUTPUT_QUBITS):
            truth_table[(format(x, '#07b').split('b')[1][::−1] + ''.join(['0' for _ in
                    range(ANCILLA_QUBITS)]))] = (
                format(x, '#07b').split('b')[1][::−1] + format((x*(l**(2**i))%N), '#07b').split('b')[1][::−1])

        U = []
        U_fallback = False
        solns_exist = np.array([False for _ in range(OUTPUT_QUBITS + ANCILLA_QUBITS)])

        for op_index, op_row in enumerate(range(OUTPUT_QUBITS + ANCILLA_QUBITS)):
            b = []
            a = []
            for input in truth_table.keys():
                b.append(int(truth_table[input][op_row]))
                a.append([int(input[input_entry]) for input_entry in
                        range(OUTPUT_QUBITS+ANCILLA_QUBITS)])

            # row = np.linalg.lstsq(np.array(a), np.array(b))[0]

            soln_exists = False

            # solve linear system with binary variables
            for m, n, o, p, q, r, s, t, u, v in product([1, 0],
                    repeat=OUTPUT_QUBITS+ANCILLA_QUBITS):
                row = [m, n, o, p, q, r, s, t, u, v]
                row_works = True
                for index, input in enumerate(a):
                    if np.array(row) @ np.array(input).transpose() != b[index] or row in U or
                            np.all(np.array(row) == 0):
                        row_works = False
                        break
                if row_works:
                    soln_exists = True
                    break

            solns_exist[op_index] = soln_exists

            U.append(row)

        if not np.all(solns_exist) or not is_non_singular(U):
            # we will fall back on the ripple carry adder solution
            U_fallback = True

        # inverse operation
        truth_table = { }

        for x in range(2 ** OUTPUT_QUBITS):
            truth_table[format(x, '#07b').split('b')[1][::−1] + ''.join(['0' for _ in
                    range(ANCILLA_QUBITS)])] = (
                    format(x, '#07b').split('b')[1][::−1] + format(int(x*((l**(2**i))**−1)%N),
                        '#07b').split('b')[1][::−1])

        V = []
        V_fallback = False
        solns_exist = np.array([False for _ in range(OUTPUT_QUBITS + ANCILLA_QUBITS)])

        for op_index, op_row in enumerate(range(OUTPUT_QUBITS + ANCILLA_QUBITS)):
            b = []
            a = []
            for input in truth_table.keys():
```

```
                b.append(int(truth_table[input][op_row]))
                a.append([int(input[input_entry]) for input_entry in range(OUTPUT_QUBITS +
                        ANCILLA_QUBITS)])

            # row = np.linalg.lstsq(np.array(a), np.array(b))[0]

            soln_exists = False

            # solve linear system with binary variables
            for m, n, o, p, q, r, s, t, u, v in product([1, 0], repeat=OUTPUT_QUBITS +
                    ANCILLA_QUBITS):
                row = [m, n, o, p, q, r, s, t, u, v]
                row_works = True
                for index, input in enumerate(a):
                    if np.array(row) @ np.array(input).transpose() != b[index] or row in U or
                            np.all(np.array(row) == 0):
                        row_works = False
                        break
                if row_works:
                    soln_exists = True
                    break

            solns_exist[op_index] = soln_exists

            V.append(row)

        if not np.all(solns_exist) or not is_non_singular(V):
            V_fallback = True

        def in_place_modular_exponentiation_via_ancillas(c):

            # apply out-of-place modular multiplication
            if not U_fallback:
                # we automatically solved for and synthesize an operation
                partial(CNOT_synth, A=np.array(U), n=len(row), m=2)(c=c)
            else:
                # otherwise, we use the known ripple carry solution
                partial(build_adder_circuit, a=(l**(2**i)))(c=c)

            # swap registers
            for s in range(OUTPUT_QUBITS):
                # qml.SWAP((INPUT_QUBITS+s, OUTPUT_QUBITS+INPUT_QUBITS+s))

                # CNOT based SWAPs used to enable control by input register
                compileMultiControlledX(wires=(c, INPUT_QUBITS + s, OUTPUT_QUBITS +
                        INPUT_QUBITS + s,))
                compileMultiControlledX(wires=(c, OUTPUT_QUBITS + INPUT_QUBITS + s,
                        INPUT_QUBITS + s,))
                compileMultiControlledX(wires=(c, INPUT_QUBITS + s, OUTPUT_QUBITS +
                        INPUT_QUBITS + s,))

            # perform inverse out-of-place multiplication
            if not V_fallback:
                # we automatically solved for and synthesize an operation
                def to_adjoint():
                    partial(CNOT_synth, A=np.array(V), n=len(row), m=2)(c=c)
                qml.adjoint(to_adjoint)
            else:
                # otherwise, we use the known ripple carry solution
                def to_adjoint():
                    partial(build_adder_circuit, a=(l**(2**i))**-1)(c=c)
                qml.adjoint(to_adjoint)

        return in_place_modular_exponentiation_via_ancillas

    else:
        print(f"Cannot handle unknown exception: {e}")
```

```
def MAJ(wires, c):
    """

    The majority gate used in the construction of the ripple carry adder:

    [1] S. A. Cuccaro, T. G. Draper, S. A. Kutin, and D. P. Moulton, ''A
    new quantum ripple-carry addition circuit,'' Oct. 22, 2004, arXiv:
    arXiv:quant-ph/0410184. doi: 10.48550/arXiv.quant-ph/0410184.

    :param wires: The wires to which the gates apply.
    :param c: The controlling input qubit.
    :return:
    """
    compileMultiControlledX((c, wires[2], wires[1]))
    compileMultiControlledX((c, wires[2], wires[0]))
    compileMultiControlledX((c, *wires))


def UMA(wires, c):
    """

    The unmajority and add gate used in the construction of the ripple carry adder:

    [1] S. A. Cuccaro, T. G. Draper, S. A. Kutin, and D. P. Moulton, ''A
    new quantum ripple-carry addition circuit,'' Oct. 22, 2004, arXiv:
    arXiv:quant-ph/0410184. doi: 10.48550/arXiv.quant-ph/0410184.

    :param wires: The wires to which the gates apply.
    :param c: The controlling input qubit.
    :return:
    """
    compileMultiControlledX((c, *wires))
    compileMultiControlledX((c, wires[2], wires[0]))
    compileMultiControlledX((c, wires[0], wires[1]))


def compileMultiControlledX(wires, debug=False):
    """

    Compile a Toffoli to Clifford+T using the approach in:

    [1] J. Welch, A. Bocharov, and K. Svore, ''Efficient Approximation of
    Diagonal Unitaries over the Clifford+T Basis,'' Quantum information & computation,
    vol. 16, Dec. 2014, doi: 10.26421/QIC16.1-2-6.

    :param wires: The wires the gate acts on, with the target wire last.
    :param debug: Whether to skip tghe compilation to Clifford + T for debugging.
    :return:
    """

    if debug:
        # use this line to generate uncompiled circuit for easy inspection
        qml.MultiControlledX(wires=wires)

    else:
        qml.T(wires=wires[0])
        qml.CNOT(wires=(wires[0], wires[1]))
        qml.adjoint(qml.T(wires=wires[1]))
        qml.CNOT(wires=(wires[0], wires[1]))
        qml.T(wires=wires[1])
        qml.H(wires=wires[2])
        qml.CNOT(wires=(wires[1], wires[2]))
        qml.adjoint(qml.T(wires=wires[2]))
        qml.CNOT(wires=(wires[0], wires[2]))
        qml.T(wires=wires[2])
```

```
        qml.CNOT(wires=(wires[1], wires[2]))
        qml.adjoint(qml.T(wires=wires[2]))
        qml.CNOT(wires=(wires[0], wires[2]))
        qml.T(wires=wires[2])
        qml.H(wires=wires[2])


def CNOT_synth(A, n, m, c):
    """

    Performs the CNOT circuit synthesis using the efficient approach in
    [1] K. N. Patel, I. L. Markov, and J. P. Hayes, ''Efficient Synthesis
    of Linear Reversible Circuits,'' Feb. 03, 2003, arXiv: arXiv:quant-ph/0302002.
    doi: 10.48550/arXiv.quant-ph/0302002.

    :param A: The initial matrix to row-reduce.
    :param n: The dimension of the square matrix.
    :param m: The size of the partitions of the square matrix.
    :param c: The qubit that controls this operation.
    :return: The circuit.
    """
    # synthesize upper and lower triangular parts
    [A, circuit_lower] = lwr_CNOT_synth(A, n, m)
    A = np.transpose(A)
    [A, circuit_upper] = lwr_CNOT_synth(A, n, m)

    # switch control / target of CNOT in upper part.
    for i in range(len(circuit_upper)):
        new = (circuit_upper[i][1], circuit_upper[i][0])
        circuit_upper[i] = new

    # combine upper, lower parts
    circuit = circuit_upper + circuit_lower

    # convert to pennylane circuit
    for j in range(len(circuit)):
        #qml.MultiControlledX(wires=(c, circuit[j][0] + INPUT_QUBITS, circuit[j][1] + INPUT_QUBITS),)
        compileMultiControlledX(wires=(c, circuit[j][0] + INPUT_QUBITS, circuit[j][1] + INPUT_QUBITS),)


def hash_pattern(arr, precision=10):
    """

    Hashes a sub-row pattern.
    :param arr: The sub-row pattern.
    :param precision: The float precision up to which to maintain uniqueness between tensors.
    :return: The hash.
    """
    hash = ''
    if arr.shape == ():
        return str(arr)
    for elem in arr:
        if not isinstance(elem, np.tensor) and not isinstance(elem, np.ndarray) and not isinstance(elem, list):
            hash += str(elem).split('.')[0] + '.'
            if len(str(elem).split('.')) > 1:
                dec = str(elem).split('.')[1]
                i = 0
                while i < len(dec) and i < precision:
                    hash += dec[i]
                    i += 1
        else:
            hash += hash_pattern(elem, precision)
    return hash


def lwr_CNOT_synth(A, n, m):
    """
```

```
    Helper function for the efficient CNOT circuit synthesis which synthesizes
    the lower triangular part.

    :param A: The square matrix to row-reduce.
    :param n: The size of the matrix.
    :param m: The size of the partitions of the square matrix.
    :return: The row-reduced matrix and the circuit.
    """
    circuit = []
    for sec in range(1, int(np.ceil(n/m))): # iterate over column sections
        patt = dict()
        # remove duplicate sub-rows in section
        for row_ind in range((sec-1)*m, n):
            sub_row_patt = copy.deepcopy(A[row_ind, (sec-1)*m: sec*m-1])
            if hash_pattern(sub_row_patt) not in patt:
                patt[hash_pattern(sub_row_patt)] = row_ind
            else:
                A[row_ind, :] += A[patt[hash_pattern(sub_row_patt)], :]
                circuit = [(patt[hash_pattern(sub_row_patt)], row_ind)] + circuit

        # use Gaussian elimination for remaining entries in column section
        for col_ind in range((sec-1)*m, sec*m-1):
            # check for 1 on diagonal
            diag_one = (A[col_ind, col_ind] == 1)
            # remove ones in rows below col_ind
            for row_ind in range(col_ind+1, n):
                if A[row_ind, col_ind] != 0:
                    if not diag_one:
                        A[col_ind, :] += A[row_ind, :]
                        circuit = [(row_ind, col_ind)] + circuit
                        diag_one = 1
                    A[row_ind, :] += A[col_ind, :]
                    circuit = [(col_ind, row_ind)] + circuit

        return [A, circuit]


shor_machine = qml.device('default.qubit',
        wires=INPUT_QUBITS+OUTPUT_QUBITS+ANCILLA_QUBITS+CARRY_QUBITS, shots=None)


def get_period(U, e):
    """

    Calls the quantum subroutine and calculates the period r from the measurement outcome histogram.
    :param U: The modular exponentiation operator.
    :param N: The number to factor.
    :return: The period r.
    """
    # measure y_m = m (2^5 / r) with high probability
    probs = shor_circuit(U)

    # compile Clifford + T circuit for error e
    # qml.clifford_t_decomposition(shor_circuit(U), e)

    # draw circuit with QFT expanded
    sys.setrecursionlimit(15000)
    print(qml.draw(shor_circuit)(U))

    # draw circuit with matplotlib
    # qml.draw_mpl(shor_circuit, decimals=2, style="pennylane")(U)
    # plt.show()

    # Calculate the period
    ys = []
    index = 0
```

```
    for prob in probs:
        if not np.isclose(prob, 0.):
            ys.append(index)
        index += 1

    rs = list(map(lambda m_y: (2**OUTPUT_QUBITS / (m_y[1] / m_y[0])) if m_y[0] != 0 else 0,
            enumerate(ys)))

    for r in rs:
        assert np.isclose(r, np.sum(rs) / len(rs))

    return np.sum(rs) / len(rs)


@partial(qml.transforms.decompose, max_expansion=1)
@qml.qnode(shor_machine)
def shor_circuit(U):
    """

    Computes the minimum r such that U^r |1> = |1>.
    :param U: The unitary that encodes the function l^x (mod N).
    :param N: The modulus, or public key, N.
    :return: The period.
    """

    # put input register in superposition
    for i in range(INPUT_QUBITS):
        qml.Hadamard(i)

    # apply modular exponentiation function
    U()

    # measure output register
    for j in range(INPUT_QUBITS, INPUT_QUBITS + OUTPUT_QUBITS):
        qml.measure(j)

    # apply QFT to input register
    qml.adjoint(qml.QFT(wires=[k for k in range(INPUT_QUBITS)]))

    # measure input register
    return qml.probs(wires=[k for k in range(INPUT_QUBITS)])


if __name__ == '__main__':
    # solve the assigned problem
    [p, q] = shor(32, 3, 10**-7)
    print(f"p: {p}, q: {q}")
```

References

1. Cardona, M., Pollak, F.H.: Energy-band structure of germanium and silicon: the k p method. Phys. Rev. **142**(2), 530–543 (1966)
2. Griffiths, D.J.: Introduction to quantum mechanics, 2nd edn. Pearson Prentice Hall, Upper Saddle River, NJ (2005)
3. Nielsen, M.A., Chuang, I.L.: Quantum computation and quantum information 10th, Anniversary Cambridge University Press, Cambridge; New York (2010)

Appendix E
Further Reading

References

1. Association, Q.W.: Quantum Computing languages landscape (2018). https://medium.com/@quantum_wa/quantum-computing-languages-landscape-1bc6dedb2a35
2. Bacon, D., Kempe, J., Lidar, D.A., Whaley, K.B.: Universal fault-tolerant quantum computation on decoherence-free subspaces. Phys. Rev. Lett. **85**, 1758–1761 (2000). https://doi.org/10.1103/PhysRevLett.85.1758.
3. Baeyer, H.C.v.: QBism: The Future of Quantum Physics. Harvard University Press (2016). Google-Books-ID: PB4_DQAAQBAJ
4. Blais, A., Huang, R.S., Wallraff, A., Girvin, S.M., Schoelkopf, R.J.: Cavity quantum electrodynamics for superconducting electrical circuits: an architecture for quantum computation. Phys. Rev. A **69**(6), 062320 (2004). https://doi.org/10.1103/PhysRevA.69.062320. ArXiv:cond-mat/0402216
5. Bogan, A., Studenikin, S., Korkusinski, M., Gaudreau, L., Zawadzki, P., Sachrajda, A., Tracy, L., Reno, J., Hargett, T.: Single hole spin relaxation probed by fast single-shot latched charge sensing. Commun. Phys. **2**(1), 17 (2019). https://doi.org/10.1038/s42005-019-0113-0.
6. Borsoi, F., Hendrickx, N.W., John, V., Motz, S., van Riggelen, F., Sammak, A., de Snoo, S.L., Scappucci, G., Veldhorst, M.: Shared control of a 16 semiconductor quantum dot crossbar array. ArXiv (2022)
7. Brodie, I., Muray, J.J.: The Physics of Micro/Nano Fabrication. Plenum Press (1992)
8. Cai, J.F., Candès, E.J., Shen, Z.: A singular value thresholding algorithm for matrix completion. SIAM J. Opt. **20**(4), 1956–1982 (2010). https://doi.org/10.1137/080738970.
9. Cardona, M., Pollak, F.H.: Energy-band structure of germanium and silicon: The kp Method. Phys. Rev. **142**(2), 530–543 (1966). https://doi.org/10.1103/PhysRev.142.530.
10. Chatterjee, A.: Auto-tuning a quantum computer. DelftX: development and applications of germanium quantum technologies (2024)

© The Editor(s) (if applicable) and The Author(s), under exclusive license to Springer Nature Switzerland AG 2026

M. S. Edwards, *An Introduction to Quantum Computing for Computer Engineers*,
https://doi.org/10.1007/978-3-032-03650-6

11. Ciocoiu, A., Khalifa, M., Salfi, J.: Towards computer-assisted design of hole spin qubits in quantum dot devices (2022). http://arxiv.org/abs/2209.12026. [cond-mat]
12. Di Matteo, O., Gamble, J., Granade, C., Rudinger, K., Wiebe, N.: Operational, gauge-free quantum tomography. Quantum **4**, 364 (2020). https://doi.org/10.22331/q-2020-11-17-364. http://arxiv.org/abs/2007.01470. [quant-ph]
13. Dirac, P.A.M.: The Principles of Quantum Mechanics. Clarendon Press (1981). Google-Books-ID: XehUpGiM6FIC
14. Edwards, M.: Three Quantum Programming Language Parser Implementations for the Web (2023). https://doi.org/10.48550/arXiv.2310.10802. http://arxiv.org/abs/2310.10802.
15. Fedele, F., Chatterjee, A., Fallahi, S., Gardner, G.C., Manfra, M.J., Kuemmeth, F.: Simultaneous operations in a two-dimensional array of singlet-triplet qubits. PRX Quantum **2**, 040306 (2021). https://doi.org/10.1103/PRXQuantum.2.040306.
16. Fujii, K.: Introduction to the Rotating Wave Approximation (RWA) : Two Coherent Oscillations (2014). ArXiv: 1301.3585 [math-ph, physics:quant-ph]
17. Greplova, E.: What is auto-tuning? DelftX: development and applications of germanium quantum technologies (2024)
18. Griffiths, D.J.: Introduction to Quantum Mechanics, 2nd edn. Cambridge University Press (2017)
19. Hagiwara, K., Hikasa, K., Nakamura, K., Tanabashi, M., Aguilar-Benitez, M., Amsler, C., Barnett, R.M., Burchat, P.R., Carone, C.D., Caso, C., Conforto, G., Dahl, O., Doser, M., Eidelman, S., Feng, J.L., Gibbons, L., Goodman, M., Grab, C., Groom, D.E., Gurtu, A., Hayes, K.G., Herna'ndez-Rey, J.J., Honscheid, K., Kolda, C., Mangano, M.L., Manley, D.M., Manohar, A.V., March-Russell, J., Masoni, A., Miquel, R., Mönig, K., Murayama, H., Navas, S., Olive, K.A., Pape, L., Patrignani, C., Piepke, A., Roos, M., Terning, J., Törnqvist, N.A., Trippe, T.G., Vogel, P., Wohl, C.G., Workman, R.L., Yao, W.M., Armstrong, B., Gee, P.S., Lugovsky, K.S., Lugovsky, S.B., Lugovsky, V.S., Artuso, M., Asner, D., Babu, K.S., Barberio, E., Battaglia, M., Bichsel, H., Biebel, O., Bloch, P., Cahn, R.N., Cattai, A., Chivukula, R.S., Cousins, R.D., Cowan, G., Damour, T., Desler, K., Donahue, R.J., Edwards, D.A., Elvira, V.D., Erler, J., Ezhela, V.V., Fasso', A., Fetscher, W., Fields, B.D., Foster, B., Froidevaux, D., Fukugita, M., Gaisser, T.K., Garren, L., Gerber, H.J., Gilman, F.J., Haber, H.E., Hagmann, C., Hewett, J., Hinchliffe, I., Hogan, C.J., Höhler, G., Igo-Kemenes, P., Jackson, J.D., Johnson, K.F., Karlen, D., Kayser, B., Klein, S.R., Kleinknecht, K., Knowles, I.G., Kreitz, P., Kuyanov, Y.V., Landua, R., Langacker, P., Littenberg, L., Martin, A.D., Nakada, T., Narain, M., Nason, P., Peacock, J.A., Quinn, H.R., Raby, S., Raffelt, G., Razuvaev, E.A., Renk, B., Rolandi, L., Ronan, M.T., Rosenberg, L.J., Sachrajda, C.T., Sanda, A.I., Sarkar, S., Schmitt, M., Schneider, O., Scott, D., Seligman, W.G., Shaevitz, M.H., Sjöstrand, T., Smoot, G.F., Spanier, S., Spieler, H., Spooner, N.J.C., Srednicki, M., Stahl, A., Stanev, T., Suzuki, M., Tkachenko, N.P., Valencia, G., van Bibber, K., Vincter, M.G., Ward, D.R., Webber, B.R., Whalley, M., Wolfenstein, L., Womersley, J., Woody, C.L., Zenin,

O.V.: Review of particle properties. Phys. Rev. D **66**, 010001 (2002). https://doi.org/10.1103/PhysRevD.66.010001.

20. Hendrickx, N.W., Franke, D.P., Sammak, A., Scappucci, G., Veldhorst, M.: Fast two-qubit logic with holes in germanium. Nature **577**(7791), 487–491, 491A–491K (2020). https://doi.org/10.1038/s41586-019-1919-3. https://www.proquest.com/docview/2350114222/abstract/F552DEAE0F1E4CF2PQ/1. Num Pages: 487–491,491A–491K Place: London, United States Publisher: Nature Publishing Group Section: Article
21. Kempe, J., Bacon, D., Lidar, D.A., Whaley, K.B.: Theory of decoherence-free fault-tolerant universal quantum computation. Phys. Rev. A **63**, 042307 (2001). https://doi.org/10.1103/PhysRevA.63.042307.
22. Kim, Y., Morvan, A., Nguyen, L.B., Naik, R.K., Jünger, C., Chen, L., Kreikebaum, J.M., Santiago, D.I., Siddiqi, I.: High-fidelity three-qubit iToffoli gate for fixed-frequency superconducting qubits. Nature Physics **18**(7), 783–788 (2022). https://doi.org/10.1038/s41567-022-01590-3
23. Knill, E., Laflamme, R.: Assumptions for fault tolerant quantum computing. Tech. Rep. LA-UR–96-2718, 373736, Office of Scientific and Technical Information (OSTI) (1996). https://doi.org/10.2172/373736. http://www.osti.gov/servlets/purl/373736-uj5Dh3/webviewable/
24. Kobayashi, T., Salfi, J., van der Heijden, J., Chua, C., House, M.G., Culcer, D., Hutchison, W.D., Johnson, B.C., McCallum, J.C., Riemann, H., Abrosimov, N.V., Becker, P., Pohl, H.J., Simmons, M.Y., Rogge, S.: Engineering long spin coherence times of spin-orbit systems. arXiv e-prints arXiv:1809.10859 (2018). https://doi.org/10.48550/arXiv.1809.10859
25. Kurkjian, A.T.K., Higginbottom, D.B., Chartrand, C., MacQuarrie, E.R., Klein, J.R., Lee-Hone, N.R., Stacho, J., Bowness, C., Bergeron, L., DeAbreu, A., Brunelle, N.A., Harrigan, S.R., Kanaganayagam, J., Kazemi, M., Marsden, D.W., Richards, T.S., Stott, L.A., Roorda, S., Morse, K.J., Thewalt, M.L.W., Simmons, S.: Optical observation of single spins in silicon. arXiv e-prints arXiv:2103.07580 (2021). https://doi.org/10.48550/arXiv.2103.07580
26. Lang, N., Büchler, H.P.: Minimal instances for toric code ground states. **86**(2), 022336 (2012). https://doi.org/10.1103/PhysRevA.86.022336
27. LaRose, R.: Overview and Comparison of Gate Level Quantum Software Platforms. Quantum **3**, 130 (2019). https://doi.org/10.22331/q-2019-03-25-130. https://quantum-journal.org/papers/q-2019-03-25-130/. Publisher: Verein zur Förderung des Open Access Publizierens in den Quantenwissenschaften
28. Lawrie, W.I.L., Eenink, H.G.J., Hendrickx, N.W., Boter, J.M., Petit, L., Amitonov, S.V., Lodari, M., Wuetz, B.P., Volk, C., Philips, S., Droulers, G., Kalhor, N., van Riggelen, F., Brousse, D., Sammak, A., Vandersypen, L.M.K., Scappucci, G., Veldhorst, M.: Quantum Dot Arrays in Silicon and Germanium. Applied Physics Letters **116**(8), 080501 (2020). https://doi.org/10.1063/5.0002013. ArXiv: 1909.06575 [cond-mat]
29. Lennon, D.T., Moon, H., Camenzind, L.C., Yu, L., Zumbühl, D.M., Briggs, G.A., Osborne, M.A., Laird, E.A., Ares, N.: Efficiently measuring a quantum

device using machine learning. npj Quantum Inf. **5**(1) (2019). https://doi.org/10.1038/s41534-019-0193-4
30. Liao, R.: Advanced topics in deep learning (2024). Graduate course
31. Linke, N.M., Maslov, D., Roetteler, M., Debnath, S., Figgatt, C., Landsman, K.A., Wright, K., Monroe, C.: Experimental comparison of two quantum computing architectures. Proceedings of the National Academy of Sciences **114**(13), 3305–3310 (2017). https://doi.org/10.1073/pnas.1618020114. Publisher: Proceedings of the National Academy of Sciences
32. Lutkenhaus, N.: Quantum cryptography at university of waterloo (2020). Graduate course
33. Luttinger, J.M., Kohn, W.: Motion of electrons and holes in perturbed periodic fields. Phys. Rev. **97**, 869–883 (1955). https://doi.org/10.1103/PhysRev.97.869.
34. Martin, R.C.: Clean architecture (2017)
35. Nguyen, V., Orbell, S.B., Lennon, D.T., Moon, H., Vigneau, F., Camenzind, L.C., Yu, L., Zumbühl, D.M., Briggs, G.A., Osborne, M.A., et al.: Deep reinforcement learning for efficient measurement of quantum devices. npj Quantum Information **7**(1) (2021). https://doi.org/10.1038/s41534-021-00434-x
36. Nowack, K.C., Koppens, F.H.L., Nazarov, Y.V., Vandersypen, L.M.K.: Coherent Control of a Single Electron Spin with Electric Fields (2007). https://doi.org/10.1126/science.1148092
37. Particle Data Group, Hagiwara, K., Hikasa, K., Nakamura, K., Tanabashi, M., Aguilar-Benitez, M., Amsler, C., Barnett, R.M., Burchat, P.R., Carone, C.D., Caso, C., Conforto, G., Dahl, O., Doser, M., Eidelman, S., Feng, J.L., Gibbons, L., Goodman, M., Grab, C., Groom, D.E., Gurtu, A., Hayes, K.G., Herna'ndez-Rey, J.J., Honscheid, K., Kolda, C., Mangano, M.L., Manley, D.M., Manohar, A.V., March-Russell, J., Masoni, A., Miquel, R., Mönig, K., Murayama, H., Navas, S., Olive, K.A., Pape, L., Patrignani, C., Piepke, A., Roos, M., Terning, J., Törnqvist, N.A., Trippe, T.G., Vogel, P., Wohl, C.G., Workman, R.L., Yao, W.M., Armstrong, B., Gee, P.S., Lugovsky, K.S., Lugovsky, S.B., Lugovsky, V.S., Artuso, M., Asner, D., Babu, K.S., Barberio, E., Battaglia, M., Bichsel, H., Biebel, O., Bloch, P., Cahn, R.N., Cattai, A., Chivukula, R.S., Cousins, R.D., Cowan, G., Damour, T., Desler, K., Donahue, R.J., Edwards, D.A., Elvira, V.D., Erler, J., Ezhela, V.V., Fasso', A., Fetscher, W., Fields, B.D., Foster, B., Froidevaux, D., Fukugita, M., Gaisser, T.K., Garren, L., Gerber, H.J., Gilman, F.J., Haber, H.E., Hagmann, C., Hewett, J., Hinchliffe, I., Hogan, C.J., Höhler, G., Igo-Kemenes, P., Jackson, J.D., Johnson, K.F., Karlen, D., Kayser, B., Klein, S.R., Kleinknecht, K., Knowles, I.G., Kreitz, P., Kuyanov, Y.V., Landua, R., Langacker, P., Littenberg, L., Martin, A.D., Nakada, T., Narain, M., Nason, P., Peacock, J.A., Quinn, H.R., Raby, S., Raffelt, G., Razuvaev, E.A., Renk, B., Rolandi, L., Ronan, M.T., Rosenberg, L.J., Sachrajda, C.T., Sanda, A.I., Sarkar, S., Schmitt, M., Schneider, O., Scott, D., Seligman, W.G., Shaevitz, M.H., Sjöstrand, T., Smoot, G.F., Spanier, S., Spieler, H., Spooner, N.J.C., Srednicki, M., Stahl, A., Stanev, T., Suzuki, M., Tkachenko, N.P., Valencia, G., van Bibber, K., Vincter, M.G., Ward, D.R., Webber, B.R., Whalley, M., Wolfenstein, L., Womersley, J., Woody, C.L., Zenin, O.V.: Review of Particle Properties. Physical

Review D **66**(1), 010001 (2002). https://doi.org/10.1103/PhysRevD.66.010001. Publisher: American Physical Society

38. Piot, N., Brun, B., Schmitt, V., Zihlmann, S., Michal, V.P., Apra, A., Abadillo-Uriel, J.C., Jehl, X., Bertrand, B., Niebojewski, H., Hutin, L., Vinet, M., Urdampilleta, M., Meunier, T., Niquet, Y.M., Maurand, R., Franceschi, S.D.: A single hole spin with enhanced coherence in natural silicon. Nat. Nanotech. **17**(10), 1072–1077 (2022). https://doi.org/10.1038/s41565-022-01196-z
39. Pons Viver, M., Puente, A.: Generation of Coulomb matrix elements for the 2D quantum harmonic oscillator. J. Math. Phys. **60**(8), 081905 (2019). https://doi.org/10.1063/1.5108599.
40. List of nmr isotopes. https://web.archive.org/web/20180201154413/http://www-lcs.ensicaen.fr/pyPulsar/index.php/List_of_NMR_isotopes
41. qosf/awesome-quantum-software (2024). https://github.com/qosf/awesome-quantum-software. Original-date: 2018-04-25T16:35:33Z
42. Ramasesha, S.: k p perturbation method (2023)
43. Rashba, E.I.: Theory of electric dipole spin resonance in quantum dots: Mean field theory with Gaussian fluctuations and beyond. **78**(19), 195302 (2008). https://doi.org/10.1103/PhysRevB.78.195302
44. Razeghi, M.: D luttinger-kohn hamiltonian. The MOCVD Challenge (1995)
45. Reiher, M., Wiebe, N., Svore, K.M., Wecker, D., Troyer, M.: Elucidating reaction mechanisms on quantum computers. Proc. Nat. Acad. Sci. **114**(29), 7555–7560 (2017). https://doi.org/10.1073/pnas.1619152114
46. Resch, K.: Quantum computing one ion at a time (2023). https://uwaterloo.ca/entangler/summer-2023/recent-research/quantum-computing-one-ion-time. Section: Recent research
47. van Riggelen, F., Lawrie, W.I.L., Russ, M., Hendrickx, N.W., Sammak, A., Rispler, M., Terhal, B.M., Scappucci, G., Veldhorst, M.: Phase flip code with semiconductor spin qubits: Supplementary Information. npj Quantum Information (2022)
48. Salfi, J., Mol, J.A., Culcer, D., Rogge, S.: Charge-Insensitive Single-Atom Spin-Orbit Qubit in Silicon. **116**(24), 246801 (2016). https://doi.org/10.1103/PhysRevLett.116.246801
49. Sammak, A., Sabbagh, D., Hendrickx, N.W., Lodari, M., Paquelet Wuetz, B., Tosato, A., Yeoh, L., Bollani, M., Virgilio, M., Schubert, M.A., Zaumseil, P., Capellini, G., Veldhorst, M., Scappucci, G.: Shallow and undoped germanium quantum wells: A playground for spin and hybrid quantum technology. Advanced Functional Materials **29**(14), 1807613 (2019). https://doi.org/https://doi.org/10.1002/adfm.201807613.
50. Senko, C.: Quantum mechanics 1 at university of waterloo (2019). Graduate course
51. Simon, S.H.: The Oxford Solid State Basics. Oxford University Press (2019)
52. Szumniak, P., Bednarek, S., Pawłowski, J., Partoens, B.: All electrically controlled quantum gates for single heavy hole spin qubits. Physical Review B **87**(19), 195307 (2013). https://doi.org/10.1103/PhysRevB.87.195307. http://arxiv.org/abs/1304.2674. ArXiv: 1304.2674 [cond-mat, physics:quant-ph]

53. Tiwari, Y., Singh Poonia, V.: Modeling of quantum dot based CNOT and Toffoli gates in a noisy environment. arXiv e-prints arXiv:2105.07021 (2021). https://doi.org/10.48550/arXiv.2105.07021
54. Tougaw, D., Will, J.D.: Designing a Turing-complete cellular automata system using quantum-dot cellular automata. Journal of Computational Electronics **19**(3), 1337–1343 (2020). https://doi.org/10.1007/s10825-020-01518-1.
55. Troyer, M.: State-of-the-art algorithm accelerates path for quantum computers to address climate change (2020). https://www.microsoft.com/en-us/research/blog/state-of-the-art-algorithm-accelerates-path-for-quantum-computers-to-address-climate-change/
56. Veldhorst, M., Yang, C.H., Hwang, J.C.C., Huang, W., Dehollain, J.P., Muhonen, J.T., Simmons, S., Laucht, A., Hudson, F.E., Itoh, K.M., Morello, A., Dzurak, A.S.: A two-qubit logic gate in silicon. Nature **526**(7573), 410–414 (2015). https://doi.org/10.1038/nature15263. https://www.nature.com/articles/nature15263. Number: 7573 Publisher: Nature Publishing Group
57. Verdon, G., McCourt, T., Luzhnica, E., Singh, V., Leichenauer, S., Hidary, J.: Quantum graph neural networks (2019). https://arxiv.org/abs/1909.12264
58. Vijay, S., Fu, L.: Quantum Error Correction for Complex and Majorana Fermion Qubits. arXiv e-prints arXiv: 1703.00459 (2017). https://doi.org/10.48550/arXiv.1703.00459
59. Wang, C.A., Scappucci, G., Veldhorst, M., Russ, M.: Modelling of planar germanium hole qubits in electric and magnetic fields (2022). http://arxiv.org/abs/2208.04795. ArXiv:2208.04795 [cond-mat, physics:quant-ph]
60. Wang, Z., Mao, L., Xue, N., Lu, W.: Suppression of shot noise in a spin–orbit coupled quantum dot. Royal Society Open Science **8**(4), 201432 (2021). https://doi.org/10.1098/rsos.201432. https://royalsocietypublishing.org/doi/10.1098/rsos.201432. Publisher: Royal Society
61. Zhu, X.Y., Tu, T., Guo, A.L., Zhou, Z.q., Li, C.F., Guo, G.C.: Dynamics of probing a quantum-dot spin qubit with superconducting resonator photons. Sci. Rep. **8**(1), 15761 (2018). https://doi.org/10.1038/s41598-018-34108-0. Number: 1 Publisher: Nature Publishing Group
62. Zinner, M., Dahlhausen, F., Boehme, P., Ehlers, J., Bieske, L., Fehring, L.: Quantum computing's potential for drug discovery: Early stage industry dynamics. Drug Discovery Today **26**(7), 1680–1688 (2021). https://doi.org/https://doi.org/10.1016/j.drudis.2021.06.003. https://www.sciencedirect.com/science/article/pii/S1359644621002750
63. Zmuidzinas, J.: Superconducting Microresonators: Physics and Applications. Annual Review of Condensed Matter Physics **3**(1), 169–214 (2012). https://doi.org/10.1146/annurev-conmatphys-020911-125022. _eprint: https://doi.org/10.1146/annurev-conmatphys-020911-125022
64. Zwolak, J.P., Taylor, J.M.: Colloquium: Advances in automation of quantum dot devices control. Rev. Mod. Phys. **95**, 011006 (2023). https://doi.org/10.1103/RevModPhys.95.011006. https://link.aps.org/doi/10.1103/RevModPhys.95.011006

GPSR Compliance
The European Union's (EU) General Product Safety Regulation (GPSR) is a set of rules that requires consumer products to be safe and our obligations to ensure this.

If you have any concerns about our products, you can contact us on

ProductSafety@springernature.com

In case Publisher is established outside the EU, the EU authorized representative is:

Springer Nature Customer Service Center GmbH
Europaplatz 3
69115 Heidelberg, Germany

www.ingramcontent.com/pod-product-compliance
Lightning Source LLC
Chambersburg PA
CBHW071503160226
39774CB00001B/18
* 9 7 8 3 0 3 2 0 3 6 4 9 0 *